BOOKS BY F. ALBERT COTTON:

COTTON AND WILKINSON—ADVANCED INORGANIC
CHEMISTRY, 2nd edition

Chemical Applications of Group Theory

Chemical

Applications

of Group

Theory

F. Albert Cotton

Department of Chemistry, Massachusetts Institute of Technology

Second Edition

WILEY—INTERSCIENCE

a Division of John Wiley and Sons, Inc.

New York · London · Sydney · Toronto

To Diane

Preface

In the seven years since the first edition of *Chemical Applications of Group Theory* was written, I have continued to teach a course along the lines of this book every other year. Steady, evolutionary change in the course finally led to a situation where the book and the course itself were no longer as closely related as they should be. I have, therefore, revised and augmented the book.

The new book has not lost the character or flavor of the old one—at least, I hope not. It aims to teach the use of symmetry arguments to the typical experimental chemist in a way that he will find meaningful and useful. At the same time I have tried to avoid that excessive and unnecessary superficiality (an unfortunate consequence of a misguided desire, evident in many books and articles on " theory for the chemist," to shelter the poor chemist from the rigors of mathematics) which only leads, in the end, to incompetence and its attendant frustrations. Too brief or too superficial a tuition in the use of symmetry arguments is a waste of whatever time is devoted to it. I think that the subject needs and merits a student's attention for the equivalent of a one-semester course. The student who masters this book will know *what* he is doing, *why* he is doing it, and *how* to do it. The range of subject matter is that which, in my judgment, the great majority of organic, inorganic, and physical chemists are likely to encounter in their daily research activity.

This book differs from its ancestor in three ways. First, the amount of illustrative and exercise material has been enormously increased. Since the demand for a *teaching textbook* in this field far exceeds what I had previously anticipated, I have tried now to equip the new edition with the pedagogic paraphernalia appropriate to meet this need.

Second, the treatment of certain subjects has been changed—improved, I hope—as a result of my continuing classroom experience. These improvements in presentation are neither extensions in coverage nor rigorizations; they are simply better ways of covering the ground. Such improvements will be found especially in Chapters 2 and 3, where the ideas of abstract group theory and the groups of highest symmetry are discussed.

Finally, new material and more rigorous methods have been introduced in several places. The major examples are (1) the explicit presentation of projection operators, and (2) an outline of the F and G matrix treatment of molecular vibrations. Although projection operators may seem a *trifle* forbidding at the outset, their potency and convenience and the nearly universal relevance of the symmetry-adapted linear combinations (SALC's) of basis functions which they generate justify the effort of learning about them. The student who does so frees himself forever from the tyranny and uncertainty of "intuitive" and "seat-of-the-pants" approaches. A new chapter which develops and illustrates projection operators has therefore been added, and many changes in the subsequent exposition have necessarily been made.

Because chemists seem to have become increasingly interested in employing vibration spectra quantitatively—or at least semiquantitatively—to obtain information on bond strengths, it seemed mandatory to augment the previous treatment of molecular vibrations with a description of the efficient F and G matrix method for conducting vibrational analyses. The fact that the convenient projection operator method for setting up symmetry coordinates has also been introduced makes inclusion of this material particularly feasible and desirable.

In view of the enormous impact which symmetry-based rules concerning the stereochemistry of concerted addition and cyclization reactions (Woodward-Hoffman rules) have had in recent years a detailed introduction to this subject has been added.

In conclusion, it is my pleasant duty to thank a new generation of students for their assistance. Many have been those whose questions and criticisms have stimulated me to seek better ways to present the subject. I am especially grateful to Professor David L. Weaver, Drs. Marie D. LaPrade, Barry G. DeBoer and James Smith and to Messrs. J. G. Bullitt, J. R. Pipal, C. M. Lukehart and J. G. Norman, Jr. for their generous assistance in correcting proof. Finally, Miss Marilyn Milan, by the speed and excellence of her typing, did much to lighten the task of preparing a new manuscript.

F. Albert Cotton

Cambridge, Massachusetts
May 1970

Preface to the First Edition

This book is the outgrowth of a one-semester course which has been taught for several years at the Massachusetts Institute of Technology to seniors and graduate students in chemistry. The treatment of the subject matter is unpretentious in that I have not hesitated to be mathematically unsophisticated, occasionally unrigorous, or somewhat prolix, where I felt that this really helps to make the subject more meaningful and comprehensible for the average student. By the average student, I mean one who does not aspire to be a theoretician but who wants to have a feel for the strategy used by theoreticians in treating problems in which symmetry properties are important and to have a working knowledge of the more common and well-established techniques. I feel that the great power and beauty of symmetry methods, not to mention the prime importance in all fields of chemistry of the results they give, make it very worthwhile for all chemists to be acquainted with the basic principles and main applications of group theoretical methods.

Despite the fact that there seems to be a growing desire among chemists at large to acquire this knowledge, it is still true that only a very few, other than professional theoreticians, have done so. The reason is not hard to discover. There is, so far as I know, no book available which is not likely to strike some terror into the hearts of all but those with an innate love of apparently esoteric theory. It seemed to me that ideas of the sort developed in this book would not soon be assimilated by a wide community of chemists until they were presented in as unpretentious and down-to-earth a manner as possible. That is what I have tried to do here. I have attempted to make this the kind of book which "one can read in bed without a pencil," as my colleague, John Waugh, once aptly described another textbook which has found wide favor because of its down-to-earth character.*

Perhaps the book may also serve as a first introduction for students intend-

---

* This statement is actually (and intentionally) not applicable to parts of Chapter 3 where I have made no concessions to the reader who refuses to inspect steric models in conjunction with study of the text.

ing to do theoretical work, giving them some overall perspective before they aim for depth.

I am most grateful for help I have received from many quarters in writing this book. Over the years students in the course have offered much valuable criticism and advice. In checking the final draft and the proofs I have had very welcome and efficient assistance from Dr. A. B. Blake and Messrs. R. C. Elder, T. E. Haas, and J. T. Mague. I, of course assume sole responsibility for all remaining errors. Finally, I wish to thank Mrs Nancy Blake for expert secretarial assistance.

F. ALBERT COTTON

Cambridge, Massachusetts
January 1963

Part III.   Appendices

Chemical Applications of Group Theory

Part I

Principles

Introduction

The experimental chemist in his daily work and thought is concerned with observing and, to as great an extent as possible, understanding and interpreting his observations on the nature of chemical compounds. Today, chemistry is a vast subject. In order to do thorough and productive experimental work, one must know so much descriptive chemistry and so much about experimental techniques that there is not time to be also a master of chemical theory. Theoretical work of profound and creative nature requires a vast training in mathematics and physics which is now the particular province of specialists. And yet, if one is to do more than merely *perform* experiments, one must have some theoretical framework for thought. In order to formulate experiments imaginatively and interpret them correctly, an understanding of the ideas provided by theory as to the behavior of molecules and other arrays of atoms is essential.

The problem in educating student chemists—and in educating ourselves—is to decide what kind of theory and how much of it is desirable. In other words, to what extent can the experimentalist afford to spend time on theoretical studies and at what point should he say, " Beyond this I have not the time or the inclination to go "? The answer to this question must of course vary with the special field of experimental work and with the individual. In some areas fairly advanced theory is indispensable. In others relatively little is really useful. For the most part, however, it seems fair to say that molecular quantum mechanics, that is, the theory of chemical bonding and molecular dynamics, is of general importance.

As we shall see in Chapter 5, the number and kinds of energy levels which an atom or molecule may have are rigorously and precisely determined by the symmetry of the molecule or of the environment of the atom. Thus, from symmetry considerations alone, we can always tell what the qualitative features of a problem must be. We shall know, without any quantitative calculations whatever, how many energy states there are and what interactions and transitions between them *may* occur. In other words, symmetry considerations *alone* can give us a complete and rigorous answer to the question

"What is possible and what is completely impossible?" Symmetry considerations alone *cannot*, however, tell us how likely it is that the possible things will actually take place. Symmetry can tell us that, *in principle*, two states of the system *must* differ in their energy, but only by computation or measurement can we determine how great the difference will be. Again, symmetry can tell us that only certain absorption bands in the electronic or vibrational spectrum of a molecule may occur. But to learn where they will occur and how great their intensity will be, calculations must be made.

Some illustrations of these statements may be helpful. Let us choose one illustration from each of the four major fields of application which are covered in Part II. In Chapter 7 the symmetry properties of molecular orbitals are discussed, with emphasis on the π molecular orbitals of unsaturated hydrocarbons, although other systems are also treated. It is shown how problems involving large numbers of orbitals and thus, potentially, high-order secular equations can be formulated so that symmetry considerations simplify these equations to the maximum extent possible. It is also shown how symmetry considerations permit the development of rules of great simplicity and generality (the so-called Woodward-Hoffman rules) governing certain concerted reactions. In Chapter 8, the method of constructing hybrid orbitals is explained, the molecular orbital approach to molecules of the AB_n type is outlined, and the relationship between the two treatments is explored.

In Chapter 9 the symmetry considerations underlying the main parts of the crystal and ligand field treatments of inner orbitals in complexes are developed. Finally, in Chapter 10, it is shown that by using symmetry considerations alone we may predict the number of vibrational fundamentals, their activities in the infrared and Raman, and the way in which the various bonds and interbond angles contribute to them for any molecule possessing some symmetry. The actual magnitudes of the frequencies depend on the interatomic forces in the molecule, and these cannot be predicted from symmetry properties. However, the technique of using symmetry restrictions to set up the equations required in calculations in their most amenable form (the *FG* matrix method) is presented in detail.

The main purpose of this book is to describe the methods by which we can extract the information which symmetry alone will provide. An understanding of this approach requires only a superficial knowledge of quantum mechanics. In several of the applications of symmetry methods, however, it would be artificial and stultifying to exclude religiously *all* quantitative considerations. Thus, in the chapter on molecular orbitals, it is natural to go a few steps beyond the procedure for determining the symmetries of the possible molecular orbitals and explain how the requisite linear combinations of atomic orbitals may be written down and how their energies may be estimated. It has also appeared desirable to introduce some quantitative ideas into the treatment of ligand field theory.

It has been assumed, necessarily, that the reader has some prior familiarity with the basic notions of quantum theory. He is expected to know in a general way what the wave equation is, the significance of the Hamiltonian operator, the physical meaning of a wave function, and so forth, but no detailed knowledge of mathematical intricacies is presumed. Even the contents of a rather qualitative book such as Coulson's *Valence* should be sufficient, although, of course, further background knowledge will not be amiss.

The following comments on the organization of the book may prove useful to the prospective reader. It is divided into two parts. Part I, which includes Chapters 1–6, covers the principles which are basic to all of the applications. The applications are described in Part II, embracing Chapters 7–10. The material in Part I has been written to be read sequentially; that is, each chapter deliberately builds on the material developed in all preceding chapters. In Part II, however, the aim has been to keep the chapters as independent of each other as possible without excessive repetition, although each one, of course, depends on all the material in Part I. This plan is advantageous to a reader whose immediate goal is to study only one particular area of application, since he can proceed directly to it, whichever it may be; it is also allows the teacher to select which applications to cover in a course too short to include all of them, or, if time permits, to take them all but in an order different from that chosen here.

Certain specialized points are expanded somewhat in Appendices in order not to divert the main discussion too far or for too long. Also, some useful tables are given as Appendices. Finally, Appendix IX provides a reference list for each of the four chapters in Part II, indicating where further discussion and research examples of the various applications may be found.

Definitions and Theorems
of Group Theory

2.1 The Defining Properties of a Group

A *group* is a collection of *elements* which are interrelated according to certain rules. We need not specify what the elements are or attribute any physical significance to them in order to discuss the group which they constitute. In this book, of course, we shall be concerned entirely with the groups formed by the sets of symmetry operations which may be carried out on molecules, but the basic definitions and theorems of group theory are far more general.

In order for any set of elements to form a mathematical group, the following conditions or rules must be satisfied.

1. *The product of any two elements in the group and the square of each element must be an element in the group.* In order for this condition to have meaning, we must, of course, have agreed on what we mean by the terms "multiply" and "product." They need not mean what they do in ordinary algebra and arithmetic. Perhaps we might say "combine" instead of "multiply" and "combination" instead of "product" in order to avoid unnecessary and perhaps incorrect connotations. Let us not yet commit ourselves to any particular law of combination but merely say that, if A and B are two elements of a group, we indicate that we are combining them by simply writing AB or BA. Now immediately the question arises if it makes any difference whether we write AB or BA. In ordinary algebra it does not, and we say that multiplication is commutative, that is $xy = yx$, or $3 \times 6 = 6 \times 3$. In group theory, the commutative law does not in general hold. Thus AB may give C while BA may give D, where C and D are two more elements in the group. There are some groups, however, in which combination is commutative, and such groups are called *Abelian* groups. Because of the fact that multiplication is not in general commutative, it is sometimes convenient to have a means of stating whether an element B is to be multiplied by A in the sense AB or BA. In the first case we can say that B is *left-multiplied* by A, and in the second case that B is *right-multiplied* by A.

2. *One element in the group must commute with all others and leave them*

unchanged. It is customary to designate this element with the letter E, and it is usually called the *identity element.* Symbolically we define it by writing $EX = XE = X$.

3. *The associative law of multiplication must hold.* This is expressed in the following equality:

$$A(BC) = (AB)C$$

In plain words, we may combine B with C in the order BC and then combine this product, S, with A in the order AS, or we may combine A with B in the order AB, obtaining a product, say R, which we then combine with C in the order RC and get the same final product either way. In general, of course, the associative property must hold for the continued product of any number of elements, viz.,

$$(AB)(CD)(EF)(GH) = A(BC)(DE)(FG)H = (AB)C(DE)(FG)H \cdots$$

4. *Every element must have a reciprocal, which is also an element of the group.* The element R is the reciprocal of the element S if $RS = SR = E$, where E is the identity. Obviously, if R is the reciprocal of S, then S is the reciprocal of R. Also, E is its own reciprocal.

At this point we shall prove a small theorem concerning reciprocals which will be of use later. The rule is:

The reciprocal of a product of two or more elements is equal to the product of the reciprocals, in reverse order. This means that

$$(ABC \cdots XY)^{-1} = Y^{-1}X^{-1} \cdots C^{-1}B^{-1}A^{-1}$$

PROOF. For simplicity we shall prove this for a ternary product, but it will be obvious that it is true generally. If A, B, and C are group elements, their product, say D, must also be a group element, viz.,

$$ABC = D$$

If now we right-multiply each side of this equation by $C^{-1}B^{-1}A^{-1}$, we obtain

$$ABCC^{-1}B^{-1}A^{-1} = DC^{-1}B^{-1}A^{-1}$$
$$ABEB^{-1}A^{-1} = DC^{-1}B^{-1}A^{-1}$$

$$\cdot$$
$$\cdot$$
$$\cdot$$

$$E \qquad = DC^{-1}B^{-1}A^{-1}$$

Since D times $C^{-1}B^{-1}A^{-1}$ equals E, $C^{-1}B^{-1}A^{-1}$ is the reciprocal of D, and since D equals ABC, we have

$$D^{-1} = (ABC)^{-1} = C^{-1}B^{-1}A^{-1}$$

which proves the above rule.

2.2   Some Examples of Groups

Groups may be either finite or infinite, that is, they may contain a limited or an unlimited number of elements. The symmetry groups with which we shall be concerned are mostly finite, but two, namely, those to which linear molecules may belong, are infinite. The number of elements in a finite group is called its *order*, and the conventional symbol for the order is h. To illustrate the above defining rules, we may consider an infinite group and then some finite groups.

As an infinite group we may take all of the integers, both positive and negative, and zero. If we take as our law of combination the ordinary algebraic process of addition, then rule 1 is satisfied. Clearly, any integer may be obtained by adding two others. Note that we have an Abelian group since the order of addition is immaterial. The identity of the group is 0, since $0 + n = n + 0 = n$. Also, the associative law of combination holds, since, for example, $[(+3) + (-7)] + (+1043) = (+3) + [(-7) + (+1043)]$. The reciprocal of any element, n, is $(-n)$, since $(+n) + (-n) = 0$.

Group Multiplication Tables

If we have a complete and nonredundant list of the h elements of a finite group and we know what all of the possible products (there are h^2) are, then the group is completely and uniquely defined—at least in an abstract sense. The foregoing information can be presented most conveniently in the form of the group multiplication table. This consists of h rows and h columns. Each column is labeled with a group element, and so is each row. The entry in the table under a given column and along a given row is the product of the elements which head that column and that row. Because multiplication is in general not commutative, we must have an agreed upon and consistent rule for the order of multiplication. Arbitrarily, we shall take the factors in the order (column element) × (row element). Thus at the intersection of the column labeled by X and the row labeled by Y we find the element which is the product XY.

We now prove an important theorem about group multiplication tables, called the *rearrangement theorem*.

Each row and each column in the group multiplication table lists each of the group elements once and only once. From this, it follows that no two rows may be identical nor may any two columns be identical. Thus each row and each column is a rearranged list of the group elements.

PROOF. Let the group consist of the h elements $E, A_2, A_3, \ldots, A_h$. The elements in a given row, say the nth row, are

$$EA_n, A_2 A_n, \ldots, A_n A_n, \ldots, A_h A_n$$

Since no two group elements, A_i and A_j for instance, are the same, no two products, $A_i A_n$ and $A_j A_n$, can be the same. The h entries in the nth row are all different. Since there are only h group elements, each of them must be present once and only once. The argument can obviously be adapted to the columns.

Let us now systematically examine the possible abstract groups of low order, using their multiplication tables to define them. There is, of course, formally a group of order 1, which consists of the identity element alone. There is only one possible group of order 2. It has the following multiplication table and will be designated G_2.

G_2	E	A
E	E	A
A	A	E

For a group of order 3, the multiplication table will have to be, in part, as follows:

	E	A	B
E	E	A	B
A	A		
B	B		

There is then only one way to complete the table. Either $AA = B$ or $AA = E$. If $AA = E$, then $BB = E$ and we would augment the table to give:

	E	A	B
E	E	A	B
A	A	E	
B	B		E

But then we can get no further, since we would have to accept $BA = A$ and $AB = A$ in order to complete the last column and the last row, respectively, thus repeating A in both the second column and the second row The alternative, $AA = B$, leads unambiguously to the following table:

G_3	E	A	B
E	E	A	B
A	A	B	E
B	B	E	A

Cyclic Groups

G_3 is the simplest, nontrivial member of an important set of groups, the *cyclic* groups. We note that $AA = B$, while $AB\ (=AAA) = E$. Thus we can consider the entire group to be generated by taking the element A and its powers, $A^2(=B)$ and $A^3(=E)$. In general, the cyclic group of order h is defined as an element X and all of its h powers up to $X^h = E$. We shall presently examine several other cyclic groups. An important property of cyclic groups is that they are *Abelian*, that is, all multiplications are commutative. This must be so, since the various group elements are all of the form X^n, X^m, etc., and, clearly, $X^n X^m = X^m X^n$ for all m and n.

To continue, we ask how many groups of order 4 there may be and what their multiplication table(s) will be. Obviously there will be a cyclic group of order 4. Let us employ the relations

$$X = A \qquad X^3 = C$$
$$X^2 = B \qquad X^4 = E$$

It then follows that the multiplication table, in the usual format, is as follows:

$G_4{}^{(1)}$	E	A	B	C
E	E	A	B	C
A	A	B	C	E
B	B	C	E	A
C	C	E	A	B

That there is a second type of G_4 group, $G_4{}^{(2)}$, is fairly obvious. We note that for $G_4{}^{(1)}$ only one element, namely B, is its own inverse. Suppose, instead, we assume that each of two elements, A and B, is its own inverse. We shall then have no choice but to also make C its own inverse, since each of the four E's in the table must lie in a different row and column. Thus, we would obtain

	E	A	B	C
E	E	A	B	C
A	A	E		
B	B		E	
C	C			E

A moment's consideration will show that there is only one way to complete this table:

$G_4^{(2)}$	E	A	B	C
E	E	A	B	C
A	A	E	C	B
B	B	C	E	A
C	C	B	A	E

It is also clear that there are no other possibilities.* Thus, there are two groups of order 4, namely $G_4^{(1)}$, and $G_4^{(2)}$, which may be considered to be defined by their multiplication tables.

It is left as an exercise to show that there is only one group of order 5. Similarly, a systematic examination of the possibilities for groups of order 6 is also left as an exercise. In order to have illustrative material for several topics which we shall take up next, the multiplication table for one of the groups of order 6 is given.

$G_6^{(1)}$	E	A	B	C	D	F
E	E	A	B	C	D	F
A	A	E	D	F	B	C
B	B	F	E	D	C	A
C	C	D	F	E	A	B
D	D	C	A	B	F	E
F	F	B	C	A	E	D

* If we make up a table in which only one element (other than E) is its own inverse and let that element be A or C instead of B as in the $G_4^{(1)}$ table given, we are *not* inventing a different G_4. We are only permuting the arbitrary symbols for the group elements.

2.3   Subgroups

Inspection of the multiplication table for the group $G_6^{(1)}$ will show that within this group of order 6 there are smaller groups. The identity E in itself is a group of order 1. This will, of course, be true in any group and is trivial. Of a nontrivial nature are the groups of order 2, viz., E, A; E, B; E, C; and the group of order 3, viz., E, D, F. The last should be recognized also as the cyclic group G_3, since $D^2 = F$, $D^3 = DF = FD = E$. But to return to the main point, these smaller groups which may be found within a larger group are called subgroups. There are, of course, groups which have no subgroups other than the trivial one of E itself.

Let us now consider whether there are any restrictions on the nature of subgroups, restrictions which are logical consequences of the general definition of a group and not of any additional or special characteristics of a particular group. We may note that the orders of the group $C_6^{(1)}$ and its subgroups are 6 and 1, 2, 3; in short, the orders of the subgroups are all factors of the order of the main group. We shall now prove the following theorem:

The order of any subgroup, g, of a group of order h must be a divisor of h. In other words, $h/g = k$ with k some integer.

PROOF. Suppose that the set of g elements, $A_1, A_2, A_3, \ldots, A_g$, forms a subgroup. Now let us take another element B in the group which is not a member of this subgroup and form all of the g products: $BA_1, BA_2, \ldots, BA_g$. No one of these products can be in the subgroup. If, for example,

$$BA_2 = A_4$$

then, if we take the reciprocal of A_2, perhaps A_5, and right-multiply the above equality, we obtain

$$BA_2 A_5 = A_4 A_5$$
$$BE = A_4 A_5$$
$$B = A_4 A_5$$

But this contradicts our assumption that B is not a member of the subgroup $A_1, A_2, \ldots, A_g$, since $A_4 A_5$ can only be one of the A_i. Hence, if all the products BA_i are in the large group in addition to the A_i themselves, there are at least $2g$ members of the group. If $h > 2g$, we can choose still another element of the group, namely C, which is not one of the A_i or one of the BA_i, and on multiplying the A_i by C we will obtain g more elements, all members of the main group, but none members of the A_i or of the BA_i sets.

Thus we now know that h must be at least equal to $3g$. Eventually, however, we must reach the point where there are no more elements by which we can multiply the A_i that are not among the sets A_i, BA_i, CA_i, and so forth, already obtained. Suppose after having found k such elements, we reach the point where there are no more. Then $h = kg$, where k is, of course, an integer. Then $h/g = k$, which is what we set out to prove.

Although we have shown that the order of any subgroup, g, must be a divisor of h, we have not proved the converse, namely, that there are subgroups of all orders which are divisors of h, and, indeed, this is not in general true. Moreover, as our illustrative group proves, there can be more than one subgroup of a given order.

2.4   Classes

We have seen that in a given group it may be possible to select various smaller sets of elements, each such set including E, however, which are in themselves groups. There is another way in which the elements of a group may be separated into smaller sets, and such sets are called *classes*. Before defining a class we must consider an operation known as *similarity transformation*.

If A and X are two elements of a group, then $X^{-1}AX$ will be equal to some element of the group, say B. We have

$$B = X^{-1}AX$$

We express this relation in words by saying that B is the *similarity transform* of A by X. We also say that A and B are *conjugate*. The following properties of conjugate elements are important.

(i) *Every element is conjugate with itself.* This means that if we choose any particular element, A, it must be possible to find at least one element, X, such that

$$A = X^{-1}AX$$

If we left-multiply by A^{-1} we obtain

$$A^{-1}A = E = A^{-1}X^{-1}AX = (XA)^{-1}(AX)$$

which can hold only if A and X commute. Thus the element X may always be E, and it may be any other element which commutes with the chosen element, A.

(ii) If A is conjugate with B, *then B is conjugate with A.* This means that if

$$A = X^{-1}BX$$

then there must be some element, Y, in the group such that

$$B = Y^{-1}AY$$

That this must be so is easily proved by carrying out appropriate multiplications, viz.,

$$XAX^{-1} = XX^{-1}BXX^{-1} = B$$

Thus, if $Y = X^{-1}$ (and thus also $Y^{-1} = X$), we have

$$B = Y^{-1}AY$$

and this must be possible, since any element, say X, must have an inverse, say Y.

(iii) *If A is conjugate with B and C, then B and C are conjugate with each other.* The proof of this should be easy to work out from the foregoing discussion and is left an as exercise.

We may now define a class of elements.

A complete set of elements which are conjugate to one another is called a class of the group. In order to determine the classes within any particular group we can begin with one element and work out all of its transforms, using all the elements in the group, including itself, then take a second element, which is not one of those found to be conjugate to the first, and determine all its transforms, and so on until all elements in the group have been placed in one class or another.

Let us illustrate this procedure with the group $G_6{}^{(1)}$. All of the results given below may be verified by using the multiplication table. Let us start with E.

$$E^{-1}EE = EEE = E$$
$$A^{-1}EA = A^{-1}AE = E$$
$$B^{-1}EB = B^{-1}BE = E$$
$$\cdots$$

Thus E must constitute by itself a class, of order 1, since it is not conjugate with any other element. This will, of course, be true in any group. To continue,

$$E^{-1}AE = A$$
$$A^{-1}AA = A$$
$$B^{-1}AB = C$$
$$C^{-1}AC = B$$
$$D^{-1}AD = B$$
$$F^{-1}AF = C$$

Thus the elements A, B, and C are all conjugate and are therefore members of the same class. It is left for the reader to show that all of the transforms of B and C are either A, B, or C. Thus A, B, and C are in fact the only members of the class.

Continuing, we have

$$E^{-1}DE = D$$
$$A^{-1}DA = F$$
$$B^{-1}DB = F$$
$$C^{-1}DC = F$$
$$D^{-1}DD = D$$
$$F^{-1}DF = D$$

It will also be found that every transform of F is either D or F. Hence, D and F constitute a class of order 2.

It will be noted that the classes have orders 1, 2, and 3, which are all factors of the group order, 6. It can be proved, by a method similar to that used in connection with the orders of subgroups, that the following theorem is true:

The orders of all classes must be integral factors of the order of the group.

We shall see later that in a symmetry group the classes have useful geometrical significance.

Exercises

2.1 Show that there is only one group of order 5, and give its multiplication table.

2.2 Generalize the result of Exercise 2.1 to prove that all groups of *prime* order are uniquely defined as cyclic, Abelian groups.

2.3 Derive the multiplication tables for the other group(s) of order 6.

2.4 Identify the subgroups in the groups $G_4{}^{(1)}$, $G_4{}^{(2)}$, and the cyclic group of order 6, the multiplication table for which was developed in Exercise 2.3.

2.5 Arrange the elements of the groups G_3, $G_4{}^{(1)}$, $G_4{}^{(2)}$, G_5, and the cyclic group of order 6 into classes.

2.6 Prove the theorem that in any Abelian group each element is in a class by itself.

2.7 Invent a noncyclic group of order 8.

2.8 Work out all the subgroups and classes for the group invented in Exercise 2.7.

2.9 Show that for any cyclic group, X, X^2, X^3,..., $X^h = E$, there will be one subgroup corresponding to each integral divisor of the order h.

2.10 Suppose that an element C is added to those of the group G_3. If C commutes with A and B, what new group is generated?

2.11 Prove or disprove the following theorem: *It is impossible in a group* E, A, B, C, ... , *to have* $A^2 = B^2 \neq E$.

Molecular Symmetry and the Symmetry Groups

3.1 General Remarks

It is perhaps appropriate to begin this chapter by sketching what we intend to do here. It is certainly intuitively obvious what we mean when we say that some molecules are more symmetrical than others, or that some molecules have high symmetry whereas others have low symmetry or no symmetry. But in order to make the idea of molecular symmetry as useful as possible, we must develop some rigid mathematical criteria of symmetry. To do this we shall first consider the kinds of *symmetry elements* that a molecule may have and the *symmetry operations* generated by the symmetry elements. We shall then show that a complete but nonredundant set of symmetry *operations* (not elements) constitutes a mathematical group. Finally, we shall use the general properties of groups, developed in Chapter 2, to aid in correctly and systematically determining the symmetry operations of any molecule we may care to consider. We shall also describe here the system of notation normally used by chemists for the various symmetry groups. An alternative system used primarily in crystallography is summarized in Appendix I.

It may also be worthwhile to offer the following advice to the student of this chapter. The use of three-dimensional models is extremely helpful in learning to recognize and visualize symmetry elements. Indeed, it is most unlikely that any but a person of the most exceptional gifts in this direction can fail to profit significantly from the examination of models. At the same time, it may also be said that anyone with the intelligence to master other aspects of modern chemical knowledge should, by the use of models, surely succeed in acquiring a good working knowledge of molecular symmetry.

3.2 Symmetry Elements and Operations

The two things, symmetry elements and symmetry operations, are inextricably related and therefore are easily confused by the beginner. They are, however, different *kinds* of things, and it is important to grasp and retain, from the outset, a clear understanding of the difference between them.

Definition of a Symmetry Operation

A symmetry operation is a movement of a body such that, after the movement has been carried out, every point of the body is coincident with an equivalent point (or perhaps the same point) of the body in its original orientation. In other words, if we note the position and orientation of a body before and after a movement is carried out, that movement is a symmetry operation if these two positions and orientations are indistinguishable. This would mean that, if we were to look at the body, turn away long enough for someone to carry out a symmetry operation, and then look again, we would be completely unable to tell whether or not the operation had actually been performed, because in either case the position and orientation would be indistinguishable from the original. One final way in which we can define a symmetry operation is to say that its effect is to take the body into an *equivalent configuration*—that is, one which is indistinguishable from the original, though not necessarily identical with it.

Definition of a Symmetry Element

A symmetry element is a geometrical entity such as a line, a plane, or a point, with respect to which one or more symmetry operations may be carried out.

Symmetry elements and symmetry operations are so closely interrelated because the operation can be defined only with respect to the element, and

Table 3.1 The Four Kinds of Symmetry Elements and Operations Required in Specifying Molecular Symmetry

SYMMETRY ELEMENT	SYMMETRY OPERATION(S)
1. Plane	Reflection in the plane
2. Center of symmetry or center of inversion	Inversion of all atoms through the center
3. Proper axis	One or more rotations about the axis
4. Improper axis	One or more repetitions of the sequence: rotation followed by reflection in a plane $\perp$ to the rotation axis

at the same time the existence of a symmetry element can be demonstrated only by showing that the appropriate symmetry operations exist. Thus, since the existence of the element is contingent on the existence of the operation(s) and vice versa, we shall discuss related types of elements and operations together.

In treating molecular symmetry, only four types of symmetry elements and operations need be considered. These, in the order in which they will be discussed, are listed in Table 3.1.

3.3 Symmetry Planes and Reflections

A symmetry plane must pass through a body, that is, the plane cannot be completely outside of the body. The conditions which must be fulfilled in order that a given plane be a symmetry plane can be stated as follows. Let us apply a Cartesian coordinate system to the molecule in such a way that the plane includes two of the axes (say x and y) and is therefore perpendicular to the third (i.e., z). The position of every atom in the molecule may also be specified in this same coordinate system. Suppose now, for each and every atom, we leave the x and y coordinates fixed and change the sign of the z coordinate: thus the ith atom, originally at (x_i, y_i, z_i), is moved to the point $(x_i, y_i, -z_i)$. Another way of expressing the above operation is to say, " Let us drop a perpendicular from each atom to the plane, extend that line an equal distance on the opposite side of the plane, and move the atom to this other end of the line." If, when such an operation is carried out on every atom in a molecule, an equivalent configuration is obtained, the plane used is a symmetry plane.

Clearly, atoms lying in the plane constitute special cases, since the operation of reflecting through the plane does not move them at all. Consequently, any planar molecule is bound to have at least one plane of symmetry, namely, its molecular plane. Another significant and immediate consequence of the definition is a restriction on the numbers of various kinds of atoms in a molecule having a plane of symmetry. All atoms of a given species which do not lie in the plane must occur in even numbers, since each one must have a twin on the other side of the plane. Of course, any number of atoms of a given species may be in the plane. Furthermore, if there is only one atom of a given species in a molecule, it must be in each and every symmetry plane that the molecule may have. This means that it must be on the line of intersection between two or more planes or at the point of intersection of three or more planes (if there is such a point), since this atom must lie in all of the symmetry planes simultaneously.

The standard symbol for a plane of symmetry is σ. The same symbol is also used for the operation of reflecting through the plane.

It should be explicitly noted that the existence of *one* symmetry plane gives rise to, requires, or, as usually stated, *generates one* symmetry operation. We may also note here, for future use, that the effect of applying the same reflection operation twice is to bring all atoms back to their original positions. Thus, while the operation σ produces a configuration *equivalent* to the original, the application of the same σ twice produces a configuration *identical* with the original. Now we can conveniently denote the successive application of the operation σ n times by writing σ^n. We can then also write, $\sigma^2 = E$, where we use the symbol E to represent *any* combination of operations which takes the molecule to a configuration identical with the original one. We call E, or any combination of operations equal to E, the *identity operation*. It should be obvious that $\sigma^n = E$ when n is even and $\sigma^n = \sigma$ when n is odd.

Let us now consider some illustrative examples of symmetry planes in molecules. At one extreme are molecules which have no symmetry planes at all. One such general class consists of those which are not planar and which have odd numbers of all atoms. An example is FClSO, seen at the right.

At the other extreme are molecules possessing an infinite number of symmetry planes, that is, linear molecules. For these any plane containing the molecular axis is a symmetry plane, and there are obviously an infinite number of these planes. Most small molecules fall between these extremes; that is, they have one or a few symmetry planes. If, instead of FClSO, we take F_2SO or Cl_2SO, we have a molecule with one symmetry plane, which passes through S and O and is perpendicular to the Cl, Cl, O or the F, F, O plane. The H_2O molecule has two symmetry planes. One is, of course, coextensive with the molecular plane. The other includes the oxygen atom (it must, since there is only one such atom) and is perpendicular to the molecular plane. The effect of reflection through this second plane is to leave the oxygen atom fixed but to exchange the hydrogen atoms, while reflection through the first plane leaves all atoms unshifted. A tetrahedral molecule of the type AB_2C_2 (e.g., CH_2Cl_2) also has two mutually perpendicular planes of symmetry. One contains AB_2, and reflection through it leaves these three atoms unshifted while interchanging the C atoms; the other contains AC_2, and reflection through it interchanges only the B atoms.

The molecules NH_3 and $CHCl_3$ are representative of a type containing three planes of symmetry. For NH_3, any plane of symmetry would have to include the nitrogen atom and either one or all three of the hydrogen atoms. Since NH_3 is not planar, there can be no symmetry plane including N and all three H's; hence we look for planes including N and one H and bisecting the line between the remaining two H's. There are clearly three such planes. For $CHCl_3$, the situation is quite analogous except that the hydrogen atom must also lie in the symmetry planes.

The NH_3 molecule is only one example of the general class of pyramidal AB_3 molecules. Let us see what happens as we begin flattening such a molecule by pushing the A atom down toward the plane of the three B atoms. It should be easily seen that this does not disturb the three symmetry planes, even in the limit of coplanarity. Nor does it introduce any new planes of symmetry *except* in the limit of coplanarity. Once AB_3 becomes planar there is then a fourth symmetry plane, which is the molecular plane. Molecules and ions of the planar AB_3 type possessing four symmetry planes, three perpendicular to the molecular plane, are fairly numerous and important. There are, for example, the boron halides $CO_3{}^{2-}$, $NO_3{}^-$, and SO_3.

A planar species of the type $[PtCl_4]^{2-}$ or $[AuCl_4]^-$ possesses five symmetry planes. One is the molecular plane. There are also two, perpendicular to the molecular plane and perpendicular to each other, which pass through three atoms. Finally, there are two more, also perpendicular to the molecular plane and perpendicular to each other, which bisect Cl—Pt—Cl or Cl—Au—Cl angles.

A regular tetrahedral molecule possesses six planes of symmetry. Using the numbering system illustrated, we may specify them by stating the atoms they contain. Symmetry planes contain the atoms:

$$AB_1B_2, \; AB_1B_3, \; AB_1B_4,$$

$$AB_2B_3, \; AB_2B_4, \; AB_3B_4.$$

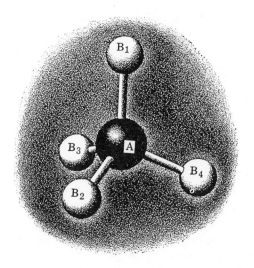

A regular octahedron possesses, in all, nine symmetry planes. Reference will be made to the accompanying numbered figure in specifying these. There are first three of the same type, viz., those including the following sets of atoms: $AB_1B_2B_3B_4$, $AB_2B_4B_5B_6$, and $AB_1B_3B_5B_6$. There are then six more of a second type, one of which includes AB_5B_6 and bisects the $B_1 — B_2$ and $B_3 — B_4$ lines, a second which includes AB_1B_3 and bisects the $B_2 — B_5$ and $B_4 — B_6$ lines, and so forth.

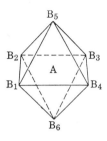

3.4   The Inversion Center

If a molecule can be brought into an equivalent configuration by changing the coordinates (x, y, z) of every atom, where the origin of coordinates lies at a point within the molecule, into $(-x, -y, -z)$, then the point at which the origin lies is said to be a center of symmetry or center of inversion. The symbol for the inversion center and for the operation of inversion is an italic i. Like a plane, the center is an element which generates only one operation.

It may be noted that, when a center of inversion exists, restrictions are placed on the numbers of all atoms, or all but one atom, in the molecule. Since the center is a point, only one atom may be at the center. If there is an atom at the center, that atom is unique, since it is the only one in the molecule which is not shifted when the inversion is performed. All other atoms must occur in pairs, since each must have a twin with which it is exchanged when the inversion is performed. From this it follows that we need not bother to look for a center of symmetry in molecules which contain an odd number of more than one species of atom.

The effect of carrying out the inversion operation n times may be expressed as i^n. It should be easily seen that $i^n = E$ when n is even, and $i^n = i$ when n is odd.

Some examples of molecules having inversion centers are octahedral AB_6, planar AB_4, planar and *trans* AB_2C_2, linear ABA, ethylene, and benzene. Two examples of otherwise fairly symmetrical molecules which do not have centers of inversion are $C_5H_5^-$ (plane pentagon) and tetrahedral AB_4 (even though A is at the "center" and B's come in even numbers).

3.5   Proper Axes and Proper Rotations

Before discussing proper axes and rotations in a general way, let us take a specific case. A line drawn perpendicular to the plane of an equilateral triangle and intersecting it at its geometric center is a proper axis of rotation for that triangle. Upon rotating the triangle by 120° ($2\pi/3$) about this axis, the triangle is brought into an equivalent configuration. It may be noted that a rotation by 240° ($2 \times 2\pi/3$) also produces an equivalent configuration.

The general symbol for a proper axis of rotation is C_n, where the subscript n denotes the *order* of the axis. By order is meant the largest value of n such that rotation through $2\pi/n$ gives an equivalent configuration. In the above example, the axis is a C_3 axis. Another way of defining the meaning of the order n of an axis is to say that it is the number of times that the smallest rotation capable of giving an equivalent configuration must be repeated in order to give a configuration not merely equivalent to the original but also identical to it. The meaning of "identical" can be amplified if we attach numbers to each apex of the triangle in our example. Then the effects of rotating by $2\pi/3$, $2 \times 2\pi/3$, and $3 \times 2\pi/3$ are seen to be:

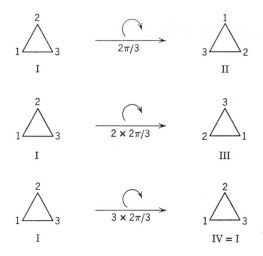

Configurations II and III are equivalent to I because without the labels (which are not real, but represent only our mental constructions) they are indistinguishable from I, although with the labels they are distinguishable. However, IV is indistinguishable from I not only without labels but also with them. Hence, it is not merely equivalent; it is *identical*.

The C_3 axis is also called a threefold axis. Moreover, we use the symbol C_3 to represent the *operation* of rotation by $2\pi/3$ around the C_3 axis. For the rotation by $2 \times 2\pi/3$ we use the symbol $C_3{}^2$, and for the rotation by $3 \times 2\pi/3$ the symbol $C_3{}^3$. Symbolically we can write $C_3{}^4 = C_3$, and hence only C_3,

$C_3{}^2$, and $C_3{}^3$ are separate and distinct operations. However, $C_3{}^3$ produces an identical configuration, and hence we may write $C_3{}^3 = E$.

After consideration of the above example, it is easy to accept some more general statements about proper axes and proper rotations. In general, an n-fold axis is denoted by C_n and a rotation by $2\pi/n$ is also represented by the symbol C_n. Rotation by $2\pi/n$ carried out successively m times is represented by the symbol $C_n{}^m$. Also, in any case, $C_n{}^n = E$, $C_n{}^{n+1} = C_n$, $C_n{}^{n+2} = C_n{}^2$, and so on.

In discussing planes of symmetry and inversion centers, attention was directed to the fact that only *one* operation, reflection, is generated by a symmetry plane, and only *one* operation, inversion, by an inversion center. A proper axis of order n, however, generates n operations, viz., $C_n, C_n{}^2, C_n{}^3, \ldots,$ $C_n{}^{n-1}, C_n{}^n (= E)$.

One last general consequence of the existence of a C_n axis concerns the requirement that there be certain numbers of each species of atom in a molecule containing the axis. Naturally, any atom which lies on a proper axis of symmetry is unshifted by any rotation about that axis. Thus there may be any number, even or odd, of each species of atom lying on an axis (unless other symmetry elements impose restrictions). However, if one atom of a certain species lies off a C_n axis, there must automatically be $n - 1$ more, or a total of n such atoms, since on applying C_n successively n times, the first atom is moved to a total of n different points. Had there not been identical atoms at all the other $n - 1$ points to begin with, the new configurations would not be equivalent configurations; this would mean that the axis would not be a C_n symmetry axis, contrary to the original assumption.

The symbol $C_n{}^m$ represents a rotation by $m \times 2\pi/n$. Let us consider the operation $C_4{}^2$, which is one of those generated by a C_4 axis. This is a rotation by $2 \times 2\pi/4 = 2\pi/2$, and can therefore be written just as well as C_2. Similarly, among the operations generated by a C_6 axis, we find $C_6{}^2$, $C_6{}^3$, and $C_6{}^4$, which may be written, respectively, as C_3, C_2, and $C_3{}^2$. It is frequent though not invariable practice to write an operation $C_n{}^m$ in what, by considering the fraction (m/n) in $(m/n)2\pi$, can be called lowest terms, and the reader should be familiar with this practice so that he immediately recognizes, for example, that the sequence $C_6, C_3, C_2, C_3{}^2, C_6{}^5, E$ is identical in meaning with $C_6, C_6{}^2, C_6{}^3, C_6{}^4, C_6{}^5, C_6{}^6$.

Let us now consider some further illustrative examples chosen from commonly encountered types of molecules. Again we may begin by considering extremes. Many molecules possess no axes of proper rotation; FClSO, for example, does not. (Actually, FClSO is a gratuitous example, since, as we saw earlier, it possesses no symmetry elements whatsoever.) Nor does Cl_2SO or F_2SO possess an axis of proper rotation. At the other extreme are linear molecules which possess ∞-fold axes of proper rotation, colinear with the

molecular axes. Since all atoms in a linear molecule lie on this axis, rotation by any angle whatever, and hence by all (∞ number) angles, leaves a
configuration indistinguishable from the original. Again, as with planes of
symmetry, most small molecules possess one axis or a few axes, generally of
low orders.

Among examples of molecules with a single axis of order 2 are H_2O and
CH_2Cl_2. No molecules possess just two twofold axes; this will be shown
later to be mathematically impossible. There are many examples of molecules
possessing three twofold axes, for example, ethylene, C_2H_4. One C_2 is colinear with the C—C axis. A second is perpendicular to the plane of the
molecule and bisects the C—C line. The third is perpendicular to the first
two and intersects both at the midpoint of the C—C line. A regular tetrahedral molecule also possesses three twofold axes, as shown in the diagram
below.

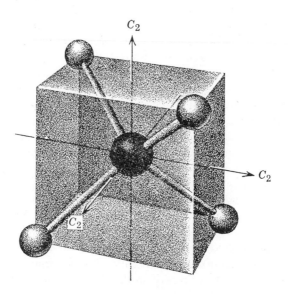

Threefold axes are quite common. Both pyramidal and planar AB_3 molecules possess threefold proper axes passing through the atom A and perpendicular to the plane of the three B atoms. A tetrahedral molecule, AB_4,
possesses four threefold axes, each passing through the atom A and one of
the B atoms. An octahedral molecule, AB_6, also possesses four threefold
axes, each passing through the centers of two opposite triangular faces and
the A atom.

The planar AB_3 molecule possesses three twofold axes perpendicular to the threefold axis as shown in the diagram. The existence of the C_3 axis and one C_2 axis perpendicular to the C_3 axis means that the other two C_2 axes, at angles of $2\pi/3$ and $4\pi/3$ to the first, *must* exist. For, on carrying out the rotation C_3, we generate the second C_2 axis from the first, and on carrying out the rotation $C_3{}^2$, we generate the third C_2 axis from the first.

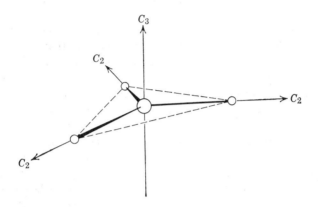

The effect of the operations C_n, $C_n{}^2$, ..., $C_n{}^{n-1}$ in replicating other symmetry elements may profitably be discussed more fully at this point. The other symmetry elements of interest are planes and axes. It will also be sufficient to limit the discussion to axes perpendicular to the axis of the replicating rotations and to planes which contain the axis of the replicating rotations. A plane perpendicular to the axis of the replicating rotations is obviously not replicated, since all rotations carry it into itself. Although a completely general discussion might be given, it seems more instructive to consider separately each of the replicating axes which can be encountered in practice (C_n, $1 < n \leqslant 8$).

An axis perpendicular to a C_2 axis or a plane containing a C_2 goes into itself on carrying out the operation C_2; hence no further axes or planes *of the same type* are required to exist in this case. We have just seen that from one axis perpendicular to a C_3 axis two similar ones are generated. The same is true for a plane of symmetry containing a C_3 axis. We may also deal with the C_5 and C_7 cases (and, indeed, any C_n where n is odd) for they all behave in the same way. One axis perpendicular to a C_5 or C_7 axis or one plane containing a C_5 or C_7 will be made to generate four or six more separate and distinct axes or planes by the operations that the C_5 or C_7 axis makes possible.

For cases where n in C_n is even, the results are less straightforward. Suppose that we have one axis $C_2(1)$ perpendicular to a C_4 axis. On carrying out the rotation C_4, $C_2(1)$ is rotated by $2\pi/4$ and a second C_2 axis, $C_2(2)$, is thus produced. On carrying out the rotation $C_4^2(=C_2)$ about the C_4 axis, however, $C_2(1)$ merely goes into itself, and $C_2(2)$ also goes into itself. The operation C_4^3 takes $C_2(1)$ into $C_2(2)$ and $C_2(2)$ into $C_2(1)$. Hence, because C_4^2 is really only C_2 and C_4^3 is only C_4 followed by C_2, the C_4 axis requires only that the axis $C_2(1)$ be accompanied by one other such axis and not three others.

A completely analogous argument holds regarding planes. In the C_6 case, using the same line of argument, it will easily be seen that, if one axis perpendicular to a C_6 or one plane containing a C_6 exists, it must be accompanied by two more of the same type. Similarly, a C_8 axis will replicate a C_2 axis perpendicular to it so as to produce a set of four such C_2 axes.

Continuing with examples of proper axes in typical molecules, we may cite the planar $PtCl_4^{2-}$ ion, which has a C_4 axis perpendicular to the plane of the ion and four C_2 axes in the plane of the ion. The cyclopentadienyl anion, $C_5H_5^-$, possesses a C_5 axis perpendicular to the molecular plane and five C_2 axes in the molecular plane. Benzene possesses a C_6 axis and two sets of three C_2 axes. Probably the only known example of a molecule with a C_7 axis is the planar $[C_7H_7]^+$, the tropylium ion. An example of a molecule with a C_8 axis is $(C_8H_8)_2U$ (uranocene).

3.6 Improper Axes and Improper Rotations

An improper rotation may be thought of as taking place in two steps: first a proper rotation and then a reflection through a plane perpendicular to the rotation axis. The axis about which this occurs is called an axis of improper rotation or, more briefly, an *improper axis*, and is denoted by the symbol S_n, where again n indicates the order. The operation of improper rotation by $2\pi/n$ is also denoted by the symbol S_n. Obviously, if an axis C_n and a perpendicular plane exist independently, then S_n exists. More important, however, is that an S_n may exist when neither the C_n nor the perpendicular σ exist separately.

Perhaps this can best be emphasized by taking an example. Let us consider ethane in its staggered configuration. The C—C line defines a C_3 axis, but certainly not a C_6 axis. Also, there is no plane of symmetry perpendicular to the C_3 axis. Yet there is an S_6, as the diagram shows. Observe that II and III are equivalent to each other but that neither is equivalent to I; that is, neither σ nor C_6 is by itself a symmetry operation. But the combination of both, in either order, which we call S_6, is a symmetry operation since it produces IV, which is equivalent to I.

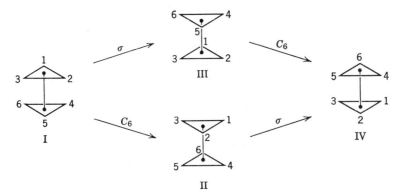

It will be shown later that the operations of rotation and reflection in a plane perpendicular to the rotation axis always give the same result regardless of the order in which they are performed. Thus the definition of improper rotation need not specify the order.

As another important example of the occurrence of improper axes and rotations, let us consider a regular tetrahedral molecule. We have already noted in Section 3.5 that the tetrahedron possesses three C_2 axes. Now each of these C_2 axes is simultaneously an S_4 axis, as can be seen in the following diagram:

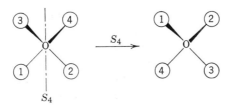

The element S_n in general generates a set of operations S_n, S_n^2, S_n^3, ... However, some important features of these operations should be noted. There are differences in the sets generated for even and odd n, so these two cases will be considered separately. Let us assume that our S_n axis is colinear with the z axis of a coordinate system and that the plane to which the reflection part of the operation S_n is referred is the xy plane.

An improper axis, S_n, of even order generates a set of operations S_n, S_n^2, S_n^3, ..., S_n^n. Let us first show that (for n even) $S_n^n = E$. S_n^n means that we carry out the operations C_n, σ, C_n, σ, ... until, in all, C_n and σ have each been carried out n times. Since n is an even number, n repetitions of σ is an identity operation, so that $S_n^n = C_n^n$; but C_n^n is also just E. Therefore, $S_n^n = E$, and $S_n^{n+1} = S_n$; $S_n^{n+2} = S_n^2$, and so on. Now by the same argument S_n^m will be equal to C_n^m whenever m is even. Thus, in any set of operations generated by an even-order S_n, certain of the S_n^m may be written in other ways. Consider, for example, the set S_6, S_6^2, S_6^3, S_6^4, S_6^5, S_6^6. S_6 can be written in no other way. $S_6^2 = C_6^2 = C_3$. $S_6^3 = S_2 = i$. $S_6^4 = C_3^2$. S_6^5 can be written in no other way. $S_6^6 = E$. Hence the complete set of operations generated by the element S_6 can, and normally would, be written: S_6, C_3, i, C_3^2, S_6^5, E. Having written the set in this way, however, we can readily make another useful observation. This set contains C_3, C_3^2, and E, which are just the operations generated by a C_3 axis. Hence the existence of the S_6 axis automatically requires that the C_3 axis exist. It should not be difficult to see that, in general, the existence of an S_n axis of even order always requires the existence of a $C_{n/2}$ axis.

Let us now turn to improper axes of odd order. Their most important property is that an odd-order S_n requires that C_n and a σ perpendicular to it must exist independently. This is easily proved. The element S_n generates operations S_n, S_n^2, S_n^3, S_n^4, ... Let us examine the operation S_n^n when n is odd. It must have the same effect as will application of C_n^n followed by $\sigma^n = \sigma$. But since $C_n^n = E$, we see that $S_n^n = \sigma$. In other words, the element S_n generates a symmetry operation σ. But if the symmetry operation σ exists, the plane to which it is referred must be a symmetry element in its own right. Now, the operation S_n requires us to reflect in the plane σ, thus carrying a configuration I into another configuration, II, and then to rotate by $2\pi/n$, thus carrying II into III. Because S_n is a symmetry operation, I and III must be equivalent configurations. However, when n is odd, σ is itself a symmetry operation, so that II is also equivalent to I. Then II is also equivalent to III, and we see that rotation by $2\pi/n$ has carried II into an equivalent configuration, III. Thus the operation C_n is also a symmetry operation in its own right.

To gain further familiarity with odd-order improper axes, let us consider how many distinct operations are generated by some such axis, say S_5. The sequence begins S_5, S_5^2, S_5^3, S_5^4, ... Using relations and conventions previously developed, we can write certain of these operations in alternative ways, as follows:

$$S_5 = C_5 \text{ then } \sigma \text{ (or } \sigma \text{ then } C_5)$$
$$S_5^2 = C_5^2$$
$$S_5^3 = C_5^3 \text{ then } \sigma$$
$$S_5^4 = C_5^4$$
$$S_5^5 = \sigma$$
$$S_5^6 = C_5$$
$$S_5^7 = C_5^2 \text{ then } \sigma$$

$$S_5^8 = C_5^3$$
$$S_5^9 = C_5^4 \text{ then } \sigma$$
$$S_5^{10} = E$$
$$S_5^{11} = C_5 \text{ then } \sigma$$

We see that for S_5 through S_5^{10} (in general, S_n through S_n^{2n}), the operations are all different ones, but commencing with S_n^{2n+1} repetition of the sequence begins. Of the ten operations, however, four plus E can be expressed as a single operation only by using symbols S_5^n, whereas the other five can be written either as C_5^n or as σ. Thus there are operations which, although they may be accomplished by using C_5^n and σ successively, cannot be represented as unit operations in any other way than S_5^n. We also see that in general the element S_n with n odd generates $2n$ operations.

3.7   Products of Symmetry Operations

In Sections 3.3–3.6 we have often discussed the question of how we can represent the net effect of applying one symmetry operation after another to a molecule, but only in a limited way. In this section we shall discuss this question with regard to a broader range of possibilities. First, we shall establish a conventional shorthand for stating that "operation X is carried out first and then operation Y, giving the same net effect as would the carrying out of the single operation Z." This we express symbolically as

$$YX = Z$$

Note that the order in which the operations are applied is the order in which they are written from *right to left*, that is, YX means X first and then Y. In general, the order makes a difference although there are cases where it does not. When the result of the sequence XY is the same as the result of the sequence YX, the two operations, X and Y, are said to *commute*. It is also normal to speak of an operation which produces the same result as does the successive application of two or more others as the *product* of the others.

One way in which we may approach the problem of finding a single operation which is the product of two others is to consider a general point with coordinates $[x_1, y_1, z_1]$. On applying a certain operation, this point will be shifted to a new position with coordinates $[x_2, y_2, z_2]$; if still another operation is applied, it will again be shifted so that its coordinates are now $[x_3, y_3, z_3]$. The net effect of applying the two operations successively is to shift the point from $[x_1, y_1, z_1]$ to $[x_3, y_3, z_3]$. We now look for a way of accomplishing this in one step. The operation which does so will be the product of the first two.

Let us illustrate this procedure by proving the statement made earlier that, if there are two twofold axes at right angles to one another, there must necessarily be a third at right angles to both. Suppose that the two given axes coincide with the x and y axes; we can designate them $C_2(x)$ and $C_2(y)$. On applying first $C_2(x)$ and then $C_2(y)$ to a general point, the following transformations of its coordinates take place:

$$[x_1, y_1, z_1] \xrightarrow{C_2(x)} [x_1, -y_1, -z_1] \xrightarrow{C_2(y)} [-x_1, -y_1, z_1]$$

That is, the value of x_3 is $-x_1$, the value of y_3 is $-y_1$, and the value of z_3 is z_1. If now we apply $C_2(z)$ to the general point, it is shifted to $[-x_1, -y_1, z_1]$. Thus we may write.

$$C_2(y)C_2(x) = C_2(z)$$

Thus, whenever $C_2(x)$ and $C_2(y)$ exist, $C_2(z)$ must also exist, because it is their product.

As a second example of how the existence of two symmetry elements may automatically require that a third one exist, we shall consider a case having a C_4 axis and one plane containing this axis. We have already seen that the operation C_4 will generate a second plane from the first one at right angles to it. It is also true, however, though less obvious, that when the C_4 axis and one of these planes exist there must then be a second plane also containing C_4 at an angle of 45° to the first one. We can prove this by the method just used. The effect of reflecting a general point $[x_1, y_1, z_1]$ through the xz plane is given by

$$\sigma(xz)[x_1, y_1, z_1] \rightarrow [x_1, -y_1, z_1]$$

whereas the effect of a clockwise C_4 rotation about the z axis upon the point is given by

$$C_4(z)[x_1, y_1, z_1] \rightarrow [y_1, -x_1, z_1]$$

From these relations we can determine the effect of applying successively $\sigma(xz)$ and then $C_4(z)$, namely,

$$C_4(z)\sigma(xz)[x_1, y_1, z_1] \rightarrow C_4(z)[x_1, -y_1, z_1] \rightarrow [-y_1, -x_1, z_1]$$

Now let us consider the effect of reflecting the point through a plane σ_d, which also contains the z axis and bisects the angles between the $+y$ and $-x$ axes and the $+x$ and $-y$ axes. This transformation is

$$\sigma_d[x_1, y_1, z_1] \rightarrow [-y_1, -x_1, z_1]$$

We see that

$$C_4(z)\sigma(xz) = \sigma_d$$

which means that the existence of $C_4(z)$ and $\sigma(xz)$ automatically requires that σ_d exist. The C_4 rotation then generates from σ_d another plane, σ_d', which passes through the first and third quadrants. The final result is that, if there is one plane containing a C_4 axis, there is automatically a set of four planes.

It may be shown in a very similar way that if $C_4(z)$ and $C_2(y)$ axes exist a C_2 axis lying in the first and third quadrants of the xy plane at $45°$ to $C_2(y)$ must also exist. This is left as an exercise.

Examination of the shifts in a general point may also be employed to show a commutative relation, for example, that $C_2(z)$ and $\sigma(xy)$ commute. Thus, we may write, in a notation which uses $\bar{x}$ instead of $-x$, $\bar{y}$ instead of $-y$, and $\bar{z}$ instead of $-z$:

$$C_2(z)[x, y, z] \rightarrow \quad [\bar{x}, \bar{y}, z]$$
$$\sigma(xy)[x, y, z] \rightarrow \quad [x, y, \bar{z}]$$
$$C_2\sigma[x, y, z] \rightarrow C_2[x, y, \bar{z}] \rightarrow [\bar{x}, \bar{y}, \bar{z}]$$

and

$$\sigma C_2[x, y, z] \rightarrow \quad \sigma[\bar{x}, \bar{y}, z] \rightarrow [\bar{x}, \bar{y}, \bar{z}]$$

We see also that the product in each case is equivalent to i.

In the above examples, where only C_2 and C_4 rotations and certain kinds of planes are concerned, the transformation of the coordinates $[x, y, z]$ to $[x, \bar{y}, \bar{z}]$ by a twofold rotation about the x axis, for example, is fairly obvious by inspection. It is also obvious that a fourfold rotation about the x axis will transform the coordinates into $[x, \bar{z}, y]$. It is also easy to see by inspection the effects of the inversion operation, an improper rotation by $2\pi/2$ or $2\pi/4$ and reflection in a plane which is the xy, xz, or yz plane or a plane rotated by $45°$ from these. However, the transformations effected by more general symmetry operations, such as rotation by $2\pi/n$ or $m2\pi/n$ and reflections in planes other than those just mentioned, are not easily handled by the simple methods and notation used above. Further discussion along this line will therefore employ some geometrical methods and also the more powerful methods of matrix algebra, which will be introduced in Chapter 4.

3.8   Equivalent Symmetry Elements and Equivalent Atoms

If a symmetry element A is carried into the element B by an operation generated by a third element X, then of course B can be carried back into A by the application of X^{-1}. The two elements A and B are said to be equivalent. If A can be carried into still a third element, C, then there will also be a way of carrying B into C, and the three elements, A, B, and C, form an equivalent set. In general, any set of symmetry elements chosen so that any member can be transformed into each and every other member of the set by application of some symmetry operation is said to be a set of equivalent symmetry elements.

For example, in a plane triangular molecule such as BF_3, each of the twofold symmetry axes lying in the plane can be carried into coincidence with each of the others by rotations of $2\pi/3$ or $2 \times 2\pi/3$, which are symmetry operations. Thus all three twofold axes are said to be equivalent to one another. In a square planar AB_4 molecule, there are four twofold axes in the molecular plane. Two of them, C_2 and C_2' lie along BAB axes, and the other two, C_2'' and C_2''', bisect BAB angles. Such a molecule also contains four symmetry planes, each of which is perpendicular to the molecular plane and intersects it along one of the twofold axes. Now it is easy to see that C_2 may be carried into C_2' and vice versa, and that C_2'' may be carried into C_2''' and vice versa, by rotations about the fourfold axis and by reflections in the symmetry planes mentioned, but there is no way to carry C_2 or C_2' into either C_2'' or C_2''' or vice versa. Thus C_2 and C_2' form one set of equivalent axes, and C_2'' and C_2''' form another. Similarly, two of the symmetry planes are equivalent to each other, but not to either of the other two, which are, however, equivalent to each other.

As other illustrations of equivalence and nonequivalence of symmetry elements, we may note that all three of the symmetry planes in BF_3 which are perpendicular to the molecular plane are equivalent, as are the three in NH_3, whereas the two planes in H_2O are not equivalent. The six twofold axes lying in the plane of the benzene molecule can be divided into two sets of equivalent axes, one set containing those which transect opposite carbon atoms and the other set containing those which bisect opposite edges of the hexagon.

Equivalent atoms in a molecule are those which may all be interchanged with one another by symmetry operations. Naturally, equivalent atoms must be of the same chemical species. Examples of equivalent atoms include all of the hydrogen atoms in methane, ethane, benzene, or cyclopropane, all of the fluorine atoms in SF_6, and all of the carbon and oxygen atoms in $Cr(CO)_6$. Examples of chemically identical atoms which are not equivalent in molecular environment are the apical and equatorial fluorine atoms in PF_5; no symmetry operation possible for this molecule ever interchanges these fluorine atoms. The α and β hydrogen and carbon atoms of naphthalene are not equivalent. All six carbon atoms of cyclohexane are equivalent in the chair configuration, but four are different from the other two in the boat configuration.

3.9   General Relations Among Symmetry Elements and Operations

We present here some very general and useful rules about how different kinds of symmetry elements and operations are related. These deal with the way in which the existence of some two symmetry elements necessitates the existence of others, and with commutation relationships. Some of the statements are presented without proof; the reader should profit by making the effort to verify them.

Products

1. The product of two proper rotations must be a proper rotation. Thus, although rotations can be created by combining reflections (see rule 2), the reverse is not possible. The special case $C_2(x)C_2(y) = C_2(z)$ has already been examined (page 28).

2. The product of two reflections, in planes A and B, intersecting at an angle ϕ_{AB}, is a rotation by $2\phi_{AB}$ about the axis defined by the line of intersection. The simplest proof of this is a geometric one, as indicated in Figure 3.1. It is clear that this rule has some far-reaching consequences. If the two

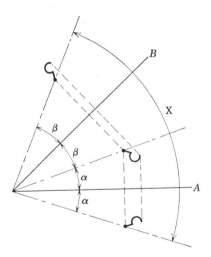

Figure 3.1   A geometric proof that the two reflection planes A and B require the existence of a C_n axis along their line of intersection with $n = 2\pi/2\phi_{AB}$. $\phi_{AB} = \alpha + \beta$, $X = \alpha + \alpha + \beta + \beta = 2(\alpha + \beta)$, $\therefore X = 2\phi_{AB}$.

planes are separated by the angle ϕ_{AB}, a C_n axis, where $n = 2\pi/2\phi_{AB}$, is required to exist. Here n must be an integer, and the C_n axis will then assure that a total of n such planes exists. Thus, the two planes imply that the entire set of operations constituting the C_{nv} group (vide infra) is present.

3. When there is a rotation axis, C_n, and a plane containing it, there must be n such planes separated by angles of $2\pi/2n$. This follows from rule 2.

4. The product of two C_2 rotations about axes which intersect at an angle θ is a rotation by 2θ about an axis perpendicular to the plane of the C_2 axes. This can be proved geometrically by a diagram similar to Figure 3.1. It also implies that a C_n axis and one perpendicular C_2 axis require the existence of a set of n C_2 axes and thus generate what we shall soon recognize as the D_n group of operations.

5. A proper rotation axis of even order and a perpendicular reflection plane generate an inversion center, i.e., $C_{2n}^n\sigma = \sigma C_{2n}^n = C_2\sigma = \sigma C_2 = i$. Similarly $C_{2n}^n i = iC_{2n}^n = C_2 i = iC_2 = \sigma$.

Commutation

The following pairs of operations always commute:

1. Two rotations about the same axis.
2. Reflections through planes perpendicular to each other.
3. The inversion and *any* reflection or rotation.
4. Two C_2 rotations about perpendicular axes.
5. Rotation and reflection in a plane perpendicular to the rotation axis.

3.10   Symmetry Elements and Optical Isomerism

Although we have followed conventional practice—and for general purposes will continue to do so—in setting out four kinds of symmetry elements and operations, σ, i, C_n, and S_n, we should note that the list can in principle be reduced to only two C_n and S_n. A reflection operation can be regarded as an S_1 operation, that is, the (trivial) rotation by $2\pi/1$ together with reflection.

The operation S_2 has the following effect on a general point, x, y, z. We suppose that the axis coincides with the z axis of a Cartesian coordinate system; the reflection component then takes place through the xy plane:

$$S_2(x, y, z) \equiv \sigma C_2(x, y, z) \rightarrow \sigma(\bar{x}, \bar{y}, z) \rightarrow (\bar{x}, \bar{y}, \bar{z})$$

But it is also true, by definition, that

$$i(x, y, z) \rightarrow (\bar{x}, \bar{y}, \bar{z})$$

Thus S_2 and i are merely two symbols for the same thing.

All symmetry operations that we wish to consider can be regarded as either proper or improper rotations.

Molecules which are *not* superimposable on their mirror images may be termed *dissymmetric*. This term is used rather than asymmetric, since the latter means, literally, having *no* symmetry. Dissymmetric molecules can and often do possess some symmetry. It is possible to give a very simple, compact rule expressing the relation between molecular symmetry and dissymmetric character:

A molecule which has no improper rotation axis will be dissymmetric.

Since improper rotation axes include $S_1 \equiv \sigma$, and $S_2 \equiv i$, the more familiar (*but incomplete!*) statement about optical isomerism existing in molecules which lack a plane or center of symmetry is subsumed in this more general one. In this connection, the tetramethylcyclooctatetraene molecule (page 50) should be examined more closely. This molecule possesses neither a center of symmetry nor any plane of symmetry. It does have an S_4 axis, and inspection will show that it *is* superimposable on its mirror image.

Proof of the above rule takes the following form.

1. A molecule has one and only one mirror image. It makes no difference where or in what orientation we place the mirror plane; we may place and orient it wherever convenient. We may therefore allow it to pass through the molecule.

2. If the molecule has an S_n axis, we may place the plane so it coincides with the plane through which the reflectional part of an S_n operation takes place. If the S_n axis is of odd order, the pure reflection operation (S_1 or S_n^n) will actually exist as a symmetry operation. The molecule is then obviously superimposable on its mirror image.

3. If the improper axis is of even order and σ_h does not exist independently, reflection will give a figure which is not superimposed on the original but needs only to be rotated by $2\pi/n$ in order to come into coincidence. This rotation of the molecule as a whole does not change its structure, and thus the molecule and its mirror image are superimposable.

It follows as an obvious corollary that dissymmetric molecules are those which either have no symmetry or have only axes of proper rotation.

3.11   The Symmetry Point Groups

Suppose that we have, by inspection, compiled a list of all of the symmetry elements possessed by a given molecule. We can then list all of the symmetry operations generated by each of these elements. Our first objective in this section is to demonstrate that such a *complete* list of symmetry *operations* satisfies the four criteria for a mathematical group. When this has been established, we shall then be free to use the theorems concerning the behavior of groups to assist in dealing with problems of molecular symmetry.

Let us first specify what we mean by a *complete* set of symmetry operations for a particular molecule. A complete set is one in which every possible product of two operations in the set is also an operation in the set. Let us consider as an example the set of operations which may be performed on a planar AB_3 molecule. These are E, C_3, $C_3{}^2$, C_2, C_2', C_2'', σ_v, σ_v', σ_v'', σ_h, S_3, and $S_3{}^2$. It should be clear that no other symmetry operations are possible. If we number the B atoms as indicated, we can systematically work through all binary products; for example:

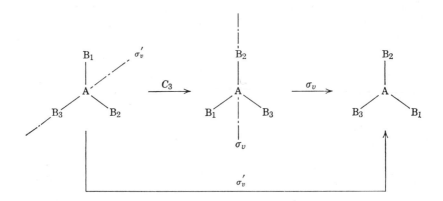

Hence we see that $\sigma_v C_3 = \sigma_v'$. Proceeding in this way, we can check all the combinations and we will find that the set given is indeed complete. This is suggested as a useful exercise.

Now, we can see that, because our set of operations is complete in the sense defined above, it satisfies the first requirement for mathematical groups, if we take as our law of combination of two symmetry operations the successive application of these operations.

The second requirement—that there must exist a group element E such that for every other element in the group, say X, $EX = XE = X$—is also seen to be satisfied. The "operation" of performing no operation at all, or that which results from a sequence of operations which sends the molecule into a configuration identical with the original (e.g., σ^2, $C_n{}^n$), is our identity, E, and we have been calling it that all along.

The associative law is obviously valid for products of symmetry operations.

The final requirement, that every element of the group have an inverse, is also satisfied. For a group composed of symmetry operations, we may define the inverse of a given operation as that second operation which will exactly undo what the given operation does. In more sophisticated terms, the reciprocal S of an operation R must be such that $RS = SR = E$. Let us consider each type of symmetry operation. For σ, reflection in a plane, the inverse is clearly σ itself: $\sigma \times \sigma = \sigma^2 = E$. For proper rotation, C_n^m, the inverse is C_n^{n-m}, for $C_n^m \times C_n^{n-m} = C_n^n = E$. For improper rotation, S_n^m, the reciprocal depends on whether m and n are even or odd, but a reciprocal exists in each of the four possible cases. When n is even, the reciprocal of S_n^m is S_n^{n-m} whether m is even or odd. When n is odd and m is even, $S_n^m = C_n^m$, the reciprocal of which is C_n^{n-m}. For S_n^m with both n and m odd we may write $S_n^m = C_n^m \sigma$. The reciprocal would be the product $C_n^{n-m} \sigma$, which is equal to $C_n^{2n-m} \sigma$, and which in turn may be written as a single operation, S_n^{2n-m}.

We have shown that complete sets of symmetry operations do constitute groups. Now we shall systematically consider what kinds of groups will be obtained from various possible collections of symmetry operations.

In the trivial case where there are no symmetry operations other than E, we have a group of order 1 called C_1.*

Let us next consider molecules whose sole symmetry element is a plane. This element generates only two operations, viz., σ and $\sigma^2 = E$. Hence the group is of order 2. The symbol normally given to this group is C_s. It is also possible to have a molecule whose sole symmetry element is an inversion center. The only operations generated by the inversion center are i and $i^2 = E$. Again we have a group of order 2; this one is conventionally designated C_i.

Let us consider now the cases where the only symmetry element is a proper axis, C_n. This generates a set of operations C_n, C_n^2, C_n^3, ..., $C_n^n = E$. Hence a molecule with C_n as its only symmetry element would belong to a group of order n, which is designated C_n. It may be noted that a C_n group is a cyclic group (see Section 2.2) and hence also Abelian.

When an improper axis is present, we must consider whether it is even or odd. When the axis, S_n, is of even order, the group of operations it generates is called S_n and consists of the n elements E, S_n, $C_{n/2}$, S_n^3, ..., S_n^{n-1}. The group S_2 is a special case because, as shown earlier, the symmetry element S_2 is equivalent to i. Thus the group which might be called the S_2 group is actually called C_i. The group of operations generated by an S_n axis when n is odd has been shown to consist of $2n$ elements, including σ_h and the

---

* This symbol and the other ones for the symmetry groups, for example, C_n, D_n, C_{nh}, C_{nv}, D_{nh}, D_{nd}, ..., which will be introduced are called the *Schoenflies symbols* after their inventor. The symmetry groups are also frequently called *point groups*, since all symmetry elements in a molecule will intersect at a common point, which is not shifted by any of the symmetry operations. There are also symmetry groups, called *space groups*, which contain operations involving translatory motions. The latter are not considered in this book.

operations generated by C_n. By convention such groups are denoted C_{nh}. This symbol emphasizes that there is a C_n axis and a horizontal plane; this combination of symmetry elements of course implies the existence of S_n just as S_n (n odd) implies the existence of C_n and σ_h. The C_{nh} groups will be discussed in more detail shortly.

Next we turn to groups which arise when two or more symmetry elements are present. In so doing, we shall divide the discussion into two parts. First, we deal with cases where there is no more than *one* axis of order higher than 2. Then, in the next section, we shall take up the groups which arise when there are several high-order ($n > 2$) axes. In each part, we shall proceed in a systematic way which should suggest strongly, if not actually prove rigorously, that all possibilities have been included.

We have already seen that, if a molecule possesses a proper axis, C_n, and also a twofold axis perpendicular to it, there must then necessarily be n such twofold axes. The n operations, $E, C_n, C_n^2, \ldots, C_n^{n-1}$, plus the n twofold rotations constitute a complete set of symmetry operations, as may be verified by actually carrying through all of the binary products. Thus such a group consists of a total of $2n$ elements. The symbol for a group of this kind is D_n.

We have now reached a point of departure in the process of adding further symmetry elements to a C_n axis. We shall consider (1) the addition of different kinds of symmetry planes to the C_n axis only, and (2) the addition of symmetry planes to a set of elements consisting of the C_n axis and the n C_2 axes perpendicular to it. In the course of this development it will be useful to have some symbols for several kinds of symmetry planes. In defining such symbols we shall consider the direction of the C_n axis, which we call the principal axis or reference axis, to be *vertical*. Hence, a symmetry plane perpendicular to this axis will be called a horizontal plane and denoted σ_h. Planes which include the C_n axis are generally called vertical planes, but there are actually two different types. In some molecules all vertical planes are equivalent and are symbolized σ_v. In others there may be two different sets of vertical planes (as in $PtCl_4^{2-}$; cf. page 30), in which case those of one set will be called σ_v and those of the other set σ_d, the d standing for dihedral. It will be best to discuss these differences more fully as we meet them.

If to the C_n axis we add a horizontal plane, we expand the original group of n operations, $C_n, C_n^2, \ldots, E$, to include all of the products $\sigma_h C_n, \sigma_h C_n^2, \sigma_h C_n^3, \ldots, \sigma_h E = \sigma_h$, making $2n$ operations in all. Now the operation $\sigma_h C_n^m = C_n^m \sigma_h$, since σ_h affects only the z coordinate of a point while C_n^m affects only its x and y coordinates, so that the order in which σ and C_n^m are performed is inconsequential. Furthermore, all of the new operations of the type σC_n^m can be expressed as single operations, viz., as improper rotations. This new set of $2n$ operations can easily be shown to be a complete set and hence to constitute a group. Such a group has the general symbol C_{nh}.

Let us look next at the consequences of adding a vertical plane to the C_n axis. First we recall (Section 3.5) that the operations generated by C_n when n is odd will require that an entire set of n such vertical planes exist. All of these planes are properly called vertical planes and symbolized σ_v. When n is even, however, we have seen (Section 3.5) that only $n/2$ planes of the same type will exist as a direct consequence of the C_n axis. However, we have also shown (Section 3.7) that another set of $n/2$ vertical planes must exist as the various products $C_n{}^m\sigma_v$. These vertical planes in this second set are usually called dihedral planes, since they bisect the dihedral angles between members of the set of σ_v's, and they are denoted σ_d. Obviously, it is completely arbitrary which set is considered vertical and which dihedral. In either case, n even or n odd, the set of operations generated by the C_n and by all of the σ's constitutes a complete set, and such a group is called C_{nv}.

We might naturally ask now about what happens when we add both the horizontal plane and the set of n vertical planes to the C_n. This gives a group called D_{nh}, which we shall develop by a different procedure.

We now consider the consequences of adding a σ_h to the group D_n. The group generated is denoted D_{nh}. We must first look at all of the products of σ_h with the operations generated by the C_2 axes and by the C_n axis. Suppose we choose a coordinate system such that the C_n axis coincides with the z axis, and one of the C_2 axes, $C_2(x)$, coincides with the x axis. We can indicate the effect of rotation about the $C_2(x)$ axis followed by σ_h on a general point $[x, y, z]$ thus:

$$[x, y, z] \xrightarrow{C_2(x)} [x, \bar{y}, \bar{z}] \xrightarrow{\sigma_h} [x, \bar{y}, z]$$

The effect of reflection in the xz plane on the same point will be

$$[x, y, z] \xrightarrow{\sigma(xz)} [x, \bar{y}, z]$$

Thus we can write

$$\sigma_h C_2(x) = \sigma(xz) = C_2(x)\sigma_h$$

where the second equality simply states that the rotation and the σ_h commute, which we have previously shown to be generally true. Of course, it now follows that, if one of the C_2 axes lies in a vertical symmetry plane, so must all of the others. There must then be a set of n operations σ_v. We may now left-multiply the above equation by σ_h, obtaining

$$\sigma_h \sigma_h C_2 = \sigma_h \sigma_v = C_2$$

and we see that all of the products of σ_h with the σ_v's are C_2's. Thus we might also take the simultaneous existence of C_n, σ_h, and the σ_v's as the criterion for the existence of the group D_{nh}. It is only by reason of convention and not because of any mathematical requirement that we take instead the simultaneous existence of C_n, n C_2's, and σ_h as the criterion.

We have now shown that the operations in a group D_{nh} include E, $(n-1)$ proper rotations about C_n, n reflections in vertical planes, σ_h, and n rotations about C_2 axes. These $3n + 1$ operations still do not constitute the complete set, however. It will be found that among the products $C_n^m\sigma_h = \sigma_h C_n^m$ are $n-1$ additional operations which are all improper rotations. For the general case where n is even, we obtain the new operations: S_n, $S_{n/2}$, ..., i ($= C_n^{n/2}\sigma_h$), ..., $S_{n/2}^{(n-2)/2}$, S_n^{n-1}. In the group D_{6h}, for example, we have S_6, S_3, i, S_3^2, and S_6^5. When n is odd we obtain in the general case the following $n-1$ improper rotations: S_n, S_n^3, S_n^5, ..., S_n^{2n-3}, S_n^{2n-1} except S_n^n ($=\sigma_h$); S_n^m where m is even being, of course, either E or one of the proper rotations which we have already recognized. Thus, we now have a total of $4n$ operations in the group D_{nh}. Systematic examination will show that the set is complete.

Our next and final task is to consider the consequences of adding to C_n and the n C_2's a set of dihedral planes, σ_d's. These are vertical planes which bisect the angles between adjacent pairs of C_2 axes. The groups generated by this combination of symmetry elements are denoted D_{nd}. The products of a σ_d with the various C_n^m operations are all other σ_d operations. However, among the various products of the type $\sigma_d C_2$ there is a set of n new operations generated by an S_{2n} axis colinear with C_n. These $4n$ operations now constitute the complete group D_{nd}.

Linear Molecules

These constitute a somewhat special case, although their possible point groups are closely related to the scheme just developed. Any linear molecule has an axis of symmetry coinciding with all the nuclei. The order of this axis is ∞; that is, rotations by any and all angles about this axis constitute symmetry operations. Also, any plane containing the molecule is a plane of symmetry. There is an infinite number of such planes, all intersecting along the molecular axis. Proceeding from this, we have just two possibilities: (1) the molecule is of the type OCO, NCCN, etc., such that it consists of two equivalent halves, or (2) it is of the type NNO, HCN, etc., and does not consist of two equivalent halves.

In the first case, the equivalence of the two halves means that any line which is a perpendicular bisector of the molecular axis is a C_2 symmetry axis; there are an infinite number of such C_2 axes. The equivalence of the two halves of the molecule also means that there is a plane of symmetry perpendicular to the molecular axis. Since there is an infinity of rotations about a unique, vertical axis, C_∞, there is also an infinity of C_2 axes perpendicular to C_∞, and there is a horizontal plane of symmetry. The group is, very reasonably, designated $D_{\infty h}$.

For linear molecules which do not consist of equivalent halves, the only symmetry operations are the rotations about C_∞ and reflections in the vertical planes. The group is called $C_{\infty v}$.

3.12   Symmetries with Multiple High-Order Axes

We have so far considered how groups of symmetry operations may be systematically built up by beginning with *one* proper axis of rotation (the reference axis) and the operations which it generates and adding to this group (a pure rotation group, C_n) the operations generated by additional symmetry elements, these being restricted to planes or *twofold* axes. We have not yet inquired about the possibility of adding to the operations generated by one higher-order ($n > 2$) axis additional high-order axes. That is the subject we must now investigate.

It turns out that there are not actually very many possibilities (only seven, to anticipate); some of these, however, are among the most important point groups we shall encounter in nature, and thus they deserve careful consideration. An interesting and systematic way to approach the subject is to recognize that groups involving several equivalent, intersecting, higher-order axes will be represented by polyhedra having faces perpendicular to such axes. For example, the tetrahedron, with four equilateral triangular faces, must have four equivalent, intersecting C_3 axes. By first making a demonstrably complete list of all such polyhedra, and then systematically considering their symmetry groups and all subgroups thereof in which the multiple axes are retained, we may expect to obtain a complete list of the symmetry groups with multiple high-order axes.

The Five Platonic Solids

To implement the above plan we consider the regular polyhedra, sometimes called the Platonic solids, of which there are five. By a *regular* polyhedron we mean a polyhedron

(1) whose faces are all some regular polygon (i.e., equilateral triangle, square, regular pentagon, hexagon, etc.) and equivalent to one another;

(2) whose vertices are all equivalent; and

(3) whose edges are all equivalent.

By "equivalent" we mean, as usual, interchangeable by symmetry operations. The five regular polyhedra are depicted and their essential characteristics listed in Table 3.2.

Table 3.2   The Five Regular Polyhedra or Platonic Solids

Tetrahedron
    Faces: 4 equilateral triangles
    Vertices: 4
    Edges: 6

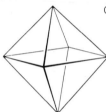

Cube
    Faces: 6 squares
    Vertices: 8
    Edges: 12

Octahedron
    Faces: 8 equilateral triangles
    Vertices: 6
    Edges: 12

Dodecahedron
    Faces: 12 regular pentagons
    Vertices: 20
    Edges: 30

Icosahedron
    Faces: 20 equilateral triangles
    Vertices: 12
    Edges: 30

Our first task is to show that the five Platonic solids do, in fact, represent all the possibilities. This is quite easy to do.

In order to construct a polyhedron, three or more of the desired faces must meet at a point so as to produce a closed, pyramidal (*not* planar) arrangement. Using equilateral triangles, we have the following possibilities:

1. Three triangles with a common vertex.
2. Four triangles with a common vertex.
3. Five triangles with a common vertex.

If six equilateral triangles share a common vertex, the sum of the angles around the vertex is $6 \times 60 = 360°$. The array is planar and cannot form part of a regular polyhedron. It is clear that the three possibilities listed give rise to the tetrahedron, the octahedron, and the icosahedron, as shown in Table 3.2.

For the next higher regular polygon, the square, there is only one possibility, viz., three squares with a common vertex, and this gives rise to the cube. Four squares having a common vertex would all lie in one plane.

With regular pentagons (internal angle, 108°), there is only one possibility, namely, three pentagons meeting at a common vertex ($3 \times 108° = 324°$), since four or more could not be fitted together ($4 \times 108° = 432°$). This single conjunction of pentagons can be replicated to produce the dodecahedron.

With hexagons, there is no way to construct a regular polyhedron, since even three hexagons sharing a vertex lie in the same plane. With all higher polyhedra not even three can be fitted together at a common vertex.

It is clear, therefore, that the five Platonic solids are the only regular polyhedra possible. Let us now examine them to see what symmetry operations may be performed on each one.

Inspection of the *tetrahedron* reveals the following symmetry elements and operations.

(i) Three S_4 axes coinciding with the x, y, and z axes. Each of these generates the operations S_4, $S_4{}^2 = C_2$, and $S_4{}^3$.

(ii) Three C_2 axes coinciding with the x, y, and z axes, each of which generates an operation C_2. These operations have already been generated, however, by the S_4's.

(iii) Four C_3 axes, each of which passes through one apex and the center of the opposite face. Each of these generates C_3 and $C_3{}^2$ operations, that is, eight operations in all.

(iv) Six planes of symmetry, each of which generates a symmetry operation.

The entire set of operations thus consists of the following 24, which are listed by classes (as will be explained in Sec. 3.15):

$$E, \ 8C_3, \ 3C_2, \ 6S_4, \ 6\sigma_d$$

This group is called T_d.

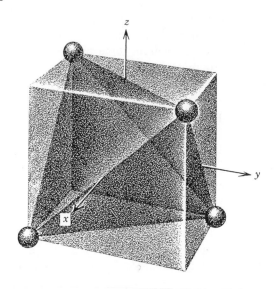

The *octahedron* has the following symmetry elements and operations.

(i) Three S_4 axes, each passing through a pair of opposite apices. Each generates the operations S_4, C_2, $S_4{}^3$.

(ii) Three C_2 axes colinear with the S_4's. The C_2 operations generated by these axes, however, are already accounted for under (i).

(iii) Three C_4 axes colinear with the S_4's and C_2's. Each generates a set of operations C_4, C_2, and $C_4{}^3$, but only C_4 and $C_4{}^3$ are new.

(iv) Six C_2' axes, which bisect opposite edges. Each generates an operation C_2'.

(v) Four S_6 axes, each passing through the centers of a pair of opposite triangular faces. Each generates a set of operations S_6, C_3, i, $C_3{}^2$, $S_6{}^5$.

(vi) Four C_3 axes colinear with the S_6's. Each generates two operations, C_3 and $C_3{}^2$, which are also generated by the colinear S_6.

(vii) An inversion center which generates an operation i, also generated by each of the S_6 axes.

(viii) Three planes of symmetry which pass through four of the six apices. Each generates an operation σ_h.

(ix) Six planes of symmetry which pass through two apices and bisect two opposite edges. Each of these generates an operation σ_d.

The entire set of operations thus consists of the following 48, grouped by classes (as will be explained in Sec. 3.15):

$$E,\ 8C_3,\ 6C_4,\ 6C_2,\ 3C_2(=C_4{}^2),\ i,\ 6S_4,\ 8S_6,\ 3\sigma_h,\ 6\sigma_d$$

This is the group called O_h.

Inspection will show that the *cube* has precisely the same set of symmetry operations as the octahedron; it, too, belongs to the point group O_h. It is worthwhile noting that the cube and the octahedron are very closely related. Each is obtainable from the other by shaving off corners, as shown in Figure 3.2 for the cube-to-octahedron conversion. In a cube the faces are penetrated

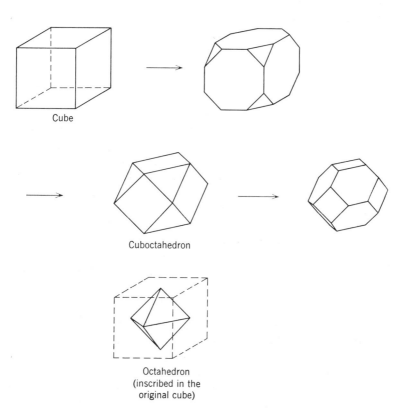

Cube

Cuboctahedron

Octahedron
(inscribed in the
original cube)

Figure 3.2   Conversion of the cube to the octahedron via the cuboctahedron.

by C_4 axes and the vertices by C_3 axes; in the octahedron the vertices lie on C_4 axes and the faces are penetrated by the C_3 axes. The nonregular polyhedron obtained as an intermediate when the triangular faces just meet is called a *cuboctahedron*. It, too, has O_h symmetry, as does every polyhedron which is transitional, in the sense of Figure 3.2, between the cube and the octahedron. The cuboctahedron occurs in nature as B_{12} cages in certain borides and has an interesting relationship to the icosahedron (cf. Exercise 3.6).

Finally, we turn to the pentagonal *dodecahedron* and the *icosahedron*. These two polyhedra have the same symmetry. They are related to each other as the cube and octahedron are related. The symmetry elements and operations are as follows.

(i) Each polyhedron has a set of six S_{10} axes. In the dodecahedron these pass through opposite pairs of pentagonal faces; in the icosahedron they pass through opposite vertices. Each S_{10} axis generates these operations: S_{10}, $S_{10}^2 = C_5$, S_{10}^3, $S_{10}^4 = C_5^2$, $S_{10}^5 = i$, $S_{10}^6 = C_5^3$, S_{10}^7, $S_{10}^8 = C_5^4$, S_{10}^9, E.

(ii) Each polyhedron has ten S_6 axes. In the dodecahedron they pass through opposite pairs of vertices; in the icosahedron they pass through pairs of opposite faces. Each of these generates these operations: S_6, $S_6^2 = C_3$, $S_6^3 = S_2 = i$, $S_6^4 = C_3^2$, S_6^5, E. Of these, i and E have already been noted.

(iii) There are six C_5 axes, colinear with the S_{10} axes. They generate C_5, C_5^2, C_5^3, C_5^4 operations, which have already been counted under S_{10}.

(iv) There are ten C_3 axes, colinear with the S_6 axes. These generate C_3 and C_3^2 operations, which have already been counted under S_6.

(v) There are fifteen C_2 axes, which in each case bisect opposite edges. These generate fifteen C_2 operations.

(vi) There are fifteen mirror planes, each containing two C_2 axes and two C_5 axes. They generate fifteen reflection operations.

Altogether there are 120 operations, which form the following classes:

$$E, 12C_5, 12C_5^2, 20C_3, 15C_2, i, 12S_{10}, 12S_{10}^3, 20S_6, 15\sigma$$

The group that they constitute is called I_h.

By direct inspection of the five regular polyhedra we have discovered three point groups: T_d, O_h, I_h. There are, however, several more, which we will now obtain straightforwardly from these. As noted earlier (page 31), products of rotations can only be rotations. Thus there are *pure rotation groups*. If from any group containing reflections we remove the reflections and all their products with the proper rotations, there will remain a subgroup consisting entirely of proper rotations.

Thus the group T_d has a pure rotational subgroup, T, of order 12. It consists of the following classes:

$$E, 4C_3, 4C_3^2, 3C_2$$

The group O_h has a pure rotational subgroup, O, of order 24. It consists of the following classes:

$$E, 6C_4, 3C_2(=C_4^2), 8C_3, 6C_2$$

The group I_h has a pure rotational subgroup, I, consisting of the following 60 operations:

$$E, 12C_5, 12C_5{}^2, 20C_3, 15C_2$$

Finally, there is one more group, called T_h. This can be derived by adding to T a set of planes, σ_h, which contain pairs of C_2 axes (as opposed to planes σ_d, which contain one C_2 axis and bisect another pair, thus giving T_d). When all the distinct products of these planes with the operations of T are enumerated and collected into classes, we have:

$$E, 4C_3, 4C_3{}^2, 3C_2, i, 4S_6, 4S_6{}^5, 3\sigma_h$$

All together, we now have the following seven groups containing multiple high-order axes:

$$
\begin{array}{ccc}
T & O & I \\
T_h & O_h & I_h \\
T_d & &
\end{array}
$$

From the systematic way in which they were obtained it should be clear that this is an exhaustive list.

3.13   A Systematic Procedure for Symmetry Classification of Molecules

In Section 3.11 we have shown that a complete and nonredundant set of symmetry operations for any molecule constitutes a mathematical group, and the various groups or kinds of groups (i.e., C_n, D_n, S_n, C_{nv}, C_{nh}, D_{nd}, T_d, ...) that we may expect to encounter among real molecules have been described. In this section we shall describe a systematic procedure for deciding to what point group any molecule belongs. This will be done in a practical, "how-to-do-it" manner, but the close relationship of this procedure to the arguments used in deriving the various groups should be evident. The following sequence of steps will lead systematically to a correct classification.

1. We determine whether the molecule belongs to one of the "special" groups, that is, $C_{\infty v}$, $D_{\infty h}$, or one of those with multiple high-order axes. Only linear molecules can belong to $C_{\infty v}$ or $D_{\infty h}$, so these cannot possibly involve any uncertainty. The specially high symmetry of the others is usually obvious. All of the cubic groups, T, T_h, T_d, O, and O_h, require four C_3 axes, while I and I_h require ten C_3's and six C_5's. These multiple C_3's and C_5's are the key things to look for. In practice only molecules built on a central

tetrahedron, octahedron, cuboctahedron, cube, or icosahedron will qualify, and these figures are usually very conspicuous.

2. If the molecule belongs to none of the special groups, we search for proper or improper axes of rotation. If no axes of either type can be found, we look for a plane or center of symmetry. If a plane only is found, the group is C_s. If a center only is found (this is *very* rare), the group is C_i. If no symmetry element at all is present, the group is the trivial one containing only the identity operation and designated C_1.

3. If an *even*-order improper axis (in practice only S_4, S_6, and S_8 are common) is found but no planes of symmetry or any proper axis except a colinear one (or more), whose presence is automatically required by the improper axis, the group is S_4, S_6, S_8, ... An S_4 axis requires a C_2 axis; an S_6 axis requires a C_3 axis; an S_8 axis requires C_4 and C_2 axes. The important point here is that the S_n (n even) groups consist exclusively of the operations generated by the S_n axis. If any additional operation is possible, we are dealing with a D_n, D_{nd}, or D_{nh} type of group. Molecules belonging to these groups are relatively rare, and the conclusion that a molecule belongs to one of these groups should be checked thoroughly before it is accepted.

4. Once it is certain that the molecule belongs to none of the groups so far considered, we look for the highest-order proper axis. It is possible that there will be no one axis of uniquely high order but instead three C_2 axes. In such a case, we look to see whether one of them is geometrically unique in some sense, for example, in being colinear with a unique molecular axis. This occurs with the molecule allene, which is one of the examples to be worked through later. If all of the axes appear quite similar to one another, then any one may be selected at random as the axis to which the vertical or horizontal character of planes will be referred. Suppose that C_n is our reference or principal axis. The crucial question now is whether there exists a set of n C_2 axes perpendicular to the C_n axis. If so, we proceed to step 5. If not, the molecule belongs to one of the groups C_n, C_{nv}, and C_{nh}. If there are no symmetry elements except the C_n axis, the group is C_n. If there are n vertical planes, the group is C_{nv}. If there is a horizontal plane, the group is C_{nh}.

5. If in addition to the principal C_n axis there are n C_2 axes lying in a plane perpendicular to the C_n axis, the molecule belongs to one of the groups D_n, D_{nh}, and D_{nd}. If there are no symmetry elements besides C_n and the n C_2 axes, the group is D_n. If there is also a horizontal plane of symmetry, the group is D_{nh}. A D_{nh} group will also, necessarily, contain n vertical planes; these planes *contain* the C_2 axes. If there is no σ_h but there is a set of n vertical planes which *pass between* the C_2 axes, the group is D_{nd}.

The five-step procedure just explained is summarized in the flow sheet of Figure 3.3.

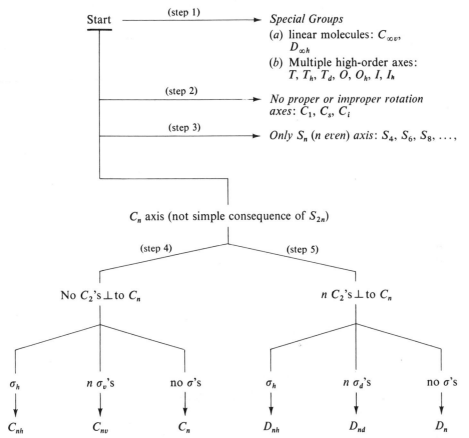

Figure 3.3   A five-stage procedure for the symmetry classification of molecules.

3.14   Illustrative Examples

The scheme just outlined for allocating molecules to their point groups will now be illustrated. We shall deal throughout with molecules which do not belong to any of the special groups, and we shall also omit molecules belonging to C_1, C_s, and C_i. Thus, each illustration will begin at step 3, the search for an even-order S_n axis.

Example 1. H_2O

3. H_2O possesses no improper axis.
4. The highest-order proper axis is a C_2 axis passing through the oxygen

atom and bisecting a line between the hydrogen atoms. There are no other C_2 axes. Therefore H_2O must belong to C_2, C_{2v}, or C_{2h}. Since it has two vertical planes, one of which is the molecular plane, it belongs to the group C_{2v}.

Example 2. NH_3

3. There is no improper axis.

4. The only proper axis is a C_3 axis; there are no C_2 axes at all. Hence, the point group must be C_3, C_{3v}, or C_{3h}. There are three vertical planes, one passing through each hydrogen atom. The group is thus C_{3v}.

Example 3. Allene

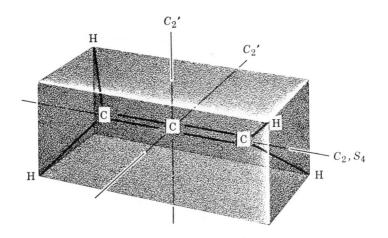

3. There is an S_4 axis coinciding with the main, molecular (C=C=C) axis. However, there are also other symmetry elements besides that C_2 axis which is a necessary consequence of the S_4. Most obvious, perhaps are the planes of symmetry passing through the $H_2C=C=C$ and $C=C=CH_2$ sets of atoms. Thus, although an S_4 axis is present, the additional symmetry rules out the point group S_4.

4. As noted, there is a C_2 axis lying along the C=C=C axis. There is no higher-order proper axis. There are two more C_2 axes perpendicular to this one, as shown in the sketch. Thus, the group must be a D type, and we proceed to step 5.

5. Taking the C_2 axis lying along the C=C=C axis of the molecule as the reference axis, we look for a σ_h. There is none, so the group D_{2h} is eliminated. There are, however, two vertical planes (which lie between C_2' axes), so the group is D_{2d}.

Example 4. H_2O_2

A. The nonplanar equilibrium configuration

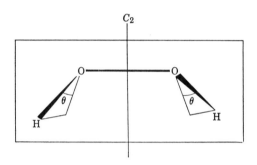

3. There is no improper axis.
4. As indicated in the sketch, there is a C_2 axis and no other proper axis. There are no planes of symmetry. The group is therefore C_2. Note that the C_2 symmetry is in no way related to the value of the angle θ except when θ equals 0° or 90°, in which case the symmetry is higher. We shall next examine these two nonequilibrium configurations of the molecule.

B. The cis-planar configuration ($\theta = 0°$)
3. Again there is no even-order S_n axis.
4. The C_2 axis, of course, remains. There are still no other proper axes. The molecule now lies in a plane, which is a plane of symmetry, and there is another plane of symmetry intersecting the molecular plane along the C_2 axis. The group is C_{2v}.

C. The trans-planar configuration ($\theta = 90°$)
3. Again, there is no even-order S_n axis. (except $S_2 \equiv i$).
4. The C_2 axis is still present, and there are no other proper axes. There is now a σ_h, which is the molecular plane. The group is C_{2h}.

Example 5. 1,3,5,7-Tetramethylcyclooctatetraene

3. There is an S_4 axis. There are no additional independent symmetry elements; the set of methyl groups destroys all the vertical planes and horizontal C_2 axes that exist in C_8H_8 itself. The group is therefore S_4.

It may be noted that this molecule contains no center of symmetry or any plane of symmetry and yet it is *not* dissymmetric. It thus provides an excellent illustration of the rule developed in Section 3.10.

Example 6. Cyclooctatetraene

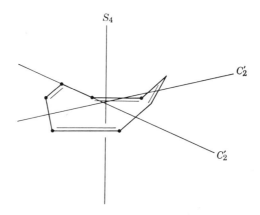

3. There is an S_4 axis. However, there are also numerous other symmetry elements which are independent of the S_4 axis. We thus proceed to step 4.

4. Coincident with the S_4 axis there is (by necessity) a C_2 axis. No proper axis of higher order can be found, but there are two more, equivalent C_2' axes in a plane perpendicular to the S_4-C_2 axis. Thus we are dealing with a D_2 type of group.

5. There is no σ_h, thus ruling out D_{2h}. There are, however, vertical planes of symmetry bisecting opposite double bonds. These pass between the C_2' axes, and the point group is D_{2d}.

Example 7. Benzene

3. There is an S_6 axis, perpendicular to the ring plane, but there are also other symmetry elements independent of the S_6 axis.

4. There is a C_6 axis perpendicular to the ring plane and six C_2 axes lying in the ring plane. Hence the group is a D_6 type.

5. Since there is a σ_h, the group is D_{6h}. Note that there are vertical planes of symmetry, but they contain the C_2 axes.

Example 8. PF₅ (Trigonal Bipyramidal)

3. There is no even-order S_n axis.

4. There is a unique C_3 axis, and there are three C_2 axes perpendicular to it.

5. There is a σ_h; the group is D_{3h}.

Example 9. Ferrocene

A. The staggered configuration

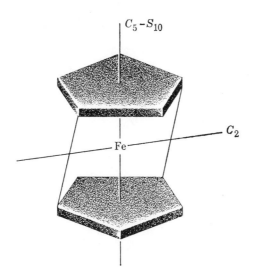

3. There is an even-order improper axis, S_{10}, as indicated in the sketch, but there are also other unrelated symmetry elements, so the group is not S_{10}.

4. The unique, high-order, proper axis is a C_5 axis, as shown. Perpendicular to this there are five C_2 axes.

5. Because of the staggered relationship of the rings there is no σ_h. There are, however, five vertical planes of symmetry which pass between the C_2 axes. The group is thus D_{5d}.

B. The eclipsed configuration

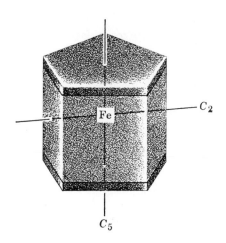

3. There is no even-order S_n axis.

4. There is a C_5 axis as shown. There are five C_2 axes perpendicular to the C_5 axis.

5. There is a σ_h, so the group is D_{5h}.

3.15   Classes of Symmetry Operations

In Section 2.4 the concept of classes of elements within a group was introduced. This concept is utilized in dealing with symmetry groups. As we shall see in Chapter 4, it is convenient and customary in writing what is called the character table of a group to consider all the elements of a given class together, since they all behave identically in the properties covered by the character table. It is the purpose of this section to explain the manner in which the symmetry operations are arranged into classes and to discuss the geometrical significance of the classes.

Of course, the general definition of a class and the method of arranging the elements of a group into classes given in Section 2.4 is perfectly applicable to a symmetry group. Let us consider, for example, the group C_{4v}. This group of operations arises when the following symmetry elements are present: C_4 and σ_v. There are eight operations in the complete set generated by these symmetry elements, viz., E, C_4, $C_4{}^2 = C_2$, $C_4{}^3$, $2\sigma_v$, $2\sigma_d$. The σ_v's are planes perpendicular to one another, intersecting along the C_4 axis, and so

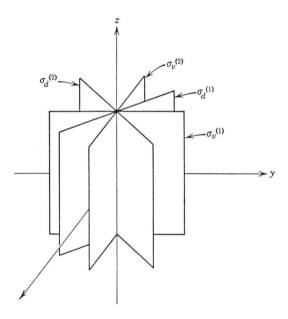

are the σ_d's. The σ_d's make $45°$ angles with the σ_v's. By methods previously explained and illustrated for determining the products of symmetry operations, a multiplication table for this group can be worked out.

Then, using this to carry out all of the possible similarity transformations, we find that there are the following classes:

$$E$$
$$C_4, C_4{}^3$$
$$C_2$$
$$\sigma_v{}^{(1)}, \sigma_v{}^{(2)}$$
$$\sigma_d{}^{(1)}, \sigma_d{}^{(2)}$$

It may be noted parenthetically that this result provides a good example of the fact that, although the orders of all classes must be integral divisors of the group order, not *all* integral divisors of the group order need be represented among the orders of the classes. Observe that, while 4 is an integral divisor of 8, there is no class of order 4 in this group.

With symmetry groups the classes have a geometrical significance which may be stated as follows: Two operations belong to the same class when one may be replaced by the other in a new coordinate system *which is accessible by a symmetry operation*. The italicized part of this prescription is quite important. Let us consider the group C_{4v} and its subgroup C_4 to see what this means. The operation $C_4{}^3$ shifts every point in the molecule by $3 \times 2\pi/4$ in, let us say, the clockwise direction. This, however is the same thing as shifting every point by $2\pi/4$ in the counterclockwise direction. Let us then for the moment think of the operation C_4 as rotation by $2\pi/4$ clockwise and $C_4' = C_4{}^3$ as rotation by $2\pi/4$ counterclockwise. Now suppose that the coordinate system in which we have been working is (*a*) such that clockwise

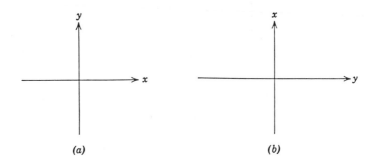

(a)   (b)

rotation by $2\pi/4$ converts a point $[x, y]$ into $[y, -x]$, while counterclockwise rotation by $2\pi/4$ converts $[x, y]$ into $[-y, x]$. Symbolically

$$C_4(z)[x, y] \rightarrow [y, -x]$$
$$C_4'(z)[x, y] \rightarrow [-y, x]$$

In coordinate system (*b*), however, the effects of C_4 (clockwise) and C_4' (counterclockwise) are

$$C_4(z)[x, y] \rightarrow [-y, x]$$
$$C_4'(z)[x, y] \rightarrow [y, -x]$$

In short, the roles of C_4 and C_4' are interchanged in coordinate system (b) from what they are in coordinate system (a). Now (and this is the important point) there is a symmetry operation in the group C_{4v} which will convert coordinate system (a) into coordinate system (b), namely, $\sigma_d^{(2)}$. Thus, in the group C_{4v}, C_4 and $C_4' = C_4{}^3$ are in the same class. However, in the group C_4 (which contains only the operations E, C_4, C_2, $C_4{}^3$) they are not in the same class because none of these four operations has the effect of transforming coordinate system (a) into system (b). Of course, since any C_n group is cyclic and hence Abelian, we can see that all operations must be in different classes, since each operation is conjugate only with itself in an Abelian group.

Returning to the group C_{4v} again, we note that C_2 is in a class by itself. This is so on geometrical grounds, because clearly there can be no way of shifting the coordinate system so that the effects of a rotation by 180° can be produced by a rotation by 90°, or by a reflection of any kind. Also, the σ_v's and σ_d's form separate classes. Only a rotation by $2\pi/8$ could change the orientation of the coordinate system to a new one in which σ_d would accomplish what σ_v did in the old one, and rotation by $2\pi/8$ is not a symmetry operation which occurs in the group.

It might be suspected intuitively that there would be a close relationship between the classes of operations and the various sets of equivalent operations in a group. In fact the classes correspond directly to the sets of equivalent operations. The reason is easy to see. The geometrical criterion for putting two operations, A and B, in the same class is that there be some third operation, C, which can be applied to the coordinate system so that operation B in the transformed coordinate system is analogous to operation A in the original coordinate system. At the same time, we say that operations A and B are equivalent if one is converted into the other (in the same coordinate system) by applying operation C to operations A and B. Now to say that operation C interchanges operations A and B when applied to them, leaving the coordinate system fixed, is perfectly equivalent to saying that operation C interchanges the functions of A and B when applied to the coordinate system, leaving the operations fixed in space. Hence *the simplest way of arranging the operations of a symmetry group into classes is to arrange them into sets of equivalent operations. These sets will be the classes.*

A practical consequence of collecting all operations in the same class when writing down the complete set, for example, at the head of a character table, is that the notation used is a little different from what we have been using thus far. This new and final form of notation will now be explained and illustrated for the four kinds of symmetry operations.

(i) *Inversion.* Only one inversion operation is possible in a molecule. If one exists it is denoted i. It will always be in a class by itself.

(ii) *Reflections.* Reflection in a horizontal plane is denoted σ_h. This operation will always be in a class by itself. When there is a set of n vertical planes all in the same class, we write simply $n\sigma_v$, and for a set of σ_d's, $n\sigma_d$. When there are some vertical planes in one class and some in another, some may be called σ_v's and the set will be indicated by $n\sigma_v$, while the second set may be denoted $n\sigma_v'$ or $n\sigma_d$ (the use of σ_v' or σ_d for the second set is somewhat arbitrary).

(iii) *Proper rotations.* In the cyclic groups each of the operations, C_n, C_n^2, C_n^3, ..., C_n^{n-1}, constitutes a class by itself and we continue to use this notation. However, in all other groups of higher symmetry, the number of classes spanned by these operations will be reduced in the following way. A C_n^m will fall into the same class with C_n^{n-m}. We have seen an example of this in the group C_{4v}, where C_4 and C_4^3 are in the same class. In these cases, we use the notation illustrated below for the various operations generated by a C_7 and a C_6 axis.

OLD NOTATION   NEW NOTATION

(Grouped by classes)

$$C_7{}^m\text{'s}\begin{cases} C_7, C_7{}^6 \\ C_7{}^2, C_7{}^5 \\ C_7{}^3, C_7{}^4 \end{cases} \qquad \begin{matrix} 2C_7 \\ 2C_7{}^2 \\ 2C_7{}^3 \end{matrix}$$

$$C_6{}^m\text{'s}\begin{cases} C_6, C_6{}^5 \\ C_6{}^2 = C_3, C_6{}^4 = C_3{}^2, \\ C_6{}^3 = C_2 \end{cases} \qquad \begin{matrix} 2C_6 \\ 2C_3 \\ C_2 \end{matrix}$$

In short, when two operations such as C_7 and $C_7{}^6$ are in the same class, one will be the same as the other only in the reverse direction, so that C_7 and $C_7{}^6$ are both called simply C_7, and so on.

(iv) *Improper rotations.* Just as with proper rotations, when two improper rotations fall in the same class it will be because one is really the same as the other except that the rotation is in the opposite sense. Thus S_6 and $S_6{}^5$ are both considered S_6's and are so written.

Exercises

3.1 What are the highest-order pure rotational subgroups of C_{8h}, D_{2d}, C_{5v}?

3.2 What group is obtained by adding to or deleting from each of the following groups the indicated symmetry operation?

C_3 plus i C_3 plus S_6

C_{3v} plus i D_{3d} minus S_6

C_{5v} plus σ_h S_4 plus i

S_6 minus i C_{3h} minus $S_6{}^5$

T_d plus i

3.3 What polyhedron is obtained from a cube if its edges are shaved away until the original cube faces disappear?

3.4 What is the conventional designation for the group of operations generated by an S_n axis when n is odd?

3.5 Write out all the operations generated by S_5 and S_8 axes, and express each one in conventional notation.

3.6 Show how a cuboctahedron may be transformed into an icosahedron by converting each square face into a pair of triangular faces with a common edge.

3.7 There are three relatively common types of dodecahedron. In addition to the pentagonal dodecahedron (point group I_h) there are the trigonal dodecahedron and the rhomboidal dodecahedron, shown at the right. To what point group does each of these belong?

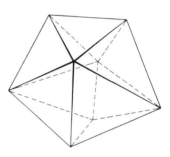

3.8 What different point groups may the biphenyl molecule belong to, depending on the rotational relationship of the two rings about the central C-C bond? Consider the same question for *m,m*-dichlorobiphenyl.

3.9 What is the point group of each of the following substituted cyclobutanes? Assume that cyclobutane itself has D_{4h} symmetry and that substituting H by X or Y changes no other structure parameters.

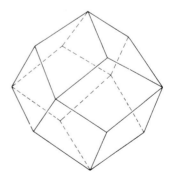

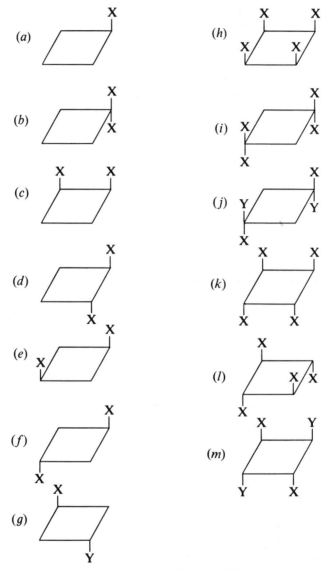

3.10 What is the point symmetry of each of the four distinct geometric isomers of an ethane-like, staggered molecule XYZC-CXYZ? Which ones are dissymmetric?

3.11 Draw structural formulae for all geometric isomers of "octahedral" complexes of the type $MA_2B_2C_2$. State the point group of each, and identify those which are dissymmetric.

3.12 Determine the point group of each of the following molecules or objects:

(*a*) 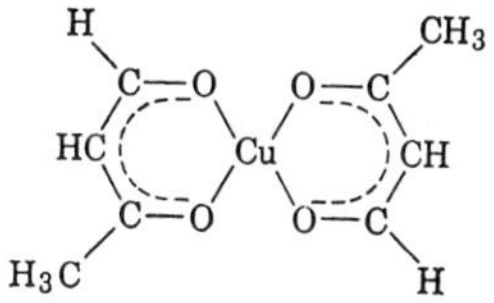

(ignore H's)

(*b*) 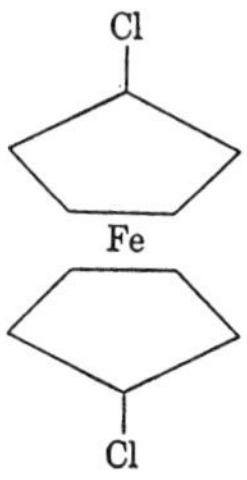

(*c*) Cl_3PO

(*d*) 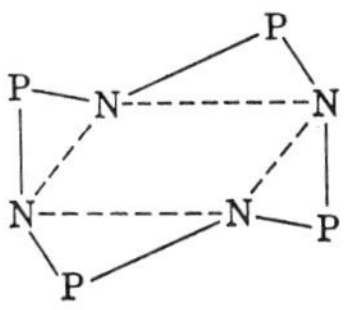

(*e*) Tennis ball (including the seam)

(*f*) *trans*–$[CrCl_2(H_2O)_4]^+$ (ignore H's)

(*g*) 1,3,5–trichlorobenzene

(*h*) *trans*–$Pt(NH_3)_2Cl_2$ (ignore H's)

(*i*) SF_5Cl

(*j*) BFClBr

(*k*) 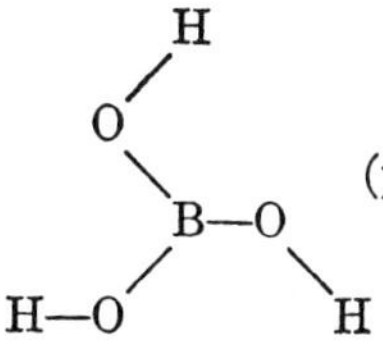

(planar)

(*l*) Spiropentane

(*m*) 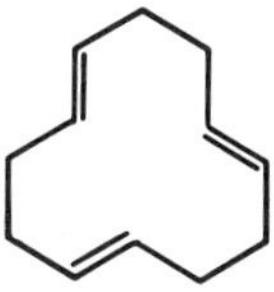

(*n*) 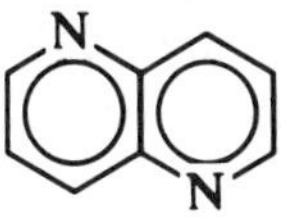

(*o*) A wineglass of the usual stemware type

(*p*)   Dibenzenechromium in each of its three rotomeric configurations

(*q*)   The chair and boat forms of cyclohexane

(*r*)   1,3-Dichloroallene, HClC=C=CHCl

(*s*)   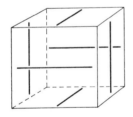 and

where the curved lines represent symmetrical, bidentate ligands.

3.13   To what symmetry is a tetrahedron with all black edges reduced if two edges which do not intersect are made red?

3.14   If you begin with an octahedron having eight black faces and paint four of them, no two of which have a common edge, white, to which symmetry group does the octahedron then belong?

3.15   If neopentane, $C(CH_3)_4$, is in a rotational conformation such that each set of methyl C-H bonds is eclipsed with a set of C-C bonds, it has T_d symmetry. Suppose that each methyl group is rotated about its C—C axis by 10° in the clockwise direction as viewed from the outside. What is the symmetry now?

3.16   What is the symmetry of a cube when a line is drawn across each of its faces in the manner shown at the left?

3.17   Suppose that the line on each face of the cube shown in Exercise 3.16 is rotated by θ, where $0 < \theta < 45°$, in the clockwise direction as seen from the outside. What is the point group now? If the angle of rotation is 45°, what is the point group?

3.18   If alternate vertices of the marked cube in Exercise 3.16 are painted black, what is the symmetry?

3.19   Suppose that we begin with an octahedron, on each face of which is drawn a smaller equilateral triangle oriented so that each vertex of the small triangle points directly toward a vertex of the larger one; the small triangles do not disturb the O_h symmetry. If each small triangle is now twisted clockwise by an angle θ, where $0 < \theta < 60°$, what symmetry does the figure possess?

3.20   Determine the highest possible point group to which each of the following belongs:

(*a*) $W_2Cl_9^{3-}$ (two octahedra sharing a face)

(*b*) Each of the two isomers of

$$\{[d\text{-}H_2NCH(CH_3)CH_2NH_2][l\text{-}H_2NCH(CH_3)CH_2NH_2]Pt\}^{2+}$$

(Pt and the four N atoms coplanar)

(*c*) A tetrahedral AB_4 molecule squashed along one of its S_4 axes (but not far enough to become entirely planar)

(*d*) Cyclooctatetraene in a crown conformation

(*e*) ϕ_4As^+

4

Representations of Groups

4.1 Some Properties of Matrices and Vectors

Because the representations of groups are in general made up of matrices, and because certain properties of representations can be advantageously formulated by using certain properties of vectors, this chapter will begin with an account of the aspects of matrix and vector algebra essential to an understanding of the following discussion of representation theory.

Definition of a Matrix

In the most general sense a matrix is a rectangular array of numbers, or symbols for numbers, which may be combined with other such arrays according to certain rules. When a matrix is written out in full, it has an appearance of which the following is typical:

$$
\begin{bmatrix}
4 & -7 & 6 & 0 \\
2 & 9 & -1 & -8 \\
2 & 0 & 5 & 4 \\
-8 & 7 & 0 & -3 \\
6 & 3 & -4 & 7
\end{bmatrix}
$$

Note the use of square brackets to enclose the array; this is a conventional way of indicating that the array is to be regarded as a matrix (instead of, perhaps, as a determinant).

In order to discuss matrices in a general way, certain general symbols are commonly used. Thus we may write a symbol for an entire matrix as a script letter, for example, $\mathscr{A}$, which stands for

$$
\begin{bmatrix}
a_{11} & a_{12} & a_{13} & \cdots & a_{1n} \\
a_{21} & a_{22} & a_{23} & \cdots & a_{2n} \\
a_{31} & a_{32} & a_{33} & \cdots & a_{3n} \\
\vdots & & & & \vdots \\
a_{m1} & a_{m2} & a_{m3} & \cdots & a_{mn}
\end{bmatrix}
$$

We may also represent the above matrix by $[a_{ij}]$. The vertical sets are called *columns*, and the horizontal ones *rows*. The symbol a_{ij} represents that element of the matrix $\mathscr{A}$ which stands in the ith *row* and the jth *column*. The m and n tell us the order of the matrix; m gives the number of rows and n the number of columns. A matrix in which $m = n$ is called a *square matrix* and will be of special importance to us. The elements in the set a_{ij} with $i = j$, that is, a_{11}, a_{22}, a_{33}, etc., in a square matrix are called the *diagonal elements* because they lie entirely on the line running diagonally from upper left to lower right corners. A square matrix in which all of the diagonal elements are equal to 1 and all of the other elements are equal to zero is called a *unit matrix* and conventionally represented by the symbol $\mathscr{E}$.

A type of matrix which is of considerable importance is the one-column matrix. To have the convenience of writing such a matrix all on one line, it is sometimes written out horizontally but enclosed in braces, { }, so as to distinguish it from a one-row matrix, which is normally written on one line in square brackets. The chief significance of the column matrix, at least for our purposes, is that it affords a way of representing a vector. Indeed it is sometimes actually called a vector.

Let us consider a vector in ordinary three-dimensional space. We can specify the length and direction of this vector in the following way. We arrange to have one end of the vector lie at the origin of a Cartesian coordinate system. The other end is then at a point which may be specified by its three Cartesian coordinates, x, y, z. In fact, these three coordinates completely specify the vector itself provided it is understood that one end of the vector is at the origin of the coordinate system. We can then write these three coordinates as a column matrix, in this case one with three rows, $\{x \quad y \quad z\}$, and say that the matrix represents the vector in question.

Obviously this notation can easily be generalized for vectors in abstract spaces of any dimension. In p-dimensional space a vector can be specified by a column vector of order $(p \times 1)$. The geometrical significance of the elements of this vector matrix is the same as in real space: they give the orthogonal (Cartesian in a general sense) coordinates of one end of the vector if the other end is at the origin of the coordinate system.

It should be noted that each of the coordinates of the outer terminus of the vector is numerically equal to the length of a projection of this vector on the axis concerned. Thus the set of numbers which define the vector in the sense discussed above may also be thought of as defining it in the sense of specifying its projections on a set of p orthogonal axes in the p-dimensional space in which it exists.

Combination of Matrices

There are certain rules for adding, subtracting, and dividing matrices; these are the rules of *matrix algebra*. It should be noted first that two matrices are equal only if they are identical. If $\mathscr{A} = \mathscr{B}$, then $a_{ij} = b_{ij}$ for all i and j.

To add or subtract two matrices, say $\mathscr{A}$ and $\mathscr{B}$, to give a sum or difference $\mathscr{C}$, the three matrices must be of the same dimensions. The elements of $\mathscr{C}$ are given by

$$c_{pq} = a_{pq} \pm b_{pq}$$

A matrix may be multiplied by a scalar number or by another matrix. For multiplication of a matrix $[c_{ij}]$ by a scalar, α, we have

$$\alpha[c_{ij}] = [\alpha c_{ij}] = [c_{ij}\alpha] = [c_{ij}]\alpha$$

Multiplication of a matrix by a matrix is somewhat more complicated. In the first place, it can be done only if the two matrices are *conformable*. This means that, if we wish to take the product $\mathscr{A}\mathscr{B} = \mathscr{C}$, the number of columns in $\mathscr{A}$ must be equal to the number of rows in $\mathscr{B}$. If this requirement is satisfied, so that $\mathscr{A}$ is of order $(n \times h)$ while $\mathscr{B}$ is of order $(h \times m)$, then $\mathscr{C}$ will be of order $(n \times m)$. Each element of the product matrix is given by the following expression:

$$c_{il} = \sum_k a_{ik} b_{kl} \tag{4.1-1}$$

This sum may be written out explicitly as follows:

$$c_{il} = a_{i1}b_{1l} + a_{i2}b_{2l} + a_{i3}b_{3l} + a_{i4}b_{4l} + \cdots + a_{ih}b_{hl}$$

where a_{ih} is the last element in the ith row of $\mathscr{A}$, and b_{hl} is the last element in the lth column of $\mathscr{B}$. Perhaps this will be still clearer if we explicitly write out the process of multiplying a 3×2 matrix into a 2×4 matrix.

$$\begin{bmatrix} a_{11} & a_{12} \\ a_{21} & a_{22} \\ a_{31} & a_{32} \end{bmatrix} \begin{bmatrix} b_{11} & b_{12} & b_{13} & b_{14} \\ b_{21} & b_{22} & b_{23} & b_{24} \end{bmatrix} = \begin{bmatrix} c_{11} & c_{12} & c_{13} & c_{14} \\ c_{21} & c_{22} & c_{23} & c_{24} \\ c_{31} & c_{32} & c_{33} & c_{34} \end{bmatrix}$$

$$
\begin{aligned}
c_{11} &= a_{11}b_{11} + a_{12}b_{21} & \quad c_{21} &= a_{21}b_{11} + a_{22}b_{21} \\
c_{12} &= a_{11}b_{12} + a_{12}b_{22} & \quad c_{22} &= a_{21}b_{12} + a_{22}b_{22} \\
c_{13} &= a_{11}b_{13} + a_{12}b_{23} & \quad c_{23} &= a_{21}b_{13} + a_{22}b_{23} \\
c_{14} &= a_{11}b_{14} + a_{12}b_{24} & \quad c_{24} &= a_{21}b_{14} + a_{22}b_{24}
\end{aligned}
$$

$$c_{31} = a_{31}b_{11} + a_{32}b_{21}$$
$$c_{32} = a_{31}b_{12} + a_{32}b_{22}$$
$$c_{33} = a_{31}b_{13} + a_{32}b_{23}$$
$$c_{34} = a_{31}b_{14} + a_{32}b_{24}$$

A mnemonically helpful way of summarizing the process is to say that the *ij*th element of the product is obtained by taking the *i*th row of the first matrix into the *j*th column of the second, with emphasis on the "row-into-column" aspect. From this discussion of the process of multiplication, the conformability requirement is readily obvious. If a row of matrix $\mathscr{A}$ is to be multiplied into a column of $\mathscr{B}$, then clearly the number of elements in that row, which is the number of columns in the matrix $\mathscr{A}$, must be equal to the number of elements in a column of $\mathscr{B}$, which is the number of rows in the matrix $\mathscr{B}$.

It should be noted specifically that matrix multiplication is not in general commutative. If the matrices $\mathscr{A}$ and $\mathscr{B}$ are conformable in the sense $\mathscr{A}\mathscr{B}$, they need not necessarily be conformable in the sense $\mathscr{B}\mathscr{A}$. Indeed they can be conformable in both ways only if both are square and of the same order. But even when the conformability requirement is satisfied, commutation is not in general possible. For example, consider the following two products:

$$\begin{bmatrix} 1 & 3 \\ 2 & 2 \end{bmatrix} \begin{bmatrix} 2 & 0 \\ 1 & 1 \end{bmatrix} = \begin{bmatrix} 5 & 3 \\ 6 & 2 \end{bmatrix}$$

$$\begin{bmatrix} 2 & 0 \\ 1 & 1 \end{bmatrix} \begin{bmatrix} 1 & 3 \\ 2 & 2 \end{bmatrix} = \begin{bmatrix} 2 & 6 \\ 3 & 5 \end{bmatrix}$$

Matrix multiplication does, however, always obey the associative law. This can easily be proved by extension of Equation 4.1-1, and working through this proof is a recommended exercise.

The quotient $\mathscr{A}/\mathscr{B}$ may be equally well regarded as the product $\mathscr{A}\mathscr{B}^{-1}$, that is, as $\mathscr{A}$ multiplied into the inverse of $\mathscr{B}$. We thus reduce the question of how to carry out a division to the question of how to find an inverse. In order to find the inverse of a matrix certain properties of the corresponding determinant must be used. The subject is treated in detail in Appendix II for the interested reader; we shall simply state the main conclusions here. The expression for the inverse of a matrix contains the corresponding determinant in the denominator. Since division by zero is not defined, only matrices with nonvanishing determinants can have inverses; and since only square determinants can be nonzero, we have the rule that only square matrices can have inverses. Of course, even some square matrices will have determinants equal to zero and hence their inverses will not be defined. A matrix $\mathscr{A}$ having a determinant $|A|$ which equals zero is said to be *singular*. We shall be interested only in matrices which have inverses, that is. in so-called *nonsingular* matrices.

The product of a matrix and its inverse is commutative and equals a unit matrix:

$$\mathcal{Q}\mathcal{Q}^{-1} = \mathcal{Q}^{-1}\mathcal{Q} = \mathcal{E}$$

A Special Case of Matrix Multiplication

A special case of matrix multiplication occurs when we deal with matrices having all nonzero elements in square blocks along the diagonal, such as the following two:

$$
\begin{bmatrix}
1 & 0 & 0 & 0 & 0 & 0 \\
1 & 2 & 0 & 0 & 0 & 0 \\
0 & 0 & 3 & 0 & 0 & 0 \\
0 & 0 & 0 & 1 & 3 & 2 \\
0 & 0 & 0 & 1 & 2 & 2 \\
0 & 0 & 0 & 4 & 0 & 1
\end{bmatrix}
\begin{bmatrix}
4 & 1 & 0 & 0 & 0 & 0 \\
2 & 3 & 0 & 0 & 0 & 0 \\
0 & 0 & 1 & 0 & 0 & 0 \\
0 & 0 & 0 & 0 & 1 & 2 \\
0 & 0 & 0 & 3 & 0 & 2 \\
0 & 0 & 0 & 2 & 1 & 1
\end{bmatrix}
$$

The product of these two matrices taken in the above order is:

$$
\begin{bmatrix}
4 & 1 & 0 & 0 & 0 & 0 \\
8 & 7 & 0 & 0 & 0 & 0 \\
0 & 0 & 3 & 0 & 0 & 0 \\
0 & 0 & 0 & 13 & 3 & 10 \\
0 & 0 & 0 & 10 & 3 & 8 \\
0 & 0 & 0 & 2 & 5 & 9
\end{bmatrix}
$$

The most conspicuous feature of this product matrix is that it is blocked out in exactly the same way as are its factors. It is not difficult to see that this sort of result must always be obtained. Moreover, it should also easily be seen that the elements of a given block in the product matrix are determined only by the elements in the corresponding blocks in the factors. Thus, when two matrices which are blocked out in the same way are to be multiplied, the corresponding blocks in each may be considered independently of the remaining blocks in each. Specifically, in the above case,

$$
\begin{bmatrix} 1 & 0 \\ 1 & 2 \end{bmatrix}
\begin{bmatrix} 4 & 1 \\ 2 & 3 \end{bmatrix}
=
\begin{bmatrix} 4 & 1 \\ 8 & 7 \end{bmatrix}
$$

$$[3] \times [1] = [3]$$

$$
\begin{bmatrix}
1 & 3 & 2 \\
1 & 2 & 2 \\
4 & 0 & 1
\end{bmatrix}
\begin{bmatrix}
0 & 1 & 2 \\
3 & 0 & 2 \\
2 & 1 & 1
\end{bmatrix}
=
\begin{bmatrix}
13 & 3 & 10 \\
10 & 3 & 8 \\
2 & 5 & 9
\end{bmatrix}
$$

A set of matrices which are all blocked out along the diagonal in the same way is said to be *block-factored*. This property will be of key importance presently.

Characters of Conjugate Matrices

An important property of a square matrix is its *character*. This is simply the sum of its diagonal elements, and it is usually given the symbol χ (Greek chi). Thus

$$\chi_{\mathscr{A}} = \sum_j a_{jj}$$

We shall now prove two important theorems concerning the behavior of characters.

If $\mathscr{C} = \mathscr{A}\mathscr{B}$ and $\mathscr{D} = \mathscr{B}\mathscr{A}$, the characters of $\mathscr{C}$ and $\mathscr{D}$ are equal.

PROOF.

$$\chi_{\mathscr{C}} = \sum_j c_{jj} = \sum_j \sum_k a_{jk} b_{kj}$$

$$\chi_{\mathscr{D}} = \sum_k d_{kk} = \sum_k \sum_j b_{kj} a_{jk}$$

$$= \sum_j \sum_k b_{kj} a_{jk} = \sum_j \sum_k a_{jk} b_{kj} = \chi_{\mathscr{C}}$$

Conjugate matrices have identical characters. Conjugate matrices are related by a similarity transformation in the same way as are conjugate elements of a group. Thus, if matrices $\mathscr{R}$ and $\mathscr{P}$ are conjugate, there is some other matrix, $\mathscr{Q}$, such that

$$\mathscr{R} = \mathscr{Q}^{-1}\mathscr{P}\mathscr{Q}$$

Since the associative law holds for matrix multiplication, the theorem is proved in the following way.

PROOF.

$$\chi \text{ of } \mathscr{R} = \chi \text{ of } \mathscr{Q}^{-1}\mathscr{P}\mathscr{Q} = \chi \text{ of } (\mathscr{Q}^{-1}\mathscr{P})\mathscr{Q}$$

$$= \chi \text{ of } \mathscr{Q}(\mathscr{Q}^{-1}\mathscr{P}) = \chi \text{ of } (\mathscr{Q}\mathscr{Q}^{-1})\mathscr{P}$$

$$= \chi \text{ of } \mathscr{P}$$

Matrix Notation for Geometric Transformations

One important application of matrix algebra is in expressing the transformations of a point—or the collection of points which define a body—in space. We have employed previously five types of operations in describing the symmetry of a molecule or other object: E, σ, i, C_n, S_n. Each of these types of operation can be described by a matrix.

The Identity. When a point with coordinates x, y, z is subjected to the identity operation, its new coordinates are the same as the initial ones, viz., x, y, z. This may be expressed in a matrix equation as follows:

$$\begin{bmatrix} 1 & 0 & 0 \\ 0 & 1 & 0 \\ 0 & 0 & 1 \end{bmatrix} \begin{bmatrix} x \\ y \\ z \end{bmatrix} = \begin{bmatrix} x \\ y \\ z \end{bmatrix}$$

Thus, the identity operation is described by a unit matrix.

Reflections. If a plane of reflection is chosen to coincide with a principal Cartesian plane (i.e., an xy, xz, or yz plane), reflection of a general point has the effect of changing the sign of the coordinate measured perpendicular to the plane while leaving unchanged the two coordinates whose axes define the plane. Thus, for reflections in the three principal planes, we may write the following matrix equations:

$$\sigma(xy): \begin{bmatrix} 1 & 0 & 0 \\ 0 & 1 & 0 \\ 0 & 0 & -1 \end{bmatrix} \begin{bmatrix} x \\ y \\ z \end{bmatrix} = \begin{bmatrix} x \\ y \\ \bar{z} \end{bmatrix}$$

$$\sigma(xz): \begin{bmatrix} 1 & 0 & 0 \\ 0 & -1 & 0 \\ 0 & 0 & 1 \end{bmatrix} \begin{bmatrix} x \\ y \\ z \end{bmatrix} = \begin{bmatrix} x \\ \bar{y} \\ z \end{bmatrix}$$

$$\sigma(yz): \begin{bmatrix} -1 & 0 & 0 \\ 0 & 1 & 0 \\ 0 & 0 & 1 \end{bmatrix} \begin{bmatrix} x \\ y \\ z \end{bmatrix} = \begin{bmatrix} \bar{x} \\ y \\ z \end{bmatrix}$$

Inversion. To simply change the signs of all the coordinates without permuting any, we clearly need a *negative* unit matrix, viz.,

$$\begin{bmatrix} -1 & 0 & 0 \\ 0 & -1 & 0 \\ 0 & 0 & -1 \end{bmatrix} \begin{bmatrix} x \\ y \\ z \end{bmatrix} = \begin{bmatrix} \bar{x} \\ \bar{y} \\ \bar{z} \end{bmatrix}$$

Proper Rotation. Defining the rotation axis as the z axis, we note first that the z coordinate will be unchanged by any rotation about the z axis. Thus, the matrix we seek must be, in part,

$$\begin{bmatrix} & & 0 \\ & & 0 \\ 0 & 0 & 1 \end{bmatrix}$$

The problem of finding the four missing elements can then be solved as a two-dimensional problem in the xy plane.

Suppose that we have a point in the xy plane with coordinates x_1 and y_1, as shown in the diagram. This point defines a vector, r_1, between itself and

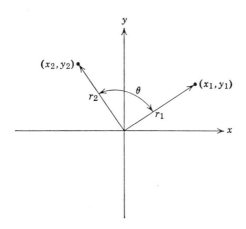

the origin. Now suppose that this vector is rotated through an angle θ so that a new vector, r_2, is produced with a terminus at the point x_2 and y_2. We now inquire about how the final coordinates, x_2 and y_2, are related to the original coordinates, x_1 and y_1, and the angle θ. This relationship is not difficult to work out. When the x component of r_1, x_1, is rotated by θ, it becomes a vector x' which has an x component of $x_1 \cos \theta$ and a y component of $x_1 \sin \theta$. Similarly, the y component of r_1, y_1, upon rotation by θ becomes a new vector y' which has an x component of $-y_1 \sin \theta$ and a y component of $y_1 \cos \theta$. Now, x_2 and y_2, the components of r_2, must be equal to the sums of the x and y components of x' and y', so we write

$$x_2 = x_1 \cos \theta - y_1 \sin \theta$$
$$y_2 = x_1 \sin \theta + y_1 \cos \theta$$

(4.1-2)

The transformation expressed by Equations 4.1-2 can be written in matrix notation in the following way:

$$\begin{bmatrix} \cos \theta & -\sin \theta \\ \sin \theta & \cos \theta \end{bmatrix} \begin{bmatrix} x_1 \\ y_1 \end{bmatrix} = \begin{bmatrix} x_2 \\ y_2 \end{bmatrix}$$

This result is for a counterclockwise rotation. Because $\cos \phi = \cos(-\phi)$ while $\sin \phi = -\sin(-\phi)$, the matrix for a clockwise rotation through the angle ϕ must be

$$\begin{bmatrix} \cos \phi & \sin \phi \\ -\sin \phi & \cos \phi \end{bmatrix}$$

Thus, finally, the total matrix equation for a clockwise rotation through ϕ about the z axis is

$$
\begin{bmatrix}
\cos \phi & \sin \phi & 0 \\
-\sin \phi & \cos \phi & 0 \\
0 & 0 & 1
\end{bmatrix}
\begin{bmatrix}
x_1 \\ y_1 \\ z_1
\end{bmatrix}
=
\begin{bmatrix}
x_2 \\ y_2 \\ z_2
\end{bmatrix}
$$

Improper Rotation. Since an improper rotation through the angle ϕ about the z axis produces the same transformation of the x and y coordinates as does a proper rotation through the same angle, but in addition changes the sign of the z coordinate, we may infer directly from the equation just derived that the matrix for clockwise rotation is

$$
\begin{bmatrix}
\cos \phi & \sin \phi & 0 \\
-\sin \phi & \cos \phi & 0 \\
0 & 0 & -1
\end{bmatrix}
$$

It will be clear that one could also have obtained this matrix by explicitly multiplying the matrices for rotation and reflection in the xy plane.

In general, the matrices which describe symmetry operations can be multiplied together so that the product of any two is the matrix for some (usually other) operation. For instance, we previously showed (page 31) somewhat tediously that the line of intersection of two perpendicular planes of symmetry must be a twofold axis of symmetry. We may employ the matrices to show the same thing very neatly. Thus, for σ_{xz}, σ_{yz}, and $C_2(z)$ we have

$$
\underset{\sigma_{xz}}{\begin{bmatrix} 1 & 0 & 0 \\ 0 & -1 & 0 \\ 0 & 0 & 1 \end{bmatrix}}
\underset{\sigma_{yz}}{\begin{bmatrix} -1 & 0 & 0 \\ 0 & 1 & 0 \\ 0 & 0 & 1 \end{bmatrix}}
=
\underset{\sigma_{yz}}{\begin{bmatrix} -1 & 0 & 0 \\ 0 & 1 & 0 \\ 0 & 0 & 1 \end{bmatrix}}
\underset{\sigma_{xz}}{\begin{bmatrix} 1 & 0 & 0 \\ 0 & -1 & 0 \\ 0 & 0 & 1 \end{bmatrix}}
=
\underset{C_2(z)}{\begin{bmatrix} -1 & 0 & 0 \\ 0 & -1 & 0 \\ 0 & 0 & 1 \end{bmatrix}}
$$

Symbolically, if a set of geometrical operations, A, B, C, D, ..., applied successively gives the same net effect as a single operation X, that is,

$$
\cdots DCBA = X
$$

then the products of the matrices representing these operations will multiply together in the same order to give a matrix corresponding to X, viz.,

$$
\cdots \mathscr{D}\mathscr{C}\mathscr{B}\mathscr{A} = \mathscr{X}
$$

The *inverse*, $\mathscr{A}^{-1}$, of a matrix, $\mathscr{A}$, is defined by the equation

$$
\mathscr{A}\mathscr{A}^{-1} = \mathscr{A}^{-1}\mathscr{A} = \mathscr{E}
$$

where $\mathscr{E}$ is the unit matrix.

All of the matrices we have just worked out, as well as all others which describe the transformations of a set of orthogonal coordinates by proper and improper rotations, are called *orthogonal* matrices. They have the convenient property that their inverses are obtained merely by transposing rows and columns. Thus, for example, the inverse of the matrix

$$
\begin{bmatrix} 0 & 1 & 0 \\ 0 & 0 & 1 \\ -1 & 0 & 0 \end{bmatrix} \quad \text{is} \quad \begin{bmatrix} 0 & 0 & -1 \\ 1 & 0 & 0 \\ 0 & 1 & 0 \end{bmatrix}
$$

as confirmed by

$$
\begin{bmatrix} 0 & 1 & 0 \\ 0 & 0 & 1 \\ -1 & 0 & 0 \end{bmatrix} \begin{bmatrix} 0 & 0 & -1 \\ 1 & 0 & 0 \\ 0 & 1 & 0 \end{bmatrix} = \begin{bmatrix} 0 & 0 & -1 \\ 1 & 0 & 0 \\ 0 & 1 & 0 \end{bmatrix} \begin{bmatrix} 0 & 1 & 0 \\ 0 & 0 & 1 \\ -1 & 0 & 0 \end{bmatrix} = \begin{bmatrix} 1 & 0 & 0 \\ 0 & 1 & 0 \\ 0 & 0 & 1 \end{bmatrix}
$$

Since rotations by ϕ clockwise and counterclockwise are inverse operations, their matrices must be inverse to each other. Thus the relation between the matrices for these two operations could now be deduced simply by saying that one must be the transpose of the other.

As a more general illustration of how matrices can be used to express symmetry operations, consider the eight C_3 operations of a tetrahedron as shown in the following sketch:

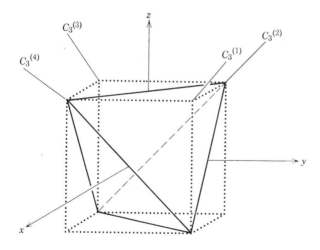

Let us first consider the effect on a general point, with coordinates x, y, and z, of a clockwise rotation by $2\pi/3$ about the axis $C_3^{(1)}$. This sends y into x, z into y, and x into z; that is, $[x, y, z]$ becomes $[y, z, x]$. Writing the two sets of coordinates as column matrices, we see that the rotation operation can be described by the matrix equation:

$$\begin{bmatrix} 0 & 1 & 0 \\ 0 & 0 & 1 \\ 1 & 0 & 0 \end{bmatrix} \begin{bmatrix} x \\ y \\ z \end{bmatrix} = \begin{bmatrix} y \\ z \\ x \end{bmatrix}$$

Similarly, counterclockwise rotation (or C_3^2 in a clockwise direction) is described by

$$\begin{bmatrix} 0 & 0 & 1 \\ 1 & 0 & 0 \\ 0 & 1 & 0 \end{bmatrix} \begin{bmatrix} x \\ y \\ z \end{bmatrix} = \begin{bmatrix} z \\ x \\ y \end{bmatrix}$$

The matrices for C_3 and C_3^2 rotations about the other C_3 axes are as follows:

$C_3^{(2)}$

$$C_3 = \begin{bmatrix} 0 & 0 & -1 \\ -1 & 0 & 0 \\ 0 & 1 & 0 \end{bmatrix} \qquad C_3^2 = \begin{bmatrix} 0 & -1 & 0 \\ 0 & 0 & 1 \\ -1 & 0 & 0 \end{bmatrix}$$

$C_3^{(3)}$

$$C_3 = \begin{bmatrix} 0 & 1 & 0 \\ 0 & 0 & -1 \\ -1 & 0 & 0 \end{bmatrix} \qquad C_3^2 = \begin{bmatrix} 0 & 0 & -1 \\ 1 & 0 & 0 \\ 0 & -1 & 0 \end{bmatrix}$$

$C_3^{(4)}$

$$C_3 = \begin{bmatrix} 0 & 0 & 1 \\ -1 & 0 & 0 \\ 0 & -1 & 0 \end{bmatrix} \qquad C_3^2 = \begin{bmatrix} 0 & -1 & 0 \\ 0 & 0 & -1 \\ 1 & 0 & 0 \end{bmatrix}$$

Since the rotations C_3 and C_3^2 in any pair about a given C_3 axis are inverse to each other, the matrices representing them should also be inverse. Moreover, since we are dealing with orthogonal matrices, it should be true that each matrix in each pair is the transpose of the other. It will be seen that this is so.

With this set of matrices it is easy to show that the product of any two threefold rotations about different axes is a twofold rotation about one of the Cartesian axes. For example, the product of C_3 about $C_3^{(1)}$ and C_3^2 about $C_3^{(3)}$ is obtained by the matrix multiplication:

$$\begin{bmatrix} 0 & 1 & 0 \\ 0 & 0 & 1 \\ 1 & 0 & 0 \end{bmatrix} \begin{bmatrix} 0 & 0 & -1 \\ 1 & 0 & 0 \\ 0 & -1 & 0 \end{bmatrix} = \begin{bmatrix} 1 & 0 & 0 \\ 0 & -1 & 0 \\ 0 & 0 & -1 \end{bmatrix}$$

The product matrix represents a twofold rotation about the x axis. The product of any matrix for a C_2 operation with any matrix for a C_3 or C_3^2

operation will be the matrix for some other C_3 or $C_3{}^2$ operation. For example, C_2 about the x axis times $C_3{}^2$ about the $C_3{}^{(2)}$ axis gives

$$\begin{bmatrix} 1 & 0 & 0 \\ 0 & -1 & 0 \\ 0 & 0 & -1 \end{bmatrix} \begin{bmatrix} 0 & -1 & 0 \\ 0 & 0 & 1 \\ -1 & 0 & 0 \end{bmatrix} = \begin{bmatrix} 0 & -1 & 0 \\ 0 & 0 & -1 \\ 1 & 0 & 0 \end{bmatrix}$$

which is the matrix for the $C_3{}^2$ rotation about the $C_3{}^{(4)}$ axis. By continuing in this way the entire set of operations whose existence depends on the presence of the four C_3 axes arranged as shown can be generated. This set, which constitutes a group, consists of the eight C_3 operations, the three C_2 operations, and the identity. The group that they form will be recognized as the pure rotation group T.

What we have just done is to substitute the algebraic process of multiplying matrices for the geometric process of successively applying symmetry operations. The matrices multiply together in the same pattern as do the symmetry operations; it is clear that they must, since they were constructed to do just that. It will be seen in the next section that this sort of relationship between a set of matrices and a group of symmetry operations has great importance and utility.

Vectors and Their Scalar Products

Since many of the basic arguments in Section 4.3 will lean heavily on the concept of orthogonal vectors in generalized, multidimensional space, a brief summary of the essentials will be included here.

A vector in p-dimensional space may be defined by the lengths of its projections on each of a set of p orthogonal axes in that space. For instance, a vector $\mathbf{A}$, in real space, with the coordinates x_1, y_1, z_1 for its outer terminus, has a projection, A_x, of length x_1 on the x axis, a projection, A_y, of length y_1 on the y axis, and a projection, A_z, of length z_1 on the z axis.

One type of product of two vectors is called the scalar product because it is merely a number, a scalar. This may be defined as the product of the lengths of the two vectors times the cosine of the angle between them. The scalar product is indicated by placing a dot between the symbols. We denote a vector as $\mathbf{A}$, its length by A, and its projections on coordinate axes by, for example, A_x, A_y, ...

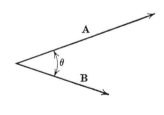

If two vectors $\mathbf{C}$ and $\mathbf{D}$ are orthogonal, their scalar or "dot" product will be zero, for

$$\mathbf{C} \cdot \mathbf{D} = CD \cos 90° = 0$$

If they are parallel or colinear, their scalar product is equal to the product of their lengths since cos $°$ = 1.

There is an equivalent but more generally useful way of writing the scalar product of two vectors. Suppose that we have two vectors **A** and **B**, both lying in the xy plane. Let **A** make an angle ϕ to the x axis and **B** a greater angle ψ. The angle between **A** and **B**, θ, is then $(\psi - \phi)$. Thus

$$\mathbf{A \cdot B} = AB \cos \theta = AB \cos (\psi - \phi) \qquad (4.1\text{-}3)$$

Now the components of **A**, that is, its projections on the x and y axes, are

$$A_x = A \cos \phi$$
$$A_y = A \sin \phi \qquad (4.1\text{-}4)$$

and similarly for **B**

$$B_x = B \cos \psi$$
$$B_y = B \sin \psi \qquad (4.1\text{-}5)$$

Using a trigonometric identity, we can write Equation 4.1-3 as follows:

$$\mathbf{A \cdot B} = AB \left(\cos \phi \cos \psi + \sin \phi \sin \psi\right)$$

which may be arranged to

$$\mathbf{A \cdot B} = A \cos \phi \, B \cos \psi + A \sin \phi \, B \sin \psi$$

Substituting the relations 4.1-4 and 4.1-5, we obtain

$$\mathbf{A \cdot B} = A_x B_x + A_y B_y$$

Thus the scalar product of vectors **A** and **B** in two-dimensional space is equal to the sum of the products of their components with no cross terms (e.g., $A_x B_y$). This result is actually only a special case of the general rule in p-dimensional space:

$$\mathbf{A \cdot B} = \sum_{i=1}^{p} A_i B_i$$

We can now restate the rule for orthogonality of two vectors in p-dimensional space as requiring that

$$\sum_{i=1}^{p} A_i B_i = 0$$

and the square of the length of a vector may be written as

$$A^2 = \sum_{i=1}^{p} A_i{}^2$$

4.2   *Representations of Groups*

A representation of a group of the type we shall be interested in may be defined as a set of matrices, each corresponding to a single operation in the group, that can be combined among themselves in a manner parallel to the way in which the group elements—in this case, the symmetry operations —combine. Thus, if two symmetry operations in a symmetry group, say C_2 and σ, combine to give a product C_2', then the matrices corresponding to C_2 and σ must multiply together to give the matrix corresponding to C_2'. But we have already seen that, if the matrices corresponding to all of the operations have been correctly written down, they will naturally have this property.

Let us, for example, work out a representation of the group C_{2v}. This group consists of the operations E, C_2, σ_v, σ_v'. Let us say that the C_2 axis coincides with the z axis of a Cartesian coordinate system, and let σ_v be the xz plane and σ_v' be the yz plane. The matrices representing the transformations effected on a general point can easily be seen to be as follows:

$$E:\begin{bmatrix} 1 & 0 & 0 \\ 0 & 1 & 0 \\ 0 & 0 & 1 \end{bmatrix} \qquad C_2:\begin{bmatrix} -1 & 0 & 0 \\ 0 & -1 & 0 \\ 0 & 0 & 1 \end{bmatrix}$$

$$\sigma_v:\begin{bmatrix} 1 & 0 & 0 \\ 0 & -1 & 0 \\ 0 & 0 & 1 \end{bmatrix} \qquad \sigma_v':\begin{bmatrix} -1 & 0 & 0 \\ 0 & 1 & 0 \\ 0 & 0 & 1 \end{bmatrix}$$

Now the group multiplication table is as follows:

	E	C_2	σ_v	σ_v'
E	E	C_2	σ_v	σ_v'
C_2	C_2	E	σ_v'	σ_v
σ_v	σ_v	σ_v'	E	C_2
σ_v'	σ_v'	σ_v	C_2	E

It can easily be shown that the matrices multiply together in the same fashion. For example:

$$\sigma_v C_2 = \sigma_v'$$

and

$$\begin{bmatrix} 1 & 0 & 0 \\ 0 & -1 & 0 \\ 0 & 0 & 1 \end{bmatrix}\begin{bmatrix} -1 & 0 & 0 \\ 0 & -1 & 0 \\ 0 & 0 & 1 \end{bmatrix} = \begin{bmatrix} -1 & 0 & 0 \\ 0 & 1 & 0 \\ 0 & 0 & 1 \end{bmatrix}$$

Again, each element in the group C_{2v} is its own inverse, so the same must be true of the matrices. This is easily shown to be so; for example:

$$\begin{bmatrix} 1 & 0 & 0 \\ 0 & -1 & 0 \\ 0 & 0 & 1 \end{bmatrix} \begin{bmatrix} 1 & 0 & 0 \\ 0 & -1 & 0 \\ 0 & 0 & 1 \end{bmatrix} = \begin{bmatrix} 1 & 0 & 0 \\ 0 & 1 & 0 \\ 0 & 0 & 1 \end{bmatrix}$$

We have now, by one procedure, namely, by considering the transformations of a general point, generated a set of matrices which form a representation for the group C_{2v}. It will be recalled that we have also done the same thing (pages 71–73) for the group T.

A question that naturally arises at this point is: How many representations can be found for any particular group, say C_{2v}, to continue with that as an example? The answer is: a very large number, limited only by our ingenuity in devising ways to generate them. There are first some very simple ones, obtained by assigning 1 or -1 to each operation, viz.,

E	C_2	σ_v	σ_v'
1	1	1	1
1	-1	1	-1
1	-1	-1	1
1	1	-1	-1

Then there are many representations of high order. For example, if we were to assign three small unit vectors directed along the x, y, and z axes to each of the atoms in H_2O and write down matrices representing the changes and interchanges of these upon applying the operations, a set of four 9×9 matrices constituting a representation of the group would be obtained. Using CH_2Cl_2 in the same way, we could obtain a representation consisting of 15×15 matrices. However, for any group, only a limited number of representations are of fundamental significance, and we shall now discuss the origin and properties of these.

Suppose that we have a set of matrices, $\mathscr{E}$, $\mathscr{A}$, $\mathscr{B}$, $\mathscr{C}$, ... which form a representation of a group. If we make the same similarity transformation on each matrix, we obtain a new set of matrices, viz.,

$$\mathscr{E}' = \mathscr{Q}^{-1}\mathscr{E}\mathscr{Q}$$
$$\mathscr{A}' = \mathscr{Q}^{-1}\mathscr{A}\mathscr{Q}$$
$$\mathscr{B}' = \mathscr{Q}^{-1}\mathscr{B}\mathscr{Q}$$

$$\cdots$$

It is easy to prove that the new set of matrices is also a representation of the group. Suppose that

$$\mathscr{A}\mathscr{B} = \mathscr{D}$$

then

$$\mathscr{A}'\mathscr{B}' = (\mathscr{2}^{-1}\mathscr{A}\mathscr{2})(\mathscr{2}^{-1}\mathscr{B}\mathscr{2}) = \mathscr{2}^{-1}\mathscr{A}(\mathscr{2}\mathscr{2}^{-1})\mathscr{B}\mathscr{2}$$
$$= \mathscr{2}^{-1}(\mathscr{A}\mathscr{B})\mathscr{2} = \mathscr{2}^{-1}\mathscr{D}\mathscr{2} = \mathscr{D}'$$

Clearly all products in the set of matrices $\mathscr{E}'$, $\mathscr{A}'$, $\mathscr{B}'$, ... will run parallel to those in the representation $\mathscr{E}$, $\mathscr{A}$, $\mathscr{B}$, ...; hence the primed set also constitutes a representation.

Let us now suppose that, when the matrix $\mathscr{A}$ is transformed to $\mathscr{A}'$ using $\mathscr{2}$ or some other matrix, we find $\mathscr{A}'$ to be a block-factored matrix, namely,

$$\mathscr{A}' = \mathscr{2}^{-1}\mathscr{A}\mathscr{2} = \begin{bmatrix} \mathscr{A}'_1 & & & & \\ & \mathscr{A}'_2 & & & \\ & & \mathscr{A}'_3 & & \\ & & & \mathscr{A}'_4 & \\ & & & & \mathscr{A}'_5 \end{bmatrix}$$

for example. If now each of the matrices $\mathscr{A}'$, $\mathscr{B}'$, $\mathscr{C}'$, and so forth is blocked out in the same way, then, as shown on page 66, corresponding blocks of each matrix can be multiplied together separately. Thus we can write such equations as:

$$\mathscr{A}'_1\mathscr{B}'_1 = \mathscr{D}'_1$$
$$\mathscr{A}'_2\mathscr{B}'_2 = \mathscr{D}'_2$$
$$\mathscr{A}'_4\mathscr{B}'_4 = \mathscr{D}'_4$$
$$\cdots$$

Therefore the various sets of matrices

$$\mathscr{E}'_1, \ \mathscr{A}'_1, \ \mathscr{B}'_1, \ \mathscr{C}'_1, \ \mathscr{D}'_1, \cdots$$
$$\mathscr{E}'_2, \ \mathscr{A}'_2, \ \mathscr{B}'_2, \ \mathscr{C}'_2, \ \mathscr{D}'_2, \cdots$$
$$\cdots$$

are in themselves representations of the group. We then call the set of matrices, $\mathscr{E}$, $\mathscr{A}$, $\mathscr{B}$, $\mathscr{C}$, $\mathscr{D}$, ..., a *reducible* representation, because it is possible, using some matrix, $\mathscr{2}$ in this case, to transform each matrix in the set into a new one so that all of the new ones can be taken apart in the same way to give two or more representations of smaller dimension. (The dimension of a representation is the order of the square matrices which constitute it.) If it is not possible to find a similarity transformation which will reduce all of the matrices of a given representation in the above manner, the representation is said to be *irreducible*. It is the irreducible representations of a group which are of fundamental importance, and their main properties will now be described.

4.3   The "Great Orthogonality Theorem" and Its Consequences

All of the properties of group representations and their characters which are important in dealing with problems in valence theory and molecular dynamics can be derived from one basic theorem concerning the elements of the matrices which constitute the irreducible representations of a group. In order to state this theorem, which we shall do without proof,* some notation must be introduced. The order of a group will, as before, be denoted by h. The dimension of the ith representation, which is the order of each of the matrices which constitute it, will be denoted by l_i. The various operations in the group will be given the generic symbol R. The element in the mth row and the nth column of the matrix corresponding to an operation R in the ith irreducible representation will be denoted $\Gamma_i(R)_{mn}$. Finally, it is necessary to take the complex conjugate of one factor on the left-hand side whenever imaginary or complex numbers are involved.

The great orthogonality theorem may then be stated as follows:

$$\sum_R [\Gamma_i(R)_{mn}][\Gamma_j(R)_{m'n'}]^* = \frac{h}{\sqrt{l_i l_j}} \delta_{ij} \delta_{mm'} \delta_{nn'} \qquad (4.3\text{-}1)$$

This means that in the set of matrices constituting any one irreducible representation any set of corresponding matrix elements, one from each matrix, behaves as the components of a vector in h-dimensional space such that all these vectors are mutually orthogonal, and each is normalized so that the square of its length equals h/l_i. This interpretation of Equation 4.3-1 will perhaps be more obvious if we, as it were, take 4.3-1 apart into three simpler equations, each of which is contained within it. We shall omit the explicit designation of complex conjugates for simplicity, but it should be remem-

---

* The proof, which is not trivial, may be found elsewhere, for example, in *Quantum Chemistry*, by H. Eyring, J. Walter, and G. E. Kimball, John Wiley, New York, 1944, p. 371.

bered that they must be used when complex numbers are involved. The three simpler equations are as follows

$$\sum_R \Gamma_i(R)_{mn} \Gamma_j(R)_{mn} = 0 \quad \text{if} \quad i \neq j \tag{4.3-2}$$

$$\sum_R \Gamma_i(R)_{mn} \Gamma_i(R)_{m'n'} = 0 \quad \text{if} \quad m \neq m' \quad \text{and/or} \quad n \neq n' \tag{4.3-3}$$

$$\sum_R \Gamma_i(R)_{mn} \Gamma_i(R)_{mn} = h/l_i \tag{4.3-4}$$

Thus, if the vectors differ by being chosen from matrices of different representations, they are orthogonal (4.3-2). If they are chosen from the same representation but from different sets of elements in the matrices of this representation, they are orthogonal (4.3-3). Finally, 4.3-4 expresses the fact that the square of the length of any such vector equals h/l_i.

We shall now discuss five important rules about irreducible representations and their characters.

1. *The sum of the squares of the dimensions of the irreducible representations of a group is equal to the order of the group, that is,*

$$\Sigma l_i^2 = l_1^2 + l_2^2 + l_3^2 + \cdots = h \tag{4.3-5}$$

PROOF. A complete proof is quite lengthy and will not be given. It is, however, easy to show that $\Sigma l_i^2 \leqslant h$. In a matrix of order l there are l^2 elements. Thus each irreducible representation, Γ_i, will provide l_i^2 h-dimensional vectors. The basic theorem requires this set of $l_1^2 + l_2^2 + l_3^2 + \cdots$ vectors to be mutually orthogonal. Since there can be no more than h orthogonal h-dimensional vectors, the sum $l_1^2 + l_2^2 + l_3^2 + \cdots$ may not exceed h. Since $\chi_i(E)$, the character of the representation of E in the ith irreducible representation, is equal to the order of the representation, we can also write this rule as

$$\sum_i [\chi_i(E)]^2 = h \tag{4.3-5a}$$

2. *The sum of the squares of the characters in any irreducible representation equals h, that is,*

$$\sum_R [\chi_i(R)]^2 = h \tag{4.3-6}$$

PROOF. From 4.3-1 we may write

$$\sum_R \Gamma_i(R)_{mm} \Gamma_i(R)_{m'm'} = \frac{h}{l_i} \delta_{mm'}$$

Summing the left side over m and m', we obtain

$$\sum_{m'}\sum_{m}\sum_{R}\Gamma_i(R)_{mm}\Gamma_i(R)_{m'm'} = \sum_{R}\left[\sum_{m}\Gamma_i(R)_{mm}\sum_{m'}\Gamma_i(R)_{m'm'}\right]$$

$$= \sum_{R}\chi_i(R)\chi_i(R)$$

$$= \sum_{R}[\chi_i(R)]^2$$

while summing the right side over m and m', we obtain

$$\frac{h}{l_i}\sum_{m'}\sum_{m}\delta_{mm'} = \frac{h}{l_i}l_i = h$$

thus proving the equality 4.3-6.

3. *The vectors whose components are the characters of two different irreducible representations are orthogonal, that is,*

$$\sum_{R}\chi_i(R)\chi_j(R) = 0 \quad \text{when} \quad i \neq j \tag{4.3-7}$$

PROOF. Setting $m = n$ in 4.3-2, we obtain

$$\sum_{R}\Gamma_i(R)_{mm}\Gamma_j(R)_{mm} = 0 \quad \text{if} \quad i \neq j$$

$$\sum_{R}\chi_i(R)\chi_j(R) = \sum_{R}\left[\sum_{m}\Gamma_i(R)_{mm}\sum_{m}\Gamma_j(R)_{mm}\right]$$

$$= \sum_{m}\left[\sum_{R}\Gamma_i(R)_{mm}\Gamma_j(R)_{mm}\right] = 0$$

4. *In a given representation (reducible or irreducible) the characters of all matrices belonging to operations in the same class are identical.*

PROOF. Since all elements in the same class are conjugate to one another, all matrices corresponding to elements in the same class in any representation must be conjugate. But we have shown on page 67 that conjugate matrices have identical characters.

5. *The number of irreducible representations of a group is equal to the number of classes in the group.*

PROOF. As for rule 1, a complete proof will not be given; we can, however, easily prove that the number of classes sets an upper limit on the number of irreducible representations. We can combine Equations 4.3-6 and 4.3-7 into one equation, viz.,

$$\sum_{R}\chi_i(R)\chi_j(R) = h\delta_{ij} \tag{4.3-8}$$

If now we denote the number of elements in the mth class by g_m, the number in the nth class by g_n, and so on, and if there are k classes altogether, 4.3-8 can be rewritten:

$$\sum_{p=1}^{k} \chi_i(R_p)\chi_j(R_p)g_p = h\delta_{ij} \tag{4.3-9}$$

where R_p refers to any one of the operations in the pth class. Equation 4.3-9 implies that the k quantities, $\chi_l(R_p)$, in each representation Γ_l behave like the components of a k-dimensional vector and that these k vectors are mutually orthogonal. Since only k k-dimensional vectors *can* be mutually orthogonal, there can be no more than k irreducible representations in a group which has k classes.

Let us now consider the irreducible representations of several typical groups to see how these rules apply. The group C_{2v} consists of four elements, and each is in a separate class. Hence (rule 5) there are four irreducible representations for this group. But it is also required (rule 1) that the sum of the squares of the dimensions of these representations equal h. Thus we are looking for a set of four positive integers, l_1, l_2, l_3, and l_4, which satisfy the relation

$$l_1{}^2 + l_2{}^2 + l_3{}^2 + l_4{}^2 = 4$$

Clearly the only solution is

$$l_1 = l_2 = l_3 = l_4 = 1$$

Thus the group C_{2v} has four one-dimensional irreducible representations.

We can actually work out the characters of these four irreducible representations—which are in this case the representations themselves because the dimensions are 1—on the basis of the vector properties of the representations and the rules derived above. One suitable vector in 4-space which has a component of 1 corresponding to E will obviously be

	E	C_2	σ_v	σ_v'
Γ_1	1	1	1	1

for

$$\sum_R [\chi_1(R)]^2 = 1^2 + 1^2 + 1^2 + 1^2 = 4$$

thus satisfying rule 2. Now all other representations will have to be such that

$$\sum_R [\chi_i(R)]^2 = 4$$

which can be true only if each $\chi_i(R) = \pm 1$. Moreover, in order for each of the other representations to be orthogonal to Γ_1 (rule 3 and Equation 4.3-7), there will have to be two $+1$'s and two -1's. Thus

$$(1)(-1) + (1)(-1) + (1)(1) + (1)(1) = 0$$

Therefore we will have

	E	C_2	σ_v	σ_v'
Γ_1	1	1	1	1
Γ_2	1	-1	-1	1
Γ_3	1	-1	1	-1
Γ_4	1	1	-1	-1

All of these representations are also orthogonal to one another. For example, taking Γ_2 and Γ_4, we have

$$(1)(1) + (-1)(1) + (-1)(-1) + (1)(-1) = 0$$

and so on. These are then the four irreducible representations of the group C_{2v}.

As another example of the working of the rules, let us consider the group C_{3v}. This consists of the following elements, listed by classes:

$$E \quad 2C_3 \quad 3\sigma_v$$

We therefore know at once that there are three irreducible representations. If we denote their dimensions by l_1, l_2, and l_3, we have (rule 1)

$$l_1{}^2 + l_2{}^2 + l_3{}^2 = h = 6$$

The only values of the l_i which will satisfy this requirement are 1, 1, and 2. Now once again, and always in any group, there will be a one-dimensional representation whose characters are all equal to 1. Thus we have

	E	$2C_3$	$3\sigma_v$
Γ_1	1	1	1

Note that (from Equation 4.3-9)

$$1^2 + 2(1)^2 + 3(1)^2 = 6$$

We now look for a second vector in 6-space all of whose components are equal to ± 1 which is orthogonal to Γ_1. The components of such a vector must consist of three $+1$'s and three -1's. Since $\chi(E)$ must always be positive and since all elements in the same class must have representations with the same character, the only possibility here is:

	E	$2C_3$	$3\sigma_v$
Γ_1	1	1	1
Γ_2	1	1	-1

Now our third representation will be of dimension 2. Hence $\chi_3(E) = 2$. In order to find out the values of $\chi_3(C_3)$ and $\chi_3(\sigma_v)$ we make use of the orthogonality relationships (rule 3, Equation 4.3-7):

$$\sum_R \chi_1(R)\chi_3(R) = [1][2] + 2[1][\chi_3(C_3)] + 3[1][\chi_3(\sigma_v)] = 0$$

$$\sum_R \chi_2(R)\chi_3(R) = [1][2] + 2[1][\chi_3(C_3)] + 3[-1][\chi_3(\sigma_v)] = 0$$

Solving these, we obtain

$$2\chi_3(C_3) + 3\chi_3(\sigma_v) = -2$$
$$-[2\chi_3(C_3) - 3\chi_3(\sigma_v) = -2]$$
$$\overline{\qquad\qquad 6\chi_3(\sigma_v) = 0}$$
$$\chi_3(\sigma_v) = 0$$

and

$$2\chi_3(C_3) + 3(0) = -2$$
$$\chi_3(C_3) = -1$$

Thus the complete set of characters of the irreducible representations is

	E	$2C_3$	$3\sigma_v$
Γ_1	1	1	1
Γ_2	1	1	-1
Γ_3	2	-1	0

We may note that there is still a check on the correctness of Γ_3: the square of the length of the vector it defines should be equal to h (rule 2), and we see that this is so:

$$2^2 + 2(-1)^2 + 3(0)^2 = 6$$

We shall conclude this section by deriving a relationship between any reducible representation of a group and the irreducible representations of that group. In terms of practical application of group theory to molecular problems, this relationship is of pivotal importance. We know already that for any reducible representation it is possible to find some similarity transformation which will reduce each matrix to one consisting of blocks along the diagonal, each of which belongs to an irreducible representation of the group. We also know that the character of a matrix is not changed by any similarity transformation. Thus we can write

$$\chi(R) = \sum_j a_j \chi_j(R) \tag{4.3-10}$$

where $\chi(R)$ is the character of the matrix corresponding to operation R in the reducible representation, and a_j represents the number of times the block constituting the jth irreducible representation will appear along the diagonal when the reducible representation is completely reduced by the necessary similarity transformation. Now we do not need to bother about the difficult question of how to find out what matrix is required to reduce completely the reducible representation in order to find the values of the a_j. We can obtain the required relationship by working only with the characters of all representations in the following way. We multiply each side of 4.3-10 by $\chi_i(R)$ and then sum each side over all operations, viz.,

$$\sum_R \chi(R)\chi_i(R) = \sum_R \sum_j a_j \chi_j(R)\chi_i(R)$$

$$= \sum_j \sum_R a_j \chi_j(R)\chi_i(R)$$

Now for each of the terms in the sum over j, we have from Equation 4.3-8

$$\sum_R a_j \chi_j(R)\chi_i(R) = a_j \sum_R \chi_j(R)\chi_i(R) = a_j h \delta_{ij}$$

since the sets of characters $\chi_j(R)$ and $\chi_i(R)$ define orthogonal vectors, the squares of whose lengths equal h. Thus, in summing over all j, only the sum over R in which $i = j$ can survive, and in that case we have

$$\sum_R \chi(R)\chi_i(R) = h a_i$$

which we rearrange to read

$$a_i = \frac{1}{h} \sum_R \chi(R)\chi_i(R) \tag{4.3-11}$$

Thus we can readily determine the number of times the ith irreducible representation occurs in a reducible representation when we know only the characters of each representation.

Let us take an example. For the group C_{3v} we give below the characters of the irreducible representations, Γ_1, Γ_2, and Γ_3, and the characters of two reducible representations, Γ_a and Γ_b.

C_{3v}	E	$2C_3$	$3\sigma_v$
Γ_1	1	1	1
Γ_2	1	1	−1
Γ_3	2	−1	0
Γ_a	5	2	−1
Γ_b	7	1	−3

Using 4.3-11, we find for Γ_a

$$a_1 = \tfrac{1}{6}[1(1)(5) + 2(1)(2) + 3(1)(-1)] = 1$$
$$a_2 = \tfrac{1}{6}[1(1)(5) + 2(1)(2) + 3(-1)(-1)] = 2$$
$$a_3 = \tfrac{1}{6}[1(2)(5) + 2(-1)(2) + 3(0)(-1)] = 1$$

and for Γ_b

$$a_1 = \tfrac{1}{6}[1(1)(7) + 2(1)(1) + 3(1)(-3)] = 0$$
$$a_2 = \tfrac{1}{6}[1(1)(7) + 2(1)(1) + 3(-1)(-3)] = 3$$
$$a_3 = \tfrac{1}{6}[1(2)(7) + 2(-1)(1) + 3(0)(-3)] = 2$$

The numbers in italics are the numbers of elements in each class. The results obtained above will be found to satisfy 4.3-10, as of course they must. For Γ_a we have

	E	$2C_3$	$3\sigma_v$
Γ_1	1	1	1
Γ_2	1	1	−1
Γ_2	1	1	−1
Γ_3	2	−1	0
Γ_a	5	2	−1

and for Γ_b

	E	$2C_3$	$3\sigma_v$
Γ_2	1	1	−1
Γ_2	1	1	−1
Γ_2	1	1	−1
Γ_3	2	−1	0
Γ_3	2	−1	0
Γ_b	7	1	−3

Indeed, in simple cases, a reducible representation may often be reduced very quickly by using Equation 4.3-10, that is, by looking for the rows of characters which add up to the correct total in each column. For more complicated cases, it is usually best to use Equation 4.3-11; however, 4.3-10 then provides a valuable check on the results.

4.4   Character Tables

Throughout all of our applications of group theory and molecular symmetry we will utilize devices called character tables. A set of these for all symmetry groups likely to be encountered among real molecules is given in Appendix IIIA.* In this section we shall explain the meaning and indicate the source of the information given in these tables. For this purpose we shall examine in detail a representative character table, one for the group C_{3v}, reproduced below. The four† main areas of the table have been assigned Roman numerals for reference in the following discussion.

C_{3v}	E	$2C_3$	$3\sigma_v$		
A_1	1	1	1	z	x^2+y^2, z^2
A_2	1	1	-1	R_z	
E	2	-1	0	$(x,y)(R_x, R_y)$	$(x^2-y^2, xy)(xz, yz)$
II		**I**		**III**	**IV**

In the top row are these entries: In the upper left corner is the Schoenflies symbol for the group. Then, along the top row of the main body of the table, are listed the elements of the group, gathered into classes; the notation is the kind explained in Section 3.15.

Area I. In area I of the table are the characters of the irreducible representations of the group. These have been fully discussed in preceding sections of this chapter and require no additional comment here.

Area II. We have previously designated the ith representation, or its set of characters, by the symbol Γ_i in a fairly arbitrary way. Although this practice is still to be found in some places and is common in older literature, most books and papers—in fact, virtually all those by English-speaking authors—now use the kind of symbols found in the C_{3v} table above and all tables in Appendix III. This nomenclature was proposed by R. S. Mulliken, and the symbols are normally called Mulliken symbols. Their meanings are as follows:

---

* Appendix IIIA will be found as a separate booklet in a pocket in the back of this book.
† In some books areas III and IV are combined.

1. All one-dimensional representations are designated either A or B; two-dimensional representations are designated E; three-dimensional species are designated T (or sometimes F).

2. One-dimensional representations which are symmetric with respect to rotation by $2\pi/n$ about the principal C_n axis [symmetric meaning: $\chi(C_n) = 1$] are designated A, while those antisymmetric in this respect [$\chi(C_n) = -1$] are designated B.

3. Subscripts 1 and 2 are usually attached to A's and B's to designate those which are, respectively, symmetric and antisymmetric with respect to a C_2 perpendicular to the principal axis or, if such a C_2 axis is lacking, to a vertical plane of symmetry.

4. Primes and double primes are attached to all letters, when appropriate, to indicate those which are, respectively, symmetric and antisymmetric with respect to σ_h.

5. In groups with a center of inversion, the subscript g (from the German *gerade*, meaning even) is attached to symbols for representations which are symmetric with respect to inversion and the subscript u (from the German *ungerade*, meaning uneven) is used for those which are antisymmetric to inversion.

6. The use of numerical subscripts for E's and T's also follows certain rules, but these cannot be easily stated precisely without some mathematical development. It will be satisfactory here to regard them as arbitrary labels.

Area III. In area III we will always find six symbols: x, y, z, R_x, R_y, R_z. The first three represent the coordinates x, y, and z, while the R's stand for rotations about the axes specified in the subscripts. We shall now show in an illustrative but by no means thorough way why these symbols are assigned to certain representations in the group C_{3v}, and this should suffice to indicate the basis for the assignments in other groups.

Any set of algebraic functions or vectors may serve as the *basis* for a representation of a group. In order to use them for a basis, we consider them to be the components of a vector and then determine the matrices which show how that vector is transformed by each symmetry operation. The resulting matrices, naturally, constitute a representation of the group. We have previously used the coordinates x, y, and z as a basis for representations of groups C_{2v} (page 75) and T (page 71). In the present case it will be easily seen that the matrices for one operation in each of the three classes are as follows:

$$
E:\quad
\begin{bmatrix}
1 & 0 & 0 \\
0 & 1 & 0 \\
0 & 0 & 1
\end{bmatrix}
\qquad
C_3:\quad
\begin{bmatrix}
\cos 2\pi/3 & -\sin 2\pi/3 & 0 \\
\sin 2\pi/3 & \cos 2\pi/3 & 0 \\
0 & 0 & 1
\end{bmatrix}
\qquad
\sigma_v:\quad
\begin{bmatrix}
1 & 0 & 0 \\
0 & -1 & 0 \\
0 & 0 & 1
\end{bmatrix}
$$

Now the first thing we can observe about these matrices is that they never mix z with x or y; that is, z' is always a function of z only. Hence z by itself forms an independent representation of the group. On the other hand C_3 mixes up x and y to give x' and y', so x and y jointly form a representation. This is equivalent to observing that the three matrices are all block-factored in the same way, namely, into the following submatrices:

$$
\begin{array}{cccc}
 & E & C_3 & \sigma_v \\
\Gamma_{x,y} & \begin{bmatrix} 1 & 0 \\ 0 & 1 \end{bmatrix} & \begin{bmatrix} \cos 2\pi/3 & -\sin 2\pi/3 \\ \sin 2\pi/3 & \cos 2\pi/3 \end{bmatrix} & \begin{bmatrix} 1 & 0 \\ 0 & -1 \end{bmatrix} \\
\Gamma_z & 1 & 1 & 1
\end{array}
$$

We see that Γ_z is the A_1 irreducible representation. This means that the coordinate z forms a basis for the A_1 representation, or, as we also say, "z transforms as (or according to) A_1." If we examine the characters of $\Gamma_{x,y}$, we find them to be those of the E representation ($2\cos 2\pi/3 = -1$), so that the coordinates x and y *together* transform as or according to the E representation. It is important to grasp that x and y are inseparable in this respect, since the representation for which they form a basis is irreducible.

A thorough treatment of how the transformation properties of the rotations are determined would be an unnecessary digression from this discussion. In simple cases we can obtain the answer in a semipictorial way by letting a curved arrow about the axis stand for a rotation. Thus such an arrow around the z axis is transformed into itself by E, it is transformed into itself by C_3, and its direction is reversed by σ_v. Thus it is the basis for a representation with the characters $1, 1, -1$, and so we see that R_z transforms as A_2.

Area IV. In this part of the table are listed all of the squares and binary products of coordinates according to their transformation properties. These results are quite easy to work out using the same procedure as for x, y, and z, except that the amount of algebra generally increases, though not always. For example, the pair of functions xz and yz must have the same transformation properties as the pair x, y, since z goes into itself under all symmetry operations in the group. Accordingly, (xz, yz) are found opposite the E representation.

4.5   Representations for Cyclic Groups

As noted earlier, a cyclic group is Abelian, and each of its h elements is in a separate class. Therefore, it must have h one-dimensional irreducible representations. To obtain these there is a perfectly general scheme which is perhaps best explained by an example. It will be evident that the example may be generalized. Let us consider the group C_5, consisting of the five commuting operations C_5, C_5^2, C_5^3, C_5^4, $C_5^5 \equiv E$; we seek a set of five one-dimensional representations, Γ^1, Γ^2, Γ^3, Γ^4, Γ^5, which are orthonormal in the sense

$$\sum_{m=1}^{m=5} [\Gamma^p(C_n^m)][\Gamma^q(C_n^m)]^* = h\delta_{pq} \tag{4.5-1}$$

We shall use the exponentials

$$\exp(2\pi i p/5) = \cos 2\pi p/5 + i \sin 2\pi p/5 \tag{4.5-2}$$

as the $\Gamma^p(C_5)$. Abbreviating these exponentials as ε^p [i.e., $\varepsilon = \exp(2\pi i/5)$], we write the first column of the following table:

	C_5	C_5^2	C_5^3	C_5^4	$C_5^5 \equiv E$
Γ^1	ε	ε^2	ε^3	ε^4	ε^5
Γ^2	ε^2	ε^4	ε^6	ε^8	ε^{10}
Γ^3	ε^3	ε^6	ε^9	ε^{12}	ε^{15}
Γ^4	ε^4	ε^8	ε^{12}	ε^{16}	ε^{20}
Γ^5	ε^5	ε^{10}	ε^{15}	ε^{20}	ε^{25}

The remaining columns follow from the group multiplications. It will now be shown that these representations satisfy the orthonormalization condition of Equation 4.5-1.

Consider any two representations, say Γ^p and Γ^q, where $q - p = r$. The left-hand side of Equation 4.5-1 takes the form

$$(\varepsilon^p)^* \varepsilon^{p+r} + (\varepsilon^{2p})^* \varepsilon^{2(p+r)} + (\varepsilon^{3p})^* \varepsilon^{3(p+r)} + (\varepsilon^{4p})^* \varepsilon^{4(p+r)} + (\varepsilon^{5p})^* \varepsilon^{5(p+r)} \tag{4.5-3a}$$

This may be rewritten as:

$$(\varepsilon^p)^* \varepsilon^p \varepsilon^r + (\varepsilon^{2p})^* \varepsilon^{2p} \varepsilon^{2r} + (\varepsilon^{3p})^* \varepsilon^{3p} \varepsilon^{3r} + (\varepsilon^{4p})^* \varepsilon^{4p} \varepsilon^{4r} + (\varepsilon^{5p})^* \varepsilon^{5p} \varepsilon^{5r} \tag{4.5-3b}$$

Since for exponentials in general

$$e^{ix}(e^{ix})^* = e^{ix}e^{-ix} = 1$$

expression 4.5-3b reduces to the relatively simple sum:

$$\varepsilon^r + \varepsilon^{2r} + \varepsilon^{3r} + \varepsilon^{4r} + \varepsilon^{5r} \tag{4.5-3c}$$

It is then clear that the representations are normalized, because if $\Gamma^p = \Gamma^q$, $r = 0$ and 4.5-3c is simply five times $e^0 = 1$, namely, 5.

If Γ^p and Γ^q are different, r is some number from 1 to 4. Because $\varepsilon^5 = 1$ (see Equation 4.5-2), we have such equalities as

$$\varepsilon^8 = \varepsilon^5 \varepsilon^3 = \varepsilon^3 \tag{4.5-4}$$

Therefore any sum of the type 4.5-3c reduces to

$$\varepsilon + \varepsilon^2 + \varepsilon^3 + \varepsilon^4 + \varepsilon^5 = \sum_{n=1}^{n=5} \exp(2\pi i n/5) \qquad (4.5\text{-}5)$$

Because of the following trigonometric identities

$$\sum_{n=1}^{n=l} \cos 2\pi n/l = 0$$

$$\sum_{n=1}^{n=l} \sin 2\pi n/l = 0$$

it follows that the sum in 4.5-5 equals 0, thus proving the orthogonality of the representations.

If we now replace all ε^n's, such as ε^5, ε^{10}, ε^{15}, ..., which are equal to unity by the number 1, and reduce all other exponents in excess of 5 to their lowest values as indicated in 4.5-4, we can rewrite our table in the following form:

	C_5	$C_5{}^2$	$C_5{}^3$	$C_5{}^4$	$C_5{}^5$
Γ^1	ε	ε^2	ε^3	ε^4	1
Γ^2	ε^2	ε^4	ε	ε^3	1
Γ^3	ε^3	ε	ε^4	ε^2	1
Γ^4	ε^4	ε^3	ε^2	ε	1
Γ^5	1	1	1	1	1

Inspection of (4.5-2) will show that two ε^i's whose exponents add up to 5 are complex conjugates of each other. For example:

$$(\varepsilon^4)^* = (\cos 2\pi\tfrac{4}{5} + i \sin 2\pi\tfrac{4}{5})^* = \cos 2\pi\tfrac{4}{5} - i \sin 2\pi\tfrac{4}{5}$$
$$= \cos 2\pi\tfrac{1}{5} + i \sin 2\pi\tfrac{1}{5} = \varepsilon$$

We now rewrite the table once more, replacing ε^3 by $(\varepsilon^2)^*$ and ε^4 by ε^* and also rearranging the rows and columns, and obtain

NEW DESIGNATION		E	C_5	$C_5{}^2$	$C_5{}^3$	$C_5{}^4$	OLD DESIGNATION
A		1	1	1	1	1	Γ^5
E_1	$\{$	1	ε	ε^2	ε^{2*}	ε^*	Γ^1
		1	ε^*	ε^{2*}	ε^2	ε	Γ^4
E_2	$\{$	1	ε^2	ε^*	ε	ε^{2*}	Γ^2
		1	ε^{2*}	ε	ε^*	ε^2	Γ^3

The new arrangement of columns is such as to place the $C_5{}^5 \equiv E$ column first. The new ordering of rows places the totally symmetric representation at the top; it is denoted A. The remaining representations are then associated in pairs, such that the elements of one row in each pair are the complex conjugates of the other. Each member of these pairs must be regarded as a separate representation in order to satisfy the requirements as to the number and dimensions of the irreducible representations. However, for certain applications to physical problems we shall prefer to add them, thus obtaining a set of characters for a representation of dimension 2. Because we are always adding a complex number to its complex conjugate, the sums are invariably pure real numbers. Because of this property of adding to give characters like those for a two-dimensional representation, the pairs of representations are jointly denoted by the Mulliken symbol E.

Let us take the group C_3 as an example. Its correct character table is, in part,

C_3	E	C_3	$C_3{}^2$	
A	1	1	1	
E	$\begin{Bmatrix} 1 & \varepsilon & \varepsilon^* \\ 1 & \varepsilon^* & \varepsilon \end{Bmatrix}$			$\varepsilon = \exp(2\pi i/3)$

and when we combine the two parts of the "E representation" we obtain

C_3	E	C_3	$C_3{}^2$
A	1	1	1
E	2	$2\cos 2\pi/3$	$2\cos 2\pi/3$

To show that the table in this form is serviceable for physical problems, let us work out the transformation properties of the x, y, and z coordinates in the same way as we did above for the group C_{3v}. Here we obtain the matrices

$$
\begin{array}{ccc}
E & C_3 & C_3{}^2 \\
\begin{bmatrix} 1 & 0 & 0 \\ 0 & 1 & 0 \\ 0 & 0 & 1 \end{bmatrix} &
\begin{bmatrix} \cos 2\pi/3 & -\sin 2\pi/3 & 0 \\ \sin 2\pi/3 & \cos 2\pi/3 & 0 \\ 0 & 0 & 1 \end{bmatrix} &
\begin{bmatrix} \cos 4\pi/3 & -\sin 4\pi/3 & 0 \\ \sin 4\pi/3 & \cos 4\pi/3 & 0 \\ 0 & 0 & 1 \end{bmatrix}
\end{array}
$$

Again we see that these matrices give a reducible representation which can be reduced on inspection to two representations having the following characters:

	E	C_3	$C_3{}^2$
Γ_z	1	1	1
$\Gamma_{x,y}$	2	$2\cos 2\pi/3$	$2\cos 2\pi/3$

Thus x and y transform according to E, and z transforms according to A_1, which are the results stated in the complete form of the mathematically correct table.

Exercises

4.1 Using matrix methods, verify all the rules concerning the products and the commutation of symmetry operations given on pages 31–32.

4.2 Derive the complete matrices for all representations of the group C_{3v}. *Hint*: Write out and reduce matrices for expressing the transformations of the general point (x, y, z).

4.3 Using matrix multiplication, carry out the required similarity transformations to determine the arrangement of the twelve operations in the group T into classes.

4.4 For the group O use the sets of functions (x, y, z), (xy, xz, yz), and $(2z^2 - x^2 - y^2, x^2 - y^2)$ to work out the complete irreducible matrix representations designated T_1, T_2, and E respectively.

4.5 Prove that all irreducible representations of Abelian groups must be one-dimensional.

4.6 Work out the character table for the group C_4. Present it in the conventional format, reducing all exponentials to their simplest form.

4.7 Consider the group D_3. Let the C_3 axis coincide with the z axis and one of the C_2 axes coincide with the x axis. Write out the complete matrices for all irreducible representations of this group. Derive from these the character table.

4.8 Write the matrices describing the effect on a point (x, y, z) of reflections in vertical planes which lie halfway between the xz and yz planes. By matrix methods determine what operations result when each of these reflections is followed by reflection in the xy plane.

Group Theory and Quantum Mechanics

5.1 Wave Functions as Bases for Irreducible Representations

It will be appropriate to give first a brief discussion of the wave equation. It is not necessary that the reader have any very extensive knowledge of wave mechanics in order to follow the development in this chapter, but the information given here is essential. The wave equation for any physical system is

$$\mathscr{H}\Psi = E\Psi \tag{5.1-1}$$

Here $\mathscr{H}$ is the Hamiltonian operator, which indicates that certain operations are to be carried out on a function written to its right. The wave equation states that, if the function is an *eigenfunction*, the result of performing the operations indicated by $\mathscr{H}$ will yield the function itself multiplied by a constant which is called an *eigenvalue*. Eigenfunctions are conventionally denoted Ψ, and the eigenvalue, which is the energy of the system, is denoted E.

The Hamiltonian operator is obtained by writing down the expression for the classical energy of the system, which is simply the sum of the potential and kinetic energies for systems of interest to us, and then replacing the momentum terms by differential operators according to the postulates of wave mechanics. We need not concern ourselves with this construction of the Hamiltonian operator in any detail. The only property of it which we shall have to use explicitly concerns its symmetry with respect to the interchange of like particles in the system to which it applies. The particles in the system will be electrons and atomic nuclei. If any two or more particles are interchanged by carrying out a symmetry operation on the system, the Hamiltonian must be *unchanged*. A symmetry operation carries the system into an equivalent configuration, which is, by definition, physically indistinguishable from the original configuration. Clearly then, the energy of the system must be the same before and after carrying out the symmetry operation. Thus we say that any symmetry operator commutes with the Hamiltonian operator, and we can write

$$R\mathscr{H} = \mathscr{H}R \tag{5.1-2}$$

The Hamiltonian operator also commutes with any constant factor, c. Thus

$$\mathscr{H}c\Psi = c\mathscr{H}\Psi = cE\Psi \qquad (5.1\text{-}3)$$

It has been more or less implied up to this point that for any eigenvalue E_i there is one appropriate eigenfunction Ψ_i. This is often true, but there are also many cases in which several eigenfunctions give the same eigenvalue, for example,

$$\mathscr{H}\Psi_{i1} = E_i\Psi_{i1}$$
$$\mathscr{H}\Psi_{i2} = E_i\Psi_{i2}$$
$$\vdots \qquad \vdots \qquad\qquad (5.1\text{-}4)$$
$$\mathscr{H}\Psi_{ik} = E_i\Psi_{ik}$$

In such cases we say that the eigenvalue is *degenerate*, and in the particular example given above we would say that the energy E_i is k-fold degenerate. Now in the case of a degenerate eigenvalue, not only does the initial set of eigenfunctions provide correct solutions to the wave equation, but also any linear combination of these is also a solution giving the same eigenvalue. This is easily shown as follows:

$$\mathscr{H}\sum_j a_{ij}\Psi_{ij} = \mathscr{H}a_{i1}\Psi_{i1} + \mathscr{H}a_{i2}\Psi_{i2} + \cdots + \mathscr{H}a_{ik}\Psi_{ik}$$
$$= E_i a_{i1}\Psi_{i1} + E_i a_{i2}\Psi_{i2} + \cdots + E_i a_{ik}\Psi_{ik}$$
$$= E_i\sum_j a_{ij}\Psi_{ij} \qquad (5.1\text{-}5)$$

One more important property of eigenfunctions must be mentioned. Eigenfunctions are so constructed as to be *orthonormal*, which means that

$$\int \Psi_i^*\Psi_j \, d\tau = \delta_{ij} \qquad (5.1\text{-}6)$$

where the integration is to be carried out over all of the coordinates, collectively represented by τ, which occur in Ψ_i and Ψ_j. When an eigenfunction belonging to the eigenvalue E_i is expressed as a linear combination of a set of eigenfunctions, we have, from 5.1-6,

$$\int \Psi_i^*\Psi_i \, d\tau = \int \left(\sum_j a_{ij}\Psi_{ij}^*\right)\left(\sum_{j'} a_{ij'}\Psi_{ij'}\right) d\tau$$

Now all products where $j \neq j'$ will vanish, for example,

$$\int a_{ij}\Psi_{ij}^* a_{ij'}\Psi_{ij'} \, d\tau = a_{ij}a_{ij'}\int \Psi_{ij}^*\Psi_{ij'} \, d\tau = 0$$

and, assuming that each Ψ_{ij} is normalized, we are left with

$$\int \sum_j a_{ij} \Psi_{ij}^* a_{ij} \Psi_{ij} \, d\tau = \sum_j a_{ij}^2 = 1 \tag{5.1-7}$$

We can now show that the eigenfunctions for a molecule are bases for irreducible representations of the symmetry group to which the molecule belongs. Let us take first the simple case of nondegenerate eigenvalues. If we take the wave equation for the molecule and carry out a symmetry operation, R, upon each side, then, from 5.1-1 and 5.1-2 we have

$$\mathcal{H} R \Psi_i = E_i R \Psi_i \tag{5.1-8}$$

Thus $R\Psi$ is itself an eigenfunction. Since Ψ is normalized, we must require, in order that $R\Psi$ also be normalized,

$$R\Psi_i = \pm 1 \Psi_i$$

Hence, by applying each of the operations in the group to an eigenfunction Ψ_i belonging to a nondegenerate eigenvalue, we generate a representation of the group with each matrix, $\Gamma_i(R)$, equal to ± 1. Since the representations are one-dimensional, they are obviously irreducible.

If we take the wave equation for the case where E_i is k-fold degenerate, then we must write, in analogy to 5.1-8,

$$\mathcal{H} R \Psi_{il} = E_i R \Psi_{il} \tag{5.1-9}$$

But here $R\Psi_{il}$ may in general be any linear combination of the Ψ_{ij}, that is,

$$R\Psi_{il} = \sum_{j=1}^{k} r_{jl} \Psi_{ij} \tag{5.1-10}$$

For some other operation, S, we have similarly,

$$S\Psi_{ij} = \sum_{m=1}^{k} s_{mj} \Psi_{im} \tag{5.1-11}$$

Because R and S are members of a symmetry group, there must be an element $T = SR$; its effect on Ψ_{il} can be expressed as

$$T\psi_{il} = \sum_{m=1}^{k} t_{ml} \psi_{im} \tag{5.1-12}$$

However, by combining the preceding expressions for the separate effects of S and R, we obtain

$$SR\psi_{il} = S \sum_{j=1}^{k} r_{jl} \psi_{ij} = \sum_{j=1}^{k} \sum_{m=1}^{k} s_{mj} r_{jl} \psi_{im} \tag{5.1-13}$$

Comparing 5.1-12 with 5.1-13, we see that

$$t_{ml} = \sum_{j=1}^{k} s_{mj} r_{jl}$$

But this is just the expression which gives the elements of a matrix, $\mathcal{T}$, which is the product, $\mathcal{SR}$, of two other matrices. Thus the matrices which describe the transformations of a set of k eigenfunctions corresponding to a k-fold degenerate eigenvalue are a k-dimensional representation for the group. Moreover, this representation is irreducible. If it were reducible we could divide the k eigenfunctions $\Psi_{i1}, \Psi_{i2}, \ldots, \Psi_{ik}$, or k linear combinations thereof, up into subsets such that the symmetry operations would send one member of the subset into a linear combination of only members of its own subset. Then the eigenvalue for members of one subset *could* be different from the eigenvalue for members of another subset. But this contradicts our original assumption that all of the Ψ_{il} *must* have the same eigenvalue.

To illustrate this explicitly, let us consider the $2p_x$ and $2p_y$ orbitals of the nitrogen atom in ammonia, which belongs to the group C_{3v}. These orbitals are represented or described by the following eigenfunctions:

$$p_x = R \sin \theta \cos \phi$$
$$p_y = R \sin \theta \sin \phi$$

where R is a constant insofar as symmetry operations are concerned, and θ and ϕ are angles in a polar coordinate system.* θ stands for an angle measured down from a reference axis, say the z axis, and ϕ denotes an angle measured in the counterclockwise direction from the x axis in the xy plane. Let us now work out the matrices which represent the transformations of these functions by each of the symmetry operations in the group C_{3v}. We consider what happens to a line whose direction is fixed initially by the angles θ_1 and ϕ_1. First of all we note that none of the operations in the group will affect θ, so that θ_2, the value of θ after application of a symmetry operation, will always equal θ_1. Hence

$$\sin \theta_2 = \sin \theta_1$$

If we rotate by $2\pi/3$ about the z axis, however, we have

$$\phi_2 = \phi_1 + 2\pi/3$$

and hence

$$\cos \phi_2 = \cos (\phi_1 + 2\pi/3) = \cos \phi_1 \cos 2\pi/3 - \sin \phi_1 \sin 2\pi/3$$
$$= -\tfrac{1}{2} \cos \phi_1 - (\sqrt{3}/2) \sin \phi_1$$
$$\sin \phi_2 = \sin (\phi_1 + 2\pi/3) = \sin \phi_1 \cos 2\pi/3 + \cos \phi_1 \sin 2\pi/3$$
$$= -\tfrac{1}{2} \sin \phi_1 + (\sqrt{3}/2) \cos \phi_1$$

* See Figure 8.1, page 195.

If we reflect in the xz plane, we have

$$\phi_2 = -\phi_1$$

and hence

$$\cos \phi_2 = \cos \phi_1$$
$$\sin \phi_2 = -\sin \phi_1$$

We can now use use this information to work out the required matrices.

E:

$$E p_x = E(R \sin \theta_1 \cos \phi_1) = R \sin \theta_2 \cos \phi_2 = R \sin \theta_1 \cos \phi_1 = p_x$$
$$E p_y = E(R \sin \theta_1 \sin \phi_1) = R \sin \theta_2 \sin \phi_2 = R \sin \theta_1 \sin \phi_1 = p_y$$

C_3:

$$
\begin{aligned}
C_3 p_x = C_3(R \sin \theta_1 \cos \phi_1) &= R \sin \theta_2 \cos \phi_2 \\
&= R (\sin \theta_1)(-\tfrac{1}{2})(\cos \phi_1 + \sqrt{3} \sin \phi_1) \\
&= -\tfrac{1}{2} R \sin \theta_1 \cos \phi_1 - (\sqrt{3}/2) R \sin \theta_1 \sin \phi_1 \\
&= -\tfrac{1}{2} p_x - (\sqrt{3}/2) p_y
\end{aligned}
$$

$$
\begin{aligned}
C_3 p_y = C_3(R \sin \theta_1 \sin \phi_1) &= R \sin \theta_2 \sin \phi_2 \\
&= R (\sin \theta_1)(-\tfrac{1}{2})(\sin \phi_1 - \sqrt{3} \cos \phi_1) \\
&= (\sqrt{3}/2) R \sin \theta_1 \cos \phi_1 - \tfrac{1}{2} R \sin \theta_1 \sin \phi_1 \\
&= (\sqrt{3}/2) p_x - \tfrac{1}{2} p_y
\end{aligned}
$$

σ_v:

$$\sigma_v p_x = \sigma_v(R \sin \theta_1 \cos \phi_1) = R \sin \theta_2 \cos \phi_2 = R \sin \theta_1 \cos \phi_1 = p_x$$
$$\sigma_v p_y = \sigma_v(R \sin \theta_1 \sin \phi_1) = R \sin \theta_2 \sin \phi_2 = -R \sin \theta_1 \sin \phi_1 = -p_y$$

Expressing these results in matrix notation, we write

$$\begin{bmatrix} 1 & 0 \\ 0 & 1 \end{bmatrix} \begin{bmatrix} p_x \\ p_y \end{bmatrix} = E \begin{bmatrix} p_x \\ p_y \end{bmatrix} \qquad \chi(E) = 2$$

$$\begin{bmatrix} -\tfrac{1}{2} & -\sqrt{3}/2 \\ \sqrt{3}/2 & -\tfrac{1}{2} \end{bmatrix} \begin{bmatrix} p_x \\ p_y \end{bmatrix} = C_3 \begin{bmatrix} p_x \\ p_y \end{bmatrix} \qquad \chi(C_3) = -1$$

$$\begin{bmatrix} 1 & 0 \\ 0 & -1 \end{bmatrix} \begin{bmatrix} p_x \\ p_y \end{bmatrix} = \sigma_v \begin{bmatrix} p_x \\ p_y \end{bmatrix} \qquad \chi(\sigma_v) = 0$$

The characters are seen to be those of the E representation of C_{3v}. Thus we see that the p_x and p_y orbitals, as a pair, provide a basis for the E

representation. It will be noted that the coordinates x and y are shown as transforming according to the E representation in the character table for the group C_{3v}. Thus the functions $\sin \theta \cos \phi$ and $\sin \theta \sin \phi$ transform in the same way as x and y. For this reason the p orbital which has an eigenfunction $\sin \theta \cos \phi$ is called p_x and the one which has an eigenfunction $\sin \theta \sin \phi$ is called p_y.

5.2   The Direct Product

Suppose that R is an operation in the symmetry group of a molecule and $X_1, X_2, \ldots, X_m$ and $Y_1, Y_2, \ldots, Y_n$ are two sets of functions (perhaps eigenfunctions of the wave equation for the molecule) which are bases for representations of the group. As shown earlier, we may write

$$RX_i = \sum_{j=1}^{m} x_{ji} X_j$$

$$RY_k = \sum_{l=1}^{n} y_{lk} Y_l$$

It is also true that

$$RX_i Y_k = \sum_{j=1}^{m} \sum_{l=1}^{n} x_{ji} y_{lk} X_j Y_l = \sum_{j} \sum_{l} z_{jl,ik} X_j Y_l$$

Thus the set of functions $X_i Y_k$, called the direct product of X_i and Y_k, also forms a basis for a representation of the group. The $z_{jl, ik}$ are the elements of a matrix $\mathscr{Z}$ of order $(mn) \times (mn)$.

We now have a very important theorem about the characters of the $\mathscr{Z}$ matrices for the various operations in the group:

The characters of the representation of a direct product are equal to the products of the characters of the representations based on the individual sets of functions.

PROOF.   This theorem is easily proved as follows:

$$\chi_{\mathscr{Z}}(R) = \sum_{jl} z_{jl, jl} = \sum_{j=1}^{m} \sum_{l=1}^{n} x_{jj} y_{ll} = \chi_{\mathscr{X}}(R)\chi_{\mathscr{Y}}(R)$$

Thus, if we want to know the characters $\chi(R)$ of a representation which is the direct product of two other representations with characters $\chi_1(R)$ and $\chi_2(R)$, these are given by

$$\chi(R) = \chi_1(R)\chi_2(R) \qquad (5.2\text{-}1)$$

for each operation R in the group.

For example, the direct products of some irreducible representations of the group C_{4v} are as follows:

C_{4v}	E	C_2	$2C_4$	$2\sigma_v$	$2\sigma_d$
A_1	1	1	1	1	1
A_2	1	1	1	-1	-1
B_1	1	1	-1	1	-1
B_2	1	1	-1	-1	1
E	2	-2	0	0	0
A_1A_2	1	1	1	-1	-1
B_1E	2	-2	0	0	0
A_1EB_2	2	-2	0	0	0
E^2	4	4	0	0	0

It should be clear from the associative property of matrix multiplication that what has been said regarding direct products of two representations can be extended to direct products of any number of representations.

In general, though not invariably, the direct product of two or more irreducible representations will be a reducible representation. For example, the direct product representations of the group given above reduce in the following way:

$$A_1A_2 = A_2 \qquad E^2 = A_1 + A_2 + B_1 + B_2$$
$$B_1E = E$$
$$A_1EB_2 = E$$

We shall now explain the importance of direct products in the solution of problems in molecular physics. Whenever we have an integral of the product of two functions, for example,

$$\int f_A f_B \, d\tau$$

the value of this integral will be equal to zero unless the integrand is invariant under all operations of the symmetry group to which the molecule belongs or unless some term in it, if it can be expressed as a sum of terms, remains invariant. This is a generalization of the familar case in which an integrand is a function of only one variable. In that case, if $y = f(x)$, the integral

$$\int_{-\infty}^{\infty} y \, dx$$

equals zero if y is an odd function, that is, if $f(x) = -f(-x)$. In this simple case we say that y is not invariant to the operation of reflecting all points in the second and third quadrants into the first and fourth quadrants and vice versa.

Now when we say that the integrand $f_A f_B$ is invariant to all symmetry operations, this means that it forms a basis for the totally symmetric representation of the group, or if some term in the expanded form of $f_A f_B$ is invariant, then that term forms a basis for the totally symmetric representation. But from what has been said above, we know how to determine the irreducible representations occurring in the representation Γ_{AB} for which $f_A f_B$ forms a basis if we know the irreducible representations for which f_A and f_B separately form bases. In general:

$$\Gamma_{AB} = \text{a sum of irreducible representations}$$

Only if one of the irreducible representations occurring in the sum is the totally symmetric one will the integral have a value other than zero. There is a theorem concerning whether the totally symmetric representation will be present in this sum:

The representation of a direct product, Γ_{AB}, will contain the totally symmetric representation only if the irreducible Γ_A = the irreducible Γ_B.

PROOF. Equation 4.3-11 tells us that the number of times, a_i, the ith irreducible representation, Γ_i, occurs in a reducible representation, say Γ_{AB}, is given by

$$a_i = \frac{1}{h} \sum_R \chi_{AB}(R) \chi_i(R) \tag{5.2-2}$$

If a_1 and χ_1 refer to the totally symmetric representation, for which all $\chi_1(R)$ equal 1, we have

$$a_1 = \frac{1}{h} \sum_R \chi_{AB}(R)$$

But from 5.2-1

$$\chi_{AB}(R) = \chi_A(R) \chi_B(R)$$

hence

$$a_1 = \frac{1}{h} \sum_R \chi_A(R) \chi_B(R) \tag{5.2-3}$$

According to the properties of characters of irreducible representations as components of vectors (Equation 4.3-8), we obtain

$$a_1 = \delta_{AB} \qquad \text{Q.E.D.}$$

We also see that if Γ_1 occurs at all it will occur only once. It is very easy to check this theorem by using the character tables in Appendix III, and doing so will perhaps help to develop familiarity with the manipulation of direct products.

From the foregoing discussion of the integrals of products of two functions it is easy to derive some important rules regarding integrands which are products of three, four, or more functions. The case of a triple product is of particular importance. In order for the integral

$$\int f_A \, f_B \, f_C \, d\tau$$

to be nonzero, the direct product of the representations of f_A, f_B, and f_C must be or contain the totally symmetric representation. That this can happen only if the representation of the direct product of any two of the functions is or contains the same representation as is given by the third function follows directly from the above theorem. This corollary is applicable chiefly in dealing with matrix elements of the type

$$\int \psi_i P \psi_j \, d\tau \tag{5.2-4}$$

where ψ_i and ψ_j are wave functions, and P is some quantum mechanical operator.

5.3   Identifying Nonzero Matrix Elements

Integrals of the general type 5.2-4 occur frequently in quantum mechanical problems. They are often termed *matrix elements*, since they occur as such in the secular equations which commonly provide the best way of formulating the problem. (See Chapters 7 and 10 for examples of secular equations.) In order to give the results just presented in Section 5.2 some concrete meaning, we shall discuss here the two commonest examples of the type of matrix element represented in 5.2-4.

Energy Elements

If we take the wave equation

$$H\psi_j = E\psi_j \tag{5.3-1}$$

left-multiply it by ψ_i, integrate both sides, and rearrange slightly, we obtain

$$\frac{\int \psi_i H \psi_j \, d\tau}{\int \psi \, \psi \, d\tau} = E \tag{5.3-2}$$

We thus have an explicit expression for an energy, which can be thought of as the energy of interaction between two states described by the wave functions ψ_i and ψ_j. If the integral which occurs in the numerator of the left-hand side of this equation is in fact required to have a value identically equal to zero, it will be helpful to know this at the earliest possible stage of a calculation so that no computational effort will be wasted on it. This information may be obtained very simply from a knowledge of the irreducible representations to which the wave functions ψ_i and ψ_j belong.

We note first that the Hamiltonian operator must have the full symmetry of the molecule; it is simply an operator expression for the energy of the molecule, and clearly the energy of the molecule cannot change in either sign or magnitude as a result of a symmetry operation. The Hamiltonian operator then belongs to the totally symmetric representation, and the symmetry of the entire integrand $\psi_i H \psi_j$ depends entirely on the representations contained in the direct product of the representations of ψ_i and ψ_j. The totally symmetric representation can occur in this direct product representation only if ψ_i and ψ_j belong to the same irreducible representation. Thus we have the completely general and enormously useful theorem that:

An energy integral, $\int \psi_i H \psi_j \, d\tau$, may be nonzero only if ψ_i and ψ_j belong to the same irreducible representation of the molecular point group.

Spectral Transition Probabilities

Perhaps the second commonest case in which the simple question of whether or not a matrix element is required by symmetry considerations to vanish occurs in connection with selection rules for various types of transition from one stationary state of a system to another with the gain or loss of a quantum of energy. If the energy difference between the states is represented by $E_i - E_j$, then radiation of frequency v will be either absorbed or emitted by the transition, if it is allowed, with v being required to satisfy the equation

$$hv = E_i - E_j$$

In general the intensity, I, of a transition from a state described by ψ_i to another described by ψ_j is given by an equation of the type

$$I \propto \int \psi_i \mu \psi_j \, d\tau \tag{5.3-3}$$

The symbol μ is a *transition moment operator*, of which there are various kinds, viz., those corresponding to changes in electric or magnetic dipoles, higher electric or magnetic multipoles, or polarizability tensors.

The commonest type of transition, and the only one to be considered right now, is the electric dipole-allowed transition. In this type the charge distributions in the two states differ in a manner corresponding to an electric dipole. Such a transition can therefore couple with electromagnetic radiation by interaction with the oscillating electric vector and thereby transfer energy to or from the electromagnetic field. Both theory and experiment show that normally this is by far the most powerful intensity-giving process, and the statement that a transition is " allowed," without further qualification, means explicitly "electric dipole allowed."

The electric dipole operator has the form

$$\mu = \sum_i e_i x_i + \sum_i e_i y_i + \sum_i e_i z_i \qquad (5.3\text{-}4)$$

where e_i represents the charge on the ith particle, and x_i, y_i, and z_i are its Cartesian coordinates. When this expression is introduced into 5.3-3, we obtain a result which is usefully expressed as three separate equations because of the orthogonality of the Cartesian coordinates:

$$I_x \propto \int \psi_i x \psi_j \, d\tau$$

$$I_y \propto \int \psi_i y \psi_j \, d\tau \qquad (5.3\text{-}5)$$

$$I_z \propto \int \psi_i z \psi_j \, d\tau$$

In these equations scalar quantities such as the e_i have been omitted, and summation over all particles is assumed.

These equations mean that the transition from the ith to the jth state (or the reverse) may acquire its intensity in any of three ways, namely, by interacting with an electric vector oscillating in the x, the y, or the z direction. If it is only the integral $\int \psi_i x \psi_j \, d\tau$ which is nonzero, we say that the transition is polarized in the x direction or that it is x-polarized. In cases of sufficiently high molecular symmetry, where x and y jointly form the basis for an irreducible representation, the integrals involving x and y will not be independent of each other. If both are nonzero, the transition is said to be xy-polarized. In problems with some kind of cubic or higher symmetry (point groups T, T_h, T_d, O, O_h, I, I_h), where the three Cartesian coordinates jointly form a basis for some three-dimensional representation, no polarization effect exists. Radiation with an electric vector in any direction will excite the transition, if it is allowed at all.

Thus the problem of deciding whether a certain transition is electric dipole allowed, and what the polarization is, reduces to that of deciding which, if any, of the three integrals in 5.3-5 are nonzero. We can always ascertain to what representations the Cartesian coordinates belong by inspection of the character table of the molecular point group. Then, according to the considerations of Section 5.2, we have the following rule:

An electric dipole transition will be allowed with x, y, or z polarization if the direct product of the representations of the two states concerned is or contains the irreducible representation to which x, y, or z, respectively, belongs.

Exercises

5.1 Write out the characters of the representations of the following direct products, and determine the irreducible representations which comprise them for group D_{6h}: $A_{1g} \times B_{1g}$; $A_{1u} \times A_{1u}$; $B_{2u} \times E_{1g}$; $E_{1g} \times E_{2u}$; $E_{1g} \times B_{2g} \times A_{2u} \times E_{1u}$.

5.2 In determining vibrational selection rules (Section 10.6) we shall need to determine whether integrals of the types $\int \psi_v{}^0 f \psi_v{}^1 \, d\tau$ are nonzero, where the function f is x, y, z, x^2, y^2, z^2, xy, yz, zx, or any combinations or sets thereof. Also, $\psi_v{}^0$ is totally symmetric and $\psi_v{}^1$ may belong to any irreducible representation. Identify the irreducible representations to which $\psi_v{}^1$ may belong in order to give nonzero integrals for molecules of symmetry C_{4v} and D_{3d}.

Symmetry-Adapted Linear Combinations

6.1 Introductory Remarks

In nearly all of the ways in which chemists employ symmetry restrictions to aid them in understanding chemical bonding and molecular dynamics—for example, constructing hybrid orbitals, constructing molecular orbitals, finding proper orbital sets under the action of a ligand field, and analyzing the vibrations of molecules, to name those subjects which will be covered explicitly in later chapters of this book—there is a common problem. This problem is to take one or more sets of orthonormal functions, which are generally either atomic orbitals or internal coordinates of a molecule, and to make orthonormal linear combinations of them in such a way that the combinations form bases for irreducible representations of the symmetry group of the molecule.

It will be obvious from the content of Chapter 5 why such combinations are desired. First, only such functions can, in themselves, constitute acceptable solutions to the wave equation or be directly combined to form acceptable solutions, as shown in Section 5.1. Second, only when the symmetry properties of wave functions are defined explicitly, in the sense of their being bases for irreducible representations, can we employ the theorems of Section 5.2 in order to determine without numerical computations which integrals or matrix elements in the problem are identically zero.

The kind of functions we need may be called *symmetry-adapted linear combinations* (SALC's). It is the purpose of this chapter to explain and illustrate the methods for constructing them in a general way. The details of adaptation to particular classes of problems will then be easy to explain as the needs arise.

6.2 Projection Operators

Let us assume that we have an orthonormal set of l_i functions $\phi_1{}^i$, $\phi_2{}^i$, ..., $\phi_{l_i}{}^i$ which form a basis for the ith irreducible representation (of dimension l_i)

of a group of order h. For any operator, $\hat{R}$, in the group we may then, by definition, write

$$\hat{R}\phi_t{}^i = \sum_s \phi_s{}^i \Gamma(R)_{st}{}^i \tag{6.2-1}$$

This equation is then multiplied by $[\Gamma(R)_{s't'}^j]^*$, and each side summed over all operations in the group, giving

$$\sum_R [\Gamma(R)_{s't'}^j]^* \hat{R}\phi_t{}^i = \sum_R \sum_s \phi_s{}^i \Gamma(R)_{st}{}^i [\Gamma(R)_{s't'}^j]^* \tag{6.2-2}$$

We note that the $\phi_s{}^i$'s are functions independent of R; hence the right side of 6.2-2 may be rewritten as

$$\sum_s \phi_s{}^i \sum_R \Gamma(R)_{st}{}^i [\Gamma(R)_{s't'}^j]^*$$

Thus, we have a series of l_i terms, each of which is a $\phi_s{}^i$ multiplied by a coefficient; each coefficient is itself expressed as a sum of products over the operations $\hat{R}$ in the group. These coefficients, however, are governed by the great orthogonality theorem (page 78), which states that

$$\sum_R \Gamma(R)_{st}{}^i [\Gamma(R)_{s't'}^j]^* = \frac{h}{\sqrt{l_i l_j}} \delta_{ij} \delta_{ss'} \delta_{tt'}$$

Thus all $\phi_s{}^i$ except $\phi_{s'}{}^i$ have coefficients of zero, and only when $i = j$ and $t = t'$ can even the term in $\phi_{s'}$ survive. Thus 6.2-2 simplifies to

$$\sum_R [\Gamma(R)_{s't'}^j]^* \hat{R}\phi_t{}^i = \left(\frac{h}{l_j}\right)\phi_{s'}{}^i \delta_{ij} \delta_{tt'} \tag{6.2-3}$$

We now introduce the symbol

$$\hat{P}_{s't'}^j = \frac{l_j}{h} \sum_R [\Gamma(R)_{s't'}^j]^* \hat{R} \tag{6.2-4}$$

and rewrite 6.2-3 as

$$\hat{P}_{s't'}^j \phi_t{}^i = \phi_{s'}{}^i \delta_{ij} \delta_{tt'} \tag{6.2-5}$$

The operator $\hat{P}_{s't'}^j$ is called a projection operator. It may be applied to an arbitrary function $\phi_t{}^i$, and only if that function itself or some term in it happens to be $\phi_{t'}{}^j$, will the result be other than zero. If $\phi_{t'}{}^j$ is a component of the arbitrary function, $\phi_{s'}{}^j$ will be "projected" out of it and the rest will be abolished. Thus, we have

$$\hat{P}^j_{s't'}\,\phi_{t'}{}^j = \phi_{s'}{}^j$$

In the very important special case where we use $\hat{P}^j_{t't'}$ we have

$$\hat{P}^j_{t't'}\,\phi_t{}^i = \phi_{t'}{}^j\,\delta_{ij}\,\delta_{tt'} \tag{6.2-6}$$

which means that $\hat{P}^j_{t't'}$ projects $\phi_{t'}{}^j$ out of an arbitrary function $\phi_t{}^i$. Thus, by using the l_j projection operators based on the l_j diagonal matrix elements, we may generate from some arbitrary function, $\phi_t{}^i$, the functions which form a basis for the jth irreducible representation.

To illustrate how 6.2-6 works, let us consider the general function, $xz + yz + z^2$, in the group C_{3v} (which is isomorphous to $G_6{}^{(1)}$). We shall use the projection operators to obtain from this arbitrary function a pair of functions which form a basis for the E representation. The matrices for this representation are given in Table 6.1. Table 6.2 shows how the arbitrary function $xz + yz + z^2$ is transformed by each of the six symmetry operators in the group. We shall first use the projection operator $\hat{P}^E_{11}$.

$$\hat{P}^E_{11}\,(xz + yz + z^2) = \tfrac{2}{6}\{(1)(xz + yz + z^2)$$

$$+ (-\tfrac{1}{2})[-\tfrac{1}{2}(1 + \sqrt{3})xz + \tfrac{1}{2}(\sqrt{3} - 1)yz + z^2]$$

$$+ (-\tfrac{1}{2})[\tfrac{1}{2}(\sqrt{3} - 1)xz - \tfrac{1}{2}(1 + \sqrt{3})yz + z^2]$$

$$+ (1)(xz - yz + z^2)$$

$$+ (-\tfrac{1}{2})[-\tfrac{1}{2}(1 + \sqrt{3})xz + \tfrac{1}{2}(1 - \sqrt{3})yz + z^2]$$

$$+ (-\tfrac{1}{2})[\tfrac{1}{2}(\sqrt{3} - 1)xz + \tfrac{1}{2}(1 + \sqrt{3})yz + z^2]\}$$

We now collect terms. The coefficients of the xz, yz, and z^2 terms are as follows:

xz: $\frac{2}{6}[1 + \frac{1}{4}(1 + \sqrt{3}) - \frac{1}{4}(\sqrt{3} - 1) + 1 + \frac{1}{4}(1 + \sqrt{3}) - \frac{1}{4}(\sqrt{3} - 1)]$

$\qquad = \frac{2}{6}[1 + \frac{1}{4} + \frac{1}{4} + 1 + \frac{1}{4} + \frac{1}{4} + \sqrt{3}(\frac{1}{4} - \frac{1}{4} + \frac{1}{4} - \frac{1}{4})]$

$\qquad = \frac{2}{6}(3 + 0) = 1$

yz: $\frac{2}{6}[1 - \frac{1}{4}(\sqrt{3} - 1) + \frac{1}{4}(1 + \sqrt{3}) - 1 - \frac{1}{4}(1 - \sqrt{3}) - \frac{1}{4}(1 + \sqrt{3})]$

$\qquad = \frac{2}{6}[1 + \frac{1}{4} + \frac{1}{4} - 1 - \frac{1}{4} - \frac{1}{4} + \sqrt{3}(-\frac{1}{4} + \frac{1}{4} + \frac{1}{4} - \frac{1}{4})]$

$\qquad = \frac{2}{6}(0) = 0$

z^2: $\frac{2}{6}(1 - \frac{1}{2} - \frac{1}{2} + 1 - \frac{1}{2} - \frac{1}{2}) = \frac{2}{6}(0) = 0$

Next we use the projection operator, $\hat{P}_{22}{}^E$.

$\hat{P}_{22}{}^E (xz + yz + z^2) = \frac{2}{6} \{(1)(xz + yz + z^2)$

$\qquad\qquad + (-\frac{1}{2})[-\frac{1}{2}(1 + \sqrt{3})xz + \frac{1}{2}(\sqrt{3} - 1)yz + z^2]$

$\qquad\qquad + (-\frac{1}{2})[\frac{1}{2}(\sqrt{3} - 1)xz - \frac{1}{2}(1 + \sqrt{3})yz + z^2]$

$\qquad\qquad + (-1)(xz - yz + z^2)$

$\qquad\qquad + (\frac{1}{2})[-\frac{1}{2}(1 + \sqrt{3})xz + \frac{1}{2}(1 - \sqrt{3})yz + z^2]$

$\qquad\qquad + (\frac{1}{2})[\frac{1}{2}(\sqrt{3} - 1)xz + \frac{1}{2}(1 + \sqrt{3})yz + z^2]\}$

Again, we collect terms and evaluate the coefficients, obtaining

xz: $\frac{2}{6}[1 + \frac{1}{4}(1 + \sqrt{3}) - \frac{1}{4}(\sqrt{3} - 1) - 1 - \frac{1}{4}(1 + \sqrt{3}) + \frac{1}{4}(\sqrt{3} - 1)]$

$\qquad = \frac{2}{6}[1 + \frac{1}{4} + \frac{1}{4} - 1 - \frac{1}{4} - \frac{1}{4} + \sqrt{3}(\frac{1}{4} - \frac{1}{4} - \frac{1}{4} + \frac{1}{4})]$

$\qquad = \frac{2}{6}(0) = 0$

yz: $\frac{2}{6}[1 - \frac{1}{4}(\sqrt{3} - 1) + \frac{1}{4}(1 + \sqrt{3}) + 1 + \frac{1}{4}(1 - \sqrt{3}) + \frac{1}{4}(1 + \sqrt{3})]$

$\qquad = \frac{2}{6}[1 + \frac{1}{4} + \frac{1}{4} + 1 + \frac{1}{4} + \frac{1}{4} + \sqrt{3}(-\frac{1}{4} + \frac{1}{4} - \frac{1}{4} + \frac{1}{4})]$

$\qquad = \frac{2}{6}(3) = 1$

z^2: $\frac{2}{6}(1 - \frac{1}{2} - \frac{1}{2} - 1 + \frac{1}{2} + \frac{1}{2}) = \frac{2}{6}(0) = 0$

Thus we have projected out of the function $xz + yz + z^2$ the two functions xz and yz, which form a basis for the E representation. The component z^2 has been abolished; it cannot, in whole or in part, contribute to a basis set for the E representation.

It will be clear from the foregoing discussion that in order to use the type of projection operator we have developed so far we need to know the individual diagonal elements of the matrices. This is inconvenient, since normally the only information readily accessible is the set of characters—the sum of all the diagonal matrix elements—for each matrix of the representation in question. For one-dimensional representations, this is a distinction without a

Table 6.1 Matrices for the E Representation of the Group C_{3v}

OPERATION	MATRIX
E	$\begin{bmatrix} 1 & 0 \\ 0 & 1 \end{bmatrix}$
C_3	$\begin{bmatrix} -\dfrac{1}{2} & \dfrac{\sqrt{3}}{2} \\ -\dfrac{\sqrt{3}}{2} & -\dfrac{1}{2} \end{bmatrix}$
C_3^2	$\begin{bmatrix} -\dfrac{1}{2} & -\dfrac{\sqrt{3}}{2} \\ \dfrac{\sqrt{3}}{2} & -\dfrac{1}{2} \end{bmatrix}$
$\sigma_v(xz)$	$\begin{bmatrix} 1 & 0 \\ 0 & -1 \end{bmatrix}$
σ_v'	$\begin{bmatrix} -\dfrac{1}{2} & \dfrac{\sqrt{3}}{2} \\ \dfrac{\sqrt{3}}{2} & \dfrac{1}{2} \end{bmatrix}$
σ_v''	$\begin{bmatrix} -\dfrac{1}{2} & -\dfrac{\sqrt{3}}{2} \\ -\dfrac{\sqrt{3}}{2} & \dfrac{1}{2} \end{bmatrix}$

Table 6.2 Transformations of Some Simple Functions of x, y, and z

OPERATOR	x	y	z	FUNCTIONS $xz + yz + z^2$
E	x	y	z	$xz + yz + z^2$
C_3	$\frac{1}{2}(-x+\sqrt{3}y)$	$\frac{1}{2}(-y-\sqrt{3}x)$	z	$\frac{1}{2}[-(1+\sqrt{3})xz + (\sqrt{3}-1)yz] + z^2$
C_3^2	$\frac{1}{2}(-x-\sqrt{3}y)$	$\frac{1}{2}(-y+\sqrt{3}x)$	z	$\frac{1}{2}[(\sqrt{3}-1)xz - (1+\sqrt{3})yz] + z^2$
$\sigma_v(x)$	x	$-y$	z	$xz - yz + z^2$
σ_v'	$\frac{1}{2}(-x-\sqrt{3}y)$	$\frac{1}{2}(y-\sqrt{3}x)$	z	$\frac{1}{2}[-(1+\sqrt{3})xz + (1-\sqrt{3})yz] + z^2$
σ_v''	$\frac{1}{2}(-x+\sqrt{3}y)$	$\frac{1}{2}(y+\sqrt{3}x)$	z	$\frac{1}{2}[(\sqrt{3}-1)xz + (1+\sqrt{3})yz] + z^2$

difference, but for two- and three-dimensional cases it is advantageous to have a projection operator which employs only the characters. It is not difficult to derive the desired operator, beginning with the explicit expression for $\hat{P}^j_{t't'}$, viz.,

$$\hat{P}^j_{t't'} = \frac{l_j}{h} \sum_R [\Gamma(R)^j_{t't'}]^* \hat{R} \tag{6.2-7}$$

If we sum each side over all values of t', we obtain

$$\hat{P}^j = \sum_{t'} \hat{P}^j_{t't'} = \frac{l_j}{h} \sum_{t'} \sum_R [\Gamma(R)^j_{t't'}]^* \hat{R}$$

$$= \frac{l_j}{h} \sum_R \left\{ \sum_{t'} [\Gamma(R)^j_{t't'}]^* \right\} \hat{R}$$

$$\hat{P}^j = \frac{l_j}{h} \sum_R \chi(R)^j \hat{R} \tag{6.2-8}$$

In this development we have employed the interchangeability of the order of the summations and the definition of the character of the matrix.

Let us now see what happens when we apply $\hat{P}^E$ to $xz + yz + z^2$.

$$\hat{P}^E(xz + yz + z^2) = \tfrac{2}{6}\{(2)(xz + yz + z^2)$$

$$+ (-1)[-\tfrac{1}{2}(1 + \sqrt{3})xz + \tfrac{1}{2}(\sqrt{3} - 1)yz + z^2]$$

$$+ (-1)[\tfrac{1}{2}(\sqrt{3} - 1)xz - \tfrac{1}{2}(1 + \sqrt{3})yz + z^2]$$

$$+ 0 + 0 + 0\}$$

$$= \tfrac{2}{6}\{[2 + \tfrac{1}{2}(1 + \sqrt{3}) - \tfrac{1}{2}(\sqrt{3} - 1)]xz$$

$$+ [2 - \tfrac{1}{2}(\sqrt{3} - 1) + \tfrac{1}{2}(1 + \sqrt{3})]yz$$

$$+ (2 - 1 - 1)z^2\}$$

$$= \tfrac{2}{6}(3xz + 3yz + 0z^2)$$

$$= xz + yz$$

We see that this operator has abolished the irrelevant part of the function and projected out a linear combination of the two separate functions, xz and yz, which we were able to obtain by employing the projection operators, $\hat{P}_{11}{}^E$ and $\hat{P}_{22}{}^E$. This should not be surprising. We obviously cannot get two separate results with only one operator. Moreover, since the operator is derived by adding the individual operators, a sum of the results given by the individual operators is what we must expect. Thus, the projection operators of the type $\hat{P}^j$ cannot be as powerful and explicit as those of the type $\hat{P}^j_{tt'}$. However, they usually suffice for solving practical problems, as we shall demonstrate in the next section.

6.3   Some Illustrations

π *Orbitals for the Cyclopropenyl Group*

The cyclopropenyl group, C_3H_3, is the simplest carbocycle with a delocalized π system and can serve as a prototype for this class of molecules. Let us see how the *p*π orbitals of the individual carbon atoms can be combined into linear combinations having the necessary symmetry properties to be π molecular orbitals—or at least the immediate precursors of actual π molecular orbitals. The subject of molecular orbitals will be discussed in detail in Chapter 8, and this illustration is intended only to demonstrate the use of projection operators in making SALC's of atomic orbitals on different atoms.

The preliminary steps before the SALC's can actually be set up are as follows.

1. Identify the point group: D_{3h}.
2. Use the three *p*π orbitals, shown and identified in the sketch, as the basis for a representation:

E	$2C_3$	$3C_2$	σ_h	$2S_3$	$3\sigma_v$
3	0	-1	-3	0	1

3. Reduce this to its irreducible components: $A_2'' + E''$.

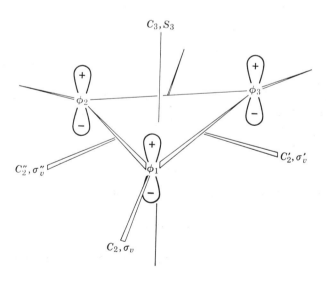

Let us first apply the projection operator $\hat{P}^{A_2''}$ to ϕ_1. This operator is, explicitly,

$$\hat{P}^{A_2''} = \frac{1}{12} \sum_R \chi(R)^{A_2''} \hat{R}$$

No complex conjugate sign appears, since all $\chi(R)$ are real, and we shall also ignore the numerical factor $\frac{1}{12}$, since the absolute as opposed to the relative values of the coefficients in the SALC can be ascertained later by a normalization procedure. We obtain

$$\hat{P}^{A_2''}\phi_1 \approx (1)\hat{E}\phi_1 + (1)\hat{C}_3\,\phi_1 + (1)\hat{C}_3{}^2\phi_1 + (-1)\hat{C}_2\,\phi_1 + (-1)\hat{C}_2'\,\phi_1$$
$$+ (-1)\hat{C}_2''\,\phi_1 + (-1)\hat{\sigma}_h\,\phi_1 + (-1)\hat{S}_3\,\phi_1 + (-1)\hat{S}_3{}^5\phi_1$$
$$+ (1)\hat{\sigma}_v\,\phi_1 + (1)\hat{\sigma}_v'\,\phi_1 + (1)\hat{\sigma}_v''\phi_1$$

where the numbers in parentheses are the characters of the A_2'' representation taken from the character table in Appendix III. Note that each operation has been written out.

Each of the operations must now be carried out, and the result written down. For example, taking rotations to be clockwise, we have

$$\hat{C}_3\phi_1 = \phi_2$$
$$\hat{C}_2'\phi_1 = -\phi_3$$

etc.

We thus obtain

$$\hat{P}^{A_2''}\,\phi_1 \approx \phi_1 + \phi_2 + \phi_3 + \phi_1 + \phi_3 + \phi_2 + \phi_1 + \phi_2 + \phi_3 + \phi_1 + \phi_3 + \phi_2$$
$$= 4(\phi_1 + \phi_2 + \phi_3) \approx \phi_1 + \phi_2 + \phi_3$$

The reader may demonstrate by applying the twelve group operations to it that this function does indeed form a basis for the A_2'' representation.

In constructing SALC's, it is customary to employ an orthonormal basis set and then to require that the SALC's be normalized. Correctly constructed SALC's must necessarily be orthogonal to each other. The orthonormality of the basis set means that the basis functions, $f_1, f_2, \ldots, f_i, \ldots, f_j, \ldots$ satisfy the condition

$$f_i f_j = \delta_{ij} \tag{6.3-1}$$

if they are vectors, and the corresponding condition

$$\int f_i f_j \, d\tau = \delta_{ij} \tag{6.3-2}$$

if they are functions of a set of coordinates collectively denoted by τ.

Assuming that the set ϕ_1, ϕ_2, and ϕ_3 satisfy 6.3-2, we may proceed to normalize the A_2'' function whose form has just been found. As it stands, it is not normalized, since

$$\int (\phi_1 + \phi_2 + \phi_3)(\phi_1 + \phi_2 + \phi_3) \, d\tau$$

$$= \int (\phi_1{}^2 + \phi_1\phi_2 + \phi_1\phi_3 + \phi_2\phi_1 + \phi_2{}^2$$

$$+ \phi_2\phi_3 + \phi_3\phi_1 + \phi_3\phi_2 + \phi_3{}^2) \, d\tau$$

$$= \int \phi_1{}^2 \, d\tau + \int \phi_1\phi_2 \, d\tau + \dots ,$$

$$= 1 + 0 + 0 + 0 + 1 + 0 + 0 + 0 + 1$$

$$= 3$$

Clearly, if we multiply the A_2'' SALC by $1/\sqrt{3}$, it will be normalized. Thus the final result for A_2'' is $(1/\sqrt{3})(\phi_1 + \phi_2 + \phi_3)$.

For the E'' representation, using again ϕ_1, we have

$$\hat{P}^{E''}\phi_1 \approx (2)\hat{E}\phi_1 + (-1)\hat{C}_3\phi_1 + (-1)\hat{C}_3{}^2\phi_1 + (0)\hat{C}_2\phi_1$$

$$+ (0)\hat{C}_2'\phi_1 + (0)\hat{C}_2''\phi_1 + (-2)\hat{\sigma}_h\phi_1 + (1)\hat{S}_3\phi_1$$

$$+ (1)\hat{S}_3{}^5\phi_2 + (0)\hat{\sigma}_v\phi_1 + (0)\hat{\sigma}_v'\phi_1 + (0)\hat{\sigma}_v''\phi_1$$

$$= 2\phi_1 - \phi_2 - \phi_3 + 2\phi_1 - \phi_2 - \phi_3 \approx 2\phi_1 - \phi_2 - \phi_3$$

It is easy to show that the normalized function is $(1/\sqrt{6})(2\phi_1 - \phi_2 - \phi_3)$.

This is *one* of *two* functions, which *together* form a basis for the E'' representation. How do we find the second one, its partner? There are several ways, but the following method is obvious and easy to apply.

If we carry out a symmetry operation on one of the two functions, it will either go into ± 1 times itself, into its partner, or into a linear combination of itself and its partner. Let us chose an operation which does *not* convert it into ± 1 times itself, viz., $\hat{C}_3$:

$$\hat{C}_3\left[\frac{1}{\sqrt{6}}(2\phi_1 - \phi_2 - \phi_3)\right] \rightarrow \frac{1}{\sqrt{6}}(2\phi_2 - \phi_3 - \phi_1)$$

It may easily be shown that the second function neither is ± 1 times the first nor is it orthogonal to it, as the partner must be. The second function must therefore be a linear combination of the first and its partner, and we can find the expression for the partner by subtracting an appropriate multiple of the first one out of the second one, leaving the partner as the remainder.

This is most easily done by ignoring, for the moment, normalization, with the idea of attending to that at the end. Thus we proceed as follows:

$$(2\phi_2 - \phi_3 - \phi_1) - (-\tfrac{1}{2})(2\phi_1 - \phi_2 - \phi_3)$$

$$= 2\phi_2 - \phi_3 - \phi_1 + \phi_1 - \tfrac{1}{2}\phi_2 - \tfrac{1}{2}\phi_3$$

$$= \tfrac{3}{2}\phi_2 - \tfrac{3}{2}\phi_3 \approx \phi_2 - \phi_3$$

This may be normalized to $(1/\sqrt{2})(\phi_2 - \phi_3)$. It is orthogonal to the first function and is thus an acceptable partner:

$$\int \frac{1}{\sqrt{6}}(2\phi_1 - \phi_2 - \phi_3)\frac{1}{\sqrt{2}}(\phi_2 - \phi_3)\,d\tau$$

$$= \frac{1}{\sqrt{12}}\int(2\phi_1\phi_2 - 2\phi_1\phi_3 - \phi_2{}^2 + \phi_2\phi_3 - \phi_3\phi_2 + \phi_3{}^2)\,d\tau$$

$$= \frac{1}{\sqrt{12}}\left(2\int\phi_1\phi_2\,d\tau - 2\int\phi_1\phi_3\,d\tau - \int\phi_2{}^2\,d\tau + \cdots\right)$$

$$= \frac{1}{\sqrt{12}}[2(0) - 2(0) - 1 + 0 - 0 + 1] = 0$$

The reader may demonstrate that the pair of functions

$$\frac{1}{\sqrt{6}}(2\phi_1 - \phi_2 - \phi_3), \qquad \frac{1}{\sqrt{2}}(\phi_2 - \phi_3)$$

do in fact form a basis for the matrices of the E'' representation and, moreover, that each of these is orthogonal to the SALC of A_2'' symmetry.

The procedure just used, although routine and reliable, is lengthy, particularly for the two-dimensional representation. The results could have been obtained with less labor by recognizing that the rotational symmetry, the behavior of the SALC's upon rotation about the principal axis, alone fixes their basic form. Their behavior under the other symmetry operations is a direct consequence of the inherent symmetry of an individual $p\pi$ orbital toward these operations, σ_h or a C_2 passing through it, being added to the symmetry properties under the pure rotations about the principal axis. This can be seen by inspecting the D_{3h} character table. For all A-type representations, A_1', A_2', A_1'', and A_2'', the characters are the same for the C_3 and $C_3{}^2$ operations; similarly, the E' and E'' representations are identical within the subgroup C_3. The thing which decides that we are dealing specifically with A_2'' and E'' SALC's is the inherent nature of the $p\pi$ basis functions.

On the strength of the above considerations, a procedure which restricts attention to the pure rotational symmetry about the principal axis may be used to construct the SALC's. For C_3H_3, we use the group C_3. This group, like all uniaxial pure rotation groups, is Abelian. Its three operations fall into three classes, and it must have three irreducible representations of dimension 1. In general, a group C_n has n one-dimensional representations (cf. Section 4.5), so that what we show here for C_3 will be generalizable to all C_n groups.

In the subgroup C_3, the set of $p\pi$ orbitals of C_3H_3 spans the A and E representations. The latter, however, appears in the character table as two associated one-dimensional representations; a projection operator may be written for each of these one-dimensional components individually. Thus, we shall be able to obtain each of the SALC's belonging to the E representation directly and routinely, using projection operators. This is the advantage of using only the principal axis rotational symmetry. Let us now work through the algebra and see how much this trick expedites our task of constructing the SALC's.

Application of the projection operators P^A, $P^{E(1)}$, and $P^{E(2)}$ to ϕ_1 (neglecting constant numerical factors) gives:

$$\hat{P}^A \phi_1 \approx (1)\hat{E}\phi_1 + (1)\hat{C}_3 \phi_1 + (1)\hat{C}_3{}^2\phi_1$$
$$= (1)\phi_1 + (1)\phi_2 + (1)\phi_3$$
$$= \phi_1 + \phi_2 + \phi_3$$

$$\hat{P}^{E(1)} \phi_1 \approx (1)\hat{E}\phi_1 + (\varepsilon)\hat{C}_3 \phi_1 + (\varepsilon^*)\hat{C}_3{}^2\phi_1$$
$$= \phi_1 + \varepsilon\phi_2 + \varepsilon^*\phi_3$$

$$\hat{P}^{E(2)} \phi_1 \approx (1)\hat{E}\phi_1 + (\varepsilon^*)\hat{C}_3 \phi_1 + (\varepsilon)\hat{C}_3{}^2\phi_1$$
$$= \phi_1 + \varepsilon^*\phi_2 + \varepsilon\phi_3$$

The A SALC has exactly the same form as we previously obtained for the A_2'' SALC, using full D_{3h} symmetry. The two E SALC's are actually satisfactory in the sense of being proper basis functions and being orthogonal to each other. However, we prefer to have real rather than complex coefficients. This change can be accomplished very simply because of the fact that the two sets of coefficients are arranged as pairs of complex conjugates (cf. Section 4.5). Thus, if we add them term by term, the imaginary component of each pair are eliminated, leaving a SALC with real coefficients. Also, if one set is subtracted, term by term, from the other, a set of pure imaginary coefficients will be obtained and the common factor i may be removed to leave another set of real coefficients. These addition and subtraction procedures are simply a case of forming new linear combinations of an initial set, and this is an entirely proper and rigorous thing to do. Thus, we add the two E SALC's obtained above:

$$
\begin{array}{r}
(\phi_1 + \varepsilon\phi_2 + \varepsilon^*\phi_3) \\
+ (\phi_1 + \varepsilon^*\phi_2 + \varepsilon\phi_3) \\
\hline
2\phi_1 + (\varepsilon + \varepsilon^*)\phi_2 + (\varepsilon + \varepsilon^*)\phi_3
\end{array}
$$

$$
\varepsilon + \varepsilon^* = (\cos 2\pi/3 + i \sin 2\pi/3) + (\cos 2\pi/3 - i \sin 2\pi/3)
$$
$$
= 2 \cos 2\pi/3 = 2(-\tfrac{1}{2}) = -1
$$

The first new SALC is therefore

$$
2\phi_1 - \phi_2 - \phi_3
$$

Next, we subtract the original E SALC's and divide out i:

$$
\begin{array}{r}
(\phi_1 + \varepsilon\phi_2 + \varepsilon^*\phi_3) \\
- (\phi_1 + \varepsilon^*\phi_2 + \varepsilon\phi_3) \\
\hline
(\varepsilon - \varepsilon^*)\phi_2 - (\varepsilon - \varepsilon^*)\phi_3
\end{array}
$$

$$
\frac{(\varepsilon - \varepsilon^*)}{i} = \frac{(\cos 2\pi/3 + i \sin 2\pi/3) - (\cos 2\pi/3 - i \sin 2\pi/3)}{i}
$$

$$
= (2i \sin 2\pi/3)/i
$$

$$
= 2 \sin 2\pi/3 = 2\left(\frac{\sqrt{3}}{2}\right) = \sqrt{3}
$$

The second new SALC, which should be orthogonal to the first and therefore its proper partner in forming a basis for the E representation, thus has the form

$$
\phi_2 - \phi_3
$$

Clearly, when the SALC's we have just obtained are properly normalized, they are identical to those previously obtained by using full D_{3h} symmetry. This second procedure, which was much simpler, may be summarized as follows:

1. An initial set of SALC's may be written down by inspection of the character table. Each one is of the form $a\phi_1 + b\phi_2 + c\phi_3$, with coefficients which are the characters for E, C_3, and $C_3{}^2$.

2. The pairs of SALC's for the E representation are added and subtracted (dividing the result by i) to get two new orthogonal SALC's which have all real coefficients.

3. The SALC's are normalized.

Symmetry Coordinates for an AB_3 Molecule

As a second illustration of the procedure for constructing SALC's we consider a problem which arises in analyzing the vibrations of a molecule. The subject of molecular vibrations will be treated fully in Chapter 10, at which time the key importance of the procedure we are about to explain will be apparent. For the moment it may be regarded as merely another illustration of the generality of the projection operator technique for constructing SALC's.

The accompanying sketch shows a pyramidal AB_3 molecule which belongs to the point group C_{3v}. Two sets of internal coordinates—the bond lengths,

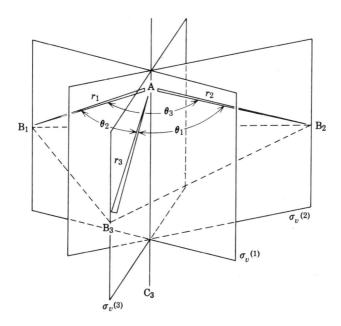

r_1, r_2, r_3, and the interbond angles, θ_1, θ_2, θ_3—are identified. The allowed vibrations of a molecule may be expressed as linear combinations of changes in the internal coordinates of the molecule. In the present case the set of bond length changes, Δr_1, Δr_2, and Δr_3, must be combined into SALC's corresponding to the A_1 and E representations of C_{3v}. The same is true of the set of bond angle changes, $\Delta\theta_1$, $\Delta\theta_2$, $\Delta\theta_3$. The reason why it is in each case the A_1 and E representations which must be matched will be explained in Chapter 10; here we merely address the question of how to do this.

As in the previous illustration dealing with the π molecular orbitals of C_3H_3, we could work directly with the true point group of the problem,

C_{3v}. However, for the same kind of reasons as before, it is advantageous to work only with the uniaxial rotational symmetry properties, which means using the subgroup C_3. In this group there are three operations: E, C_3, and $C_3{}^2$. The internal coordinate changes are affected by each of these as follows:

$$\hat{E}(\Delta r_1) = \Delta r_1 \qquad \hat{E}(\Delta \theta_1) = \Delta \theta_1$$
$$\hat{C}_3(\Delta r_1) = \Delta r_2 \qquad \hat{C}_3(\Delta \theta_1) = \Delta \theta_2$$
$$\hat{C}_3{}^2(\Delta r_1) = \Delta r_3 \qquad \hat{C}_3{}^2(\Delta_1) = \Delta \theta_3$$
$$\cdot \qquad\qquad\qquad \cdot$$
$$\cdot \qquad\qquad\qquad \cdot$$
$$\cdot \qquad\qquad\qquad \cdot$$
$$\hat{C}_3{}^2(\Delta r_2) = \Delta r_1 \qquad \hat{C}_3{}^2(\Delta \theta_2) = \Delta \theta_1$$
$$\cdot \qquad\qquad\qquad \cdot$$
$$\cdot \qquad\qquad\qquad \cdot$$
$$\text{etc.} \qquad\qquad\qquad \text{etc.}$$

The projection operators may be applied as follows:

$$\hat{P}^A(\Delta r_1) \approx (1)\hat{E}(\Delta r_1) + (1)\hat{C}_3(\Delta r_1) + (1)\hat{C}_3{}^2(\Delta r_1)$$
$$= (1)\Delta r_1 + (1)\Delta r_2 + (1)\Delta r_3$$
$$= \Delta r_1 + \Delta r_2 + \Delta r_3$$
$$\hat{P}^A(\Delta \theta_1) \approx (1)\hat{E}(\Delta \theta_1) + (1)\hat{C}_3(\Delta \theta_1) + (1)\hat{C}_3{}^2(\Delta \theta_1)$$
$$= (1)\Delta \theta_1 + (1)\Delta \theta_2 + (1)\Delta \theta_3$$
$$= \Delta \theta_1 + \Delta \theta_2 + \Delta \theta_3$$
$$\hat{P}^{E(1)}(\Delta r_1) \approx (1)\hat{E}(\Delta r_1) + \varepsilon \hat{C}_3(\Delta r_1) + \varepsilon^* \hat{C}_3{}^2(\Delta r_1)$$
$$= \Delta r_1 + \varepsilon \Delta r_2 + \varepsilon^* \Delta r_3$$
$$\hat{P}^{E(2)}(\Delta r_1) \approx (1)\hat{E}(\Delta r_1) + (\varepsilon^*)\hat{C}_3(\Delta r_1) + (\varepsilon)\hat{C}_3{}^2(\Delta r_1)$$
$$= \Delta r_1 + \varepsilon^* \Delta r_2 + \varepsilon \Delta r_3$$
$$\hat{P}^{E(1)}(\Delta \theta_1) = \Delta \theta_1 + \varepsilon \Delta \theta_2 + \varepsilon^* \Delta \theta_3$$
$$\hat{P}^{E(2)}(\Delta \theta_1) = \Delta \theta_1 + \varepsilon^* \Delta \theta^2 + \varepsilon \Delta \theta_3$$

These results are analogous to those obtained for the π orbitals of C_3H_3. The process of taking sums and differences of the pairs of E-type SALC's to eliminate complex or imaginary coefficients and the normalization procedures are carried out as before, yielding the following final results:

$$\left. \begin{aligned} S_1 &= \frac{1}{\sqrt{3}}(\Delta r_1 + \Delta r_2 + \Delta r_3) \\[2mm] S_2 &= \frac{1}{\sqrt{3}}(\Delta \theta_1 + \Delta \theta_2 + \Delta \theta_3) \end{aligned} \right\} A_1$$

$$S_{3a} = \frac{1}{\sqrt{6}} (2\Delta r_1 - \Delta r_2 - \Delta r_3)$$

$$S_{3b} = \frac{1}{\sqrt{2}} (\Delta r_2 - \Delta r_3)$$

$$\left. \begin{array}{l} \\ \\ \\ \\ \end{array} \right\} E$$

$$S_{4a} = \frac{1}{\sqrt{6}} (2\Delta\theta_1 - \Delta\theta_2 - \Delta\theta_3)$$

$$S_{4b} = \frac{1}{\sqrt{2}} (\Delta\theta_2 - \Delta\theta_3)$$

The symbol S is used to designate these *symmetry coordinates*, as they are called. They can be shown to form bases for the A_1 and E representations, as indicated by the symbols to the right, of the point group C_{3v} even though they were obtained by invoking explicitly only the symmetry of the pure rotation subgroup C_3.

Exercises

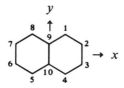

6.1 Consider the naphthalene molecule (point group D_{2h}). Each carbon atom has a p_z orbital to contribute to the π molecular orbitals of the molecule. Take these ten p_z orbitals as a basis set and answer the following questions:

(a) How do they divide into subsets of symmetry-equivalent orbitals?
(b) What representations are spanned by each subset?
(c) What are the ten normalized SALC's?

6.2 Consider the molecule $(C_5H_5)Fe(CO)_2$-$Fe(CO)_2(C_5H_5)$ in its "sawhorse" rotational configuration (C_{2v} symmetry). For the four CO distances work out the expressions for normalized symmetry coordinates.

Part II

Applications

Symmetry Aspects of

Molecular Orbital Theory

7.1 General Principles

In the valence bond theory as developed by Slater, Pauling, and others, all bonds are considered to be two-center bonds, that is, to subsist between two atoms. It is only as an afterthought, more or less, that account is taken of any interaction between such bonds. In addition, there is no a priori way to decide which pairs of atoms should be considered as bonded and which nonbonded; we make choices which are supported by chemical knowledge. The molecular orbital theory begins, at least in principle, with the idea that all orbitals in a molecule extend over the entire molecule, which means therefore that electrons occupying these orbitals may be delocalized over the entire molecule. The theory does, naturally, admit of the possibility that one or more of these *molecular orbitals* (MO's) may have significantly large values of the wave function only in certain parts of the molecule. That is, localized bonding is a special case which is adequately covered by MO theory, but localization is not "built in" as a postulate at the outset as in the valence bond treatment.

Because MO theory treats orbitals which are in general spread over the entirety of a molecule, considerations of molecular symmetry properties are extremely useful in this theory. They make it possible to determine the symmetry properties of the MO wave functions. With these known, it is often possible to draw many useful conclusions about bonding without doing any actual quantum computations at all, or by doing only very simple ones. If elaborate calculations are to be carried out, the use of MO symmetry properties can immensely alleviate the labor involved by showing that many integrals must be identically equal to zero.

The LCAO Approximation

By far the commonest approximation employed to reduce the notion of an MO to an explicit, practical form is the linear combination of atomic orbitals (LCAO) approximation. Each MO is written as a linear combination of atomic orbitals on the various atoms. Denoting the ith atomic orbital ϕ_i and the kth molecular orbital ψ_k, we write

$$\psi_k = \sum_i c_{ik} \phi_i \tag{7.1-1}$$

The ϕ_i's are a basis set, and it is convenient to choose or adjust them so that they are normalized. This property, which we shall henceforth take for granted is defined by the equation

$$\int \phi_i \phi_i \, d\tau = 1 \tag{7.1-2}$$

By using LCAO-MO's, a particular form of the wave equation, called the secular equation, is developed in the following way. The wave equation is written in the form

$$\mathcal{H}\psi - E\psi = (\mathcal{H} - E)\psi = 0 \tag{7.1-3}$$

An LCAO expression for ψ is now introduced, giving

$$\sum_i c_i(\mathcal{H} - E)\phi_i = 0 \tag{7.1-4}$$

For clarity, without loss of generality, it is easier to continue the development explicitly for the case of a two-term LCAO-MO; thus 7.1-4 takes the form

$$c_1(\mathcal{H} - E)\phi_1 + c_2(\mathcal{H} - E)\phi_2 = 0 \tag{7.1-5}$$

Equation 7.1-5 is now multiplied by ϕ_1, and the left side integrated over all spatial coordinates of the wave functions:

$$c_1 \int \phi_1(\mathcal{H} - E)\phi_1 \, d\tau + c_2 \int \phi_1(\mathcal{H} - E)\phi_2 \, d\tau = 0 \tag{7.1-6}$$

To simplify notation, the following definitions will now be introduced:

$$H_{ii} = \int \phi_i \mathcal{H} \phi_i \, d\tau \tag{7.1-7}$$

$$H_{ij} = \int \phi_i \mathcal{H} \phi_j \, d\tau \tag{7.1-8}$$

$$S_{ij} = \int \phi_i \phi_j \, d\tau \tag{7.1-9}$$

The integral H_{ii} gives the energy of the atomic orbital ϕ_i. The H_{ij} integrals give the energies of interaction between pairs of atomic orbitals. The S_{ij} are called overlap integrals. Because the energy, E, is simply a number

$$\int \phi_i E \phi_j \, d\tau = E \int \phi_i \phi_j \, d\tau = E S_{ij} \tag{7.1-10}$$

Equation 7.1-6 may now be written

$$c_1(H_{11} - E) + c_2(H_{12} - E S_{12}) = 0 \tag{7.1-11}$$

Equation 7.1-5 could also have been multiplied by ϕ_2 and integrated, leading to

$$c_1(H_{21} - E S_{21}) + c_2(H_{22} - E) = 0 \tag{7.1-12}$$

These two equations form a system of homogeneous linear equations in c_1 and c_2. They obviously have the trivial solutions $c_1 = c_2 = 0$. It is proved in the theory of homogeneous linear equations that other, nontrivial solutions can exist only if the matrix of the coefficients of the c_1's forms a determinant equal to zero (Cramer's theorem). Thus, we have the so-called secular equation:

$$\begin{vmatrix} H_{11} - E & H_{12} - E S_{12} \\ H_{21} - E S_{21} & H_{22} - E \end{vmatrix} = 0 \tag{7.1-13}$$

The numerical values of the H_{ii}'s and H_{ij}'s and S_{ij}'s can be guessed, estimated, or computed at some level of approximation, and the secular equation then solved for the values of E. From the algebra of determinants (cf. Appendix II) it follows that an $n \times n$ determinant will give rise to an nth-order polynomial equation in the energy. Equation 7.1-13, for example, gives the quadratic equation

$$(1 - S_{12}^2)E^2 - (H_{11} + H_{22} - 2H_{12} S_{12})E + H_{11}H_{22} - H_{12}^2 = 0 \tag{7.1-14}$$

where the relations $S_{ij} = S_{ji}$ and $H_{ij} = H_{ji}$ have been invoked. This equation can be solved to give two roots, E_1 and E_2, which can be shown (by the variational theorem) to be upper limits to the energies of the ground and first excited states.

If the value E_1 is inserted into Equations 7.1-11 and 7.1-12, these equations may then be solved for the coefficients c_{11} and c_{12}, which give the MO ψ_1 having the energy E_1. Similarly, E_2 may be substituted to obtain equations for the coefficients c_{21} and c_{22} giving the MO ψ_2, which has the energy E_2.

Thus, in summary, by using values of the various integrals (often called matrix elements) H_{ii}, H_{ij}, S_{ij}, the energies of MO's may be calculated in the LCAO-MO approximation without knowing the explicit form of the LCAO-MO's. After the energy values are known, it is possible to determine the coefficients, c_{ij}, and thus obtain the explicit expressions for the LCAO-MO's.

The Hückel Approximation

The LCAO-MO approach just outlined is in itself an approximation. Even so, if no further approximations are made, the evaluation of the integrals 7.1-7, 7.1-8, and 7.1-9 can be time consuming. Some simplifications and further approximations are often made, the most drastic of which are, as a group, called the Hückel approximation. For our purposes this relatively crude approximation will suffice, and it has the advantage of permitting us to carry treatments in which symmetry arguments can be employed through all stages, even to numerical results, without becoming bogged down in algebraic and computational problems.

Two points should be stressed. First, the symmetry arguments themselves are rigorous and would serve equally well in calculations of whatever degree of computational rigor one might care to make. Second, the Hückel approximation is, happily, a successful approximation; for all its apparent crudity it gives useful results.

The Hückel approximation assumes that *all* $S_{ij} = 0$ and that all $H_{ij} = 0$ unless the ith and jth orbitals are on adjacent atoms. The setting of all $S_{ij} = 0$ means that the normalizing factors for LCAO's are obtained very simply. If

$$\psi_i = N_i \sum_j a_{ij} \phi_j$$

and we require that

$$\int \psi_i \psi_i \, d\tau = 1$$

we obtain

$$\frac{1}{N_i^2} = \int \left(\sum_j a_{ij} \phi_j \right)^2 d\tau$$

$$= \sum_j a_{ij}^2 \int \phi_j \phi_j \, d\tau + \sum_{\substack{j,k \\ (j \neq k)}} a_{ij} a_{ik} \int \phi_j \phi_k \, d\tau$$

The second sum is equal to zero because overlap is assumed to be zero. The first sum is just equal to $\sum_j a_{ij}^2$, since the ϕ_j's are assumed to be normalized. Thus

$$\frac{1}{N_i^2} = \sum_j a_{ij}^2$$

or

$$N_i = \frac{1}{\sqrt{\sum_j a_{ij}^2}}$$

For the special but not uncommon case in which all a_{ij}'s are ± 1, N is just $1/\sqrt{n}$, where n is the number of atomic orbitals in the linear combination.

The Hückel approximation is applied most often to the π orbitals of hydrocarbons, in which case the following abbreviations are conventional:

$\alpha = H_{ii}$, the energy of an electron in a carbon $p\pi$ orbital before interaction with others;

$\beta = H_{ij}$, the energy of interaction between orbitals on adjacent atoms.

The appearance of the secular equations can be further simplified if α is taken as the zero of energy (i.e., set equal to zero) and β taken as the unit of energy. It can be shown that β is inherently a negative quantity. Thus an MO whose energy is positive in units of β has an *absolute* energy which is negative. An electron in such an MO is therefore more stable than an electron in an isolated $p\pi$ orbital.

Energy Level Diagrams

In many instances it is helpful to plot the calculated energies of MO's in a diagram such as the accompanying one. The energy scale is vertical, and each

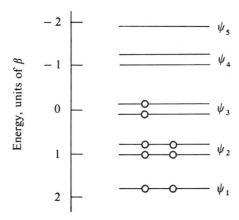

orbital is represented as a short horizontal line placed at the position corresponding to the energy of the orbital. For doubly degenerate orbitals a pair of closely spaced lines is used. The occupation of the orbitals by electrons is indicated by placing small circles on the lines.

Hund's Rule and the Exclusion Principle

The order of filling of MO's for the ground state of a molecule follows the same rules as does the filling of orbitals in the ground state of an atom. Thus an electron will go into the lowest unfilled level subject to the following restrictions: Only two electrons may occupy a single level, and their spins must be of opposite sign (exclusion principle). When electrons are to be placed in a pair of degenerate orbitals, they will (as shown for ψ_3 in the diagram) occupy each of the two degenerate orbitals singly, giving a total spin of 1 (Hund's rule).

Bonding Character of Orbitals

It is convenient, as already noted, to choose the zero of our energy scale to be that of the system in the hypothetical condition in which no interaction between the separate atomic orbitals is occurring. In the actual state of the molecule, then, some MO's will be of lower energy, some of higher energy, and, in certain cases, some of the same energy as this state which we take as zero. The MO's which are more stable than the separate, noninteracting atomic orbitals have absolute energies less than zero and are called *bonding* orbitals. Those with absolute energies greater than zero are called *antibonding* orbitals, and any orbitals having energies of precisely zero are called *nonbonding* orbitals. In the above example, we have a strongly bonding MO, ψ_1; a less strongly bonding, doubly degenerate MO, ψ_2; a nonbonding, doubly degenerate MO, ψ_3; a moderately antibonding, doubly degenerate MO, ψ_4; and a strongly antibonding, nondegenerate MO, ψ_5.

To see how the Hückel approximation simplifies the treatment of a moderately complex problem, consider the π orbitals of naphthalene. The ten $p\pi$ orbitals, numbered as in Figure 7.1, can be combined into ten linearly independent π MO's; thus, a 10×10 secular determinant can be written, as follows:

$$\begin{vmatrix} H_{11} - E & H_{12} - ES_{12} & H_{13} - ES_{13} & \cdots & H_{1,10} - ES_{1,10} \\ H_{21} - ES_{21} & H_{22} - E & & & \\ & & H_{33} - E & & \\ & & & \ddots & \\ H_{10,1} - ES_{10,1} & & \cdots & \cdots & H_{10,10} - E \end{vmatrix} = 0$$

$$(7.1\text{-}15)$$

We now introduce the Hückel approximation, employing the α, β notation, the convention that α is the zero of energy, and using β itself as the unit of energy. Explicitly, we have

$$H_{11} = H_{22} = H_{33} = \cdots \quad H_{99} = H_{10,10} = \alpha = 0$$

$$S_{ij} = \delta_{ij}$$

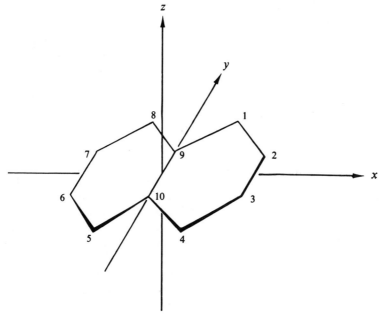

Figure 7-1 Cartesian axes and atom numbering for the naphthalene molecule.

$H_{ij} = H_{ji} = 0$, except for H_{12}, H_{23}, H_{34}, H_{56}, H_{67}, H_{78}, H_{89}, $H_{9,10}$, H_{19}, $H_{4,10}$, and $H_{5,10}$, which are all equal to β, which is equal to 1 unit of energy

The Secular equation is then:

$$
\begin{vmatrix}
-E & 1 & 0 & 0 & 0 & 0 & 0 & 0 & 1 & 0 \\
1 & -E & 1 & 0 & 0 & 0 & 0 & 0 & 0 & 0 \\
0 & 1 & -E & 1 & 0 & 0 & 0 & 0 & 0 & 0 \\
0 & 0 & 1 & -E & 0 & 0 & 0 & 0 & 0 & 1 \\
0 & 0 & 0 & 0 & -E & 1 & 0 & 0 & 0 & 1 \\
0 & 0 & 0 & 0 & 1 & -E & 1 & 0 & 0 & 0 \\
0 & 0 & 0 & 0 & 0 & 1 & -E & 1 & 0 & 0 \\
0 & 0 & 0 & 0 & 0 & 0 & 1 & -E & 1 & 0 \\
1 & 0 & 0 & 0 & 0 & 0 & 0 & 1 & -E & 1 \\
0 & 0 & 0 & 1 & 1 & 0 & 0 & 0 & 1 & -E
\end{vmatrix} = 0
$$

(7.1-16)

This is simpler than 7.1-15, in that many of the terms in the tenth-order polynomial equation which will result on expanding the determinant will now be equal to zero. Nevertheless, the basic, awkward fact is that a tenth-order equation still has to be solved. This is not a task to be confronted with pleasurable anticipation; without the use of a digital computer it would be a protracted, tedious job. Fortunately, in this case and all others in which the molecule possesses symmetry, the secular equation can be factored—that is, reduced to a collection of smaller equations—by using the symmetry properties in the right way. The method of symmetry factoring will now be explained and illustrated.

7.2   Symmetry Factoring of Secular Equations

Even with the simplifications which result from a drastic approximation such as the Hückel approximation, the secular equation for the MO's of an n-atomic molecule will, in general, involve at least an unfactored nth-order determinant, as just illustrated in the case of naphthalene. It is clearly desirable to factor such determinants, and symmetry considerations provide a systematic and rigorous means of doing this.

A secular equation such as 7.1-15 is derived from an array of the individual atomic orbitals of the basis set. Thus, in general, all of the H_{ij} and S_{ij} are nonzero. As *an approximation*, some may be set equal to zero, as in Equation 7.1-16, but still there is no logical, a priori reason why any entire classes or sets of integrals involving the individual atomic orbitals as such should systematically vanish.

Suppose that, instead of writing the secular determinant from an $n \times n$ array of atomic orbitals, we use an $n \times n$ array of n orthonormal, linear combinations of the basis set orbitals. Suppose, furthermore—and this is the key—we require these linear combinations to be SALC's, that is, each one is required to be a function which forms a basis for an irreducible representation of the point group of the molecule. Then, as shown in Chapter 5, all integrals of the types

$$\int \psi_i \psi_j \, d\tau \quad \text{and} \quad \int \psi_i \mathscr{H} \psi_j \, d\tau$$

are identically equal to zero unless ψ_i and ψ_j belong to the same irreducible representation.

The foregoing considerations lead to a three-step procedure for setting up a symmetry-factored secular equation:

1. Use the set of atomic orbitals as the basis for a representation of the group, and reduce this representation to its irreducible components.

2. Combine the basis orbitals into linear combinations corresponding to each of the irreducible representations. These SALC's can always be con-

structed systematically by using the projection operator technique developed in Chapter 6.

3. List the SALC's so that all those belonging to a given representation occur together in the list. Use this list to label the rows and columns of the secular determinant. Only the elements of the secular determinant which lie at the intersection of a row and a column belonging to the same irreducible representation can be nonzero, and these nonzero elements will lie in blocks along the principal diagonal. The secular determinant will therefore be factored.

As a very persuasive illustration of the effectiveness of symmetry factorization in reducing a computational task which would be entirely impractical without a digital computer to one which is a straightforward pencil-and-paper operation, we shall again consider the naphthalene molecule. It has been shown in Section 7.1 that the secular equation for the π MO's is the 10×10 determinantal equation, 7.1-15, if the set of ten $p\pi$ orbitals is used directly for constructing LCAO-MO's.

The naphthalene molecule belongs to the point group D_{2h}. The set of ten $p\pi$ orbitals may be used as the basis for a representation, Γ_π, of this group. This reducible representation can be decomposed into irreducible representations as follows:

$$\Gamma_\pi = 2A_u + 3B_{1u} + 2B_{2g} + 3B_{3g}$$

When the ten $p\pi$ orbitals are combined into SALC'S, ψ_1 to ψ_{10}, which are then listed in order of their symmetry types, viz.,

$$\underbrace{\psi_1, \psi_2,}_{A_u} \quad \underbrace{\psi_3, \psi_4, \psi_5,}_{B_{1u}} \quad \underbrace{\psi_6, \psi_7,}_{B_{2g}} \quad \underbrace{\psi_8, \psi_9, \psi_{10}}_{B_{3g}}$$

and these ten SALC's are used to construct the secular determinant, it takes the form shown on page 132, where all the empty, off-diagonal places must contain zeros, for rigorous, symmetry reasons.

It follows from the properties of determinants that, if the entire determinant is to have the value zero, *each block factor separately* must equal zero. Thus the 10×10 determinantal equation has been reduced to two 2×2 and two 3×3 secular equations. For example, the energies of the two MO's of A_u symmetry are given by the simple secular equation

$$\begin{vmatrix} H_{11} - E & H_{12} \\ H_{12} & H_{22} - E \end{vmatrix} = 0$$

We shall return later (page 161) to the symmetry-factored secular equation for the π MO's of naphthalene and solve for the energies, LCAO-MO coefficients, and other useful results.

This is a block-diagonal secular determinant. Column headers (top): A_u over ψ_1, ψ_2; B_{1u} over ψ_3, ψ_4, ψ_5; B_{2g} over ψ_6, ψ_7; B_{3g} over $\psi_8, \psi_9, \psi_{10}$. Row labels (left): $A_u\begin{Bmatrix}\psi_1\\\psi_2\end{Bmatrix}$, $B_{1u}\begin{Bmatrix}\psi_3\\\psi_4\\\psi_5\end{Bmatrix}$, $B_{2g}\begin{Bmatrix}\psi_6\\\psi_7\end{Bmatrix}$, $B_{3g}\begin{Bmatrix}\psi_8\\\psi_9\\\psi_{10}\end{Bmatrix}$.

A_u block:
$$\begin{matrix} H_{11}-E & H_{12} \\ H_{21} & H_{22}-E \end{matrix}$$

B_{1u} block:
$$\begin{matrix} H_{33}-E & H_{34} & H_{35} \\ H_{43} & H_{44}-E & H_{45} \\ H_{53} & H_{54} & H_{55}-E \end{matrix}$$

B_{2g} block:
$$\begin{matrix} H_{66}-E & H_{67} \\ H_{76} & H_{77}-E \end{matrix}$$

B_{3g} block:
$$\begin{matrix} H_{88}-E & H_{89} & H_{8,10} \\ H_{98} & H_{99}-E & H_{9,10} \\ H_{10,8} & H_{10,9} & H_{10,10}-E \end{matrix}$$

7.3 Carbocyclic Systems

We begin by considering the most famous and important of such systems, benzene. This molecule belongs to the point group D_{6h}. When the set of six $p\pi$ orbitals, one on each carbon atom, is taken as the basis for a representation of the group D_{6h}, we obtain the result:

D_{6h}	E	$2C_6$	$2C_3$	C_2	$3C_2'$	$3C_2''$	i	$2S_3$	$2S_6$	σ_h	$3\sigma_d$	$3\sigma_v$
Γ_π	6	0	0	0	-2	0	0	0	0	-6	2	0

We have defined C_2' axes and σ_d's as those passing through opposite carbon atoms, and C_2'' axes and σ_v's as those bisecting opposite edges of the hexagon. The above characters can be obtained at once by remembering that the matrix describing the effect of a given operation on the set of six orbitals obtains a contribution to its diagonal from each basis orbital according to the following schedule: 0 if the orbital is shifted to a different position; $+1$ if the orbital goes into itself; -1 if the orbital goes into the negative of itself, which means, simply, that it is turned upside down.

This representation is reduced (cf. Section 4.3) as follows:

$$\Gamma_\pi = A_{2u} + B_{1g} + E_{1g} + E_{2u}$$

Thus we need to form LCAO's of the indicated symmetry types, and this can be done by using projection operators for these representations of D_{6h}. However, it is advantageous to approach this task from a less direct point of view in order to arrive ultimately at an easier and more general approach to the type of problem it represents.

As pointed out in Section 6.3, for the $(CH)_3$ case, all the essential symmetry properties of the LCAO's we seek are determined by the operations of the uniaxial rotational subgroup, C_6. When the set of six $p\pi$ orbitals is used as the basis for a representation of the group C_6, the following results are obtained:

C_6	E	C_6	C_3	C_2	$C_3{}^2$	$C_6{}^5$
A	1	1	1	1	1	1
B	1	-1	1	-1	1	-1
E_1	$\begin{cases}1\\1\end{cases}$	$\begin{matrix}\varepsilon\\\varepsilon^*\end{matrix}$	$\begin{matrix}-\varepsilon^*\\-\varepsilon\end{matrix}$	$\begin{matrix}-1\\-1\end{matrix}$	$\begin{matrix}-\varepsilon\\-\varepsilon^*\end{matrix}$	$\begin{matrix}\varepsilon^*\\\varepsilon\end{matrix}$
E_2	$\begin{cases}1\\1\end{cases}$	$\begin{matrix}-\varepsilon^*\\-\varepsilon\end{matrix}$	$\begin{matrix}-\varepsilon\\-\varepsilon^*\end{matrix}$	$\begin{matrix}1\\1\end{matrix}$	$\begin{matrix}-\varepsilon^*\\-\varepsilon\end{matrix}$	$\begin{matrix}-\varepsilon\\-\varepsilon^*\end{matrix}$
Γ_ϕ	6	0	0	0	0	0

$$\Gamma_\phi = A + B + E_1 + E_2$$

Note first that $\chi(E) = 6$ while all other characters are zero. The reason is that the operation E transforms each ϕ_i into itself while *every* rotation operation necessarily shifts *every* ϕ_i to a different place. Clearly this kind of result will be obtained for any n-membered ring in a pure rotation group C_n. Second, note that the only way to add up characters of irreducible representations so as to obtain $\chi = 6$ for E and $\chi = 0$ for *every* operation other than E is to sum each column of the character table. From the basic properties of the irreducible representations of the uniaxial pure rotation groups (see Section 4.5), this is a general property for all C_n groups. Thus, the results just obtained for the benzene molecule merely illustrate the following general rule:

In a cyclic $(CII)_n$ *molecule with rotational symmetry* C_n, *there will always be* n π *molecular orbitals, one belonging to each irreducible representation of the group* C_n.

The $(CH)_3$ system discussed in Section 6.3 provides another illustration of this rule.

The A, B, E_1, and E_2 representations spanned in the pure rotation group C_6 become the A_{2u}, B_{1g}, E_{1g}, and E_{2u} representations in the group D_{6h} when the full symmetry of the benzene molecule and the inherent symmetry of the individual $p\pi$ orbitals are considered. The great advantage of working with the pure rotation group, which has already been illustrated (Section 6.3) in the very simple case of $(CH)_3$, is apparent when we come to determine the LCAO expressions for the MO's. Since each of the required linear combinations belongs uniquely to a single one-dimensional representation, the projection operator technique is extremely simple to apply—so simple, indeed, that the result can be written down as fast as one can write, simply by inspection of the character table.

Consider the following statement of the effect of applying the projection operator for *any* representation of C_6 to ϕ_1, the $p\pi$ orbital on carbon atom 1:

$$\hat{P}\phi_1 = \chi(E)\hat{E}\phi_1 + \chi(C_6)\hat{C}_6\phi_1 + \chi(C_6{}^2)\hat{C}_6{}^2\phi_1 + \chi(C_6{}^3)\hat{C}_6{}^3\phi_1$$
$$+ \chi(C_6{}^4)\hat{C}_6{}^4\phi_1 + \chi(C_6{}^5)\hat{C}_6{}^5\phi_1$$
$$= \chi(E)\phi_1 + \chi(C_6)\phi_2 + \chi(C_6{}^2)\phi_3 + \chi(C_6{}^3)\phi_4$$
$$+ \chi(C_6{}^4)\phi_5 + \chi(C_6{}^5)\phi_6$$

The second expression is simply a list of the six ϕ_i's, in numerical order, each multiplied by the character for one of the six operations, in the conventional order E, C_6, $C_6{}^2$, $\ldots$, $C_6{}^5$. This must be true for each and every representation. Hence, the sets of characters of the group *are* the coefficients of the LCAO-MO's. The argument is obviously a general one and applies to all cyclic $(CH)_n$ systems belonging to the point groups D_{nh}, each of which has a uniaxial pure rotation subgroup, C_n.

For the sake of concreteness, let us continue to use benzene as an example and write out the ψ's:

$$A: \quad \psi_1 = \phi_1 + \phi_2 + \phi_3 + \phi_4 + \phi_5 + \phi_6$$
$$B: \quad \psi_2 = \phi_1 - \phi_2 + \phi_3 - \phi_4 + \phi_5 - \phi_6$$
$$E_1: \quad \begin{cases} \psi_3 = \phi_1 + \varepsilon\phi_2 - \varepsilon^*\phi_3 - \phi_4 - \varepsilon\phi_5 + \varepsilon^*\phi_6 \\ \psi_4 = \phi_1 + \varepsilon^*\phi_2 - \varepsilon\phi_3 - \phi_4 - \varepsilon^*\phi_5 + \varepsilon\phi_6 \end{cases}$$
$$E_2: \quad \begin{cases} \psi_5 = \phi_1 - \varepsilon^*\phi_2 - \varepsilon\phi_3 + \phi_4 - \varepsilon^*\phi_5 - \varepsilon\phi_6 \\ \psi_6 = \phi_1 - \varepsilon\phi_2 - \varepsilon^*\phi_3 + \phi_4 - \varepsilon\phi_5 - \varepsilon^*\phi_6 \end{cases}$$

From a practical point of view there are two disadvantages in these LCAO-MO's. First, they contain imaginary coefficients. Second, they are not normalized to unity.

As explained in Section 6.3, we can easily convert the pairs of SALC's belonging to each pair of E-type representations into new linear combinations with real numbers as coefficients by (1) adding them, and (2) subtracting them and dividing out i. Thus, by adding ψ_3 and ψ_4 we obtain

$$\psi(E_1 a) = 2\phi_1 + (\varepsilon + \varepsilon^*)\phi_2 - (\varepsilon^* + \varepsilon)\phi_3 - 2\phi_4 - (\varepsilon + \varepsilon^*)\phi_5 + (\varepsilon^* + \varepsilon)\phi_6$$

which reduces to

$$\psi(E_1 b) = 2\phi_1 + \phi_2 - \phi_3 - 2\phi_4 - \phi_5 + \phi_6$$

The second linear combination, $(\psi_3 - \psi_4)/i$, is explicitly

$$\psi(E_1 b) = [(\varepsilon - \varepsilon^*)\phi_2 - (\varepsilon^* - \varepsilon)\phi_3 - (\varepsilon - \varepsilon^*)\phi_5 + (\varepsilon^* - \varepsilon)\phi_6]/i$$
$$= -\sqrt{3}\phi_2 - \sqrt{3}\phi_3 + \sqrt{3}\phi_5 + \sqrt{3}\phi_6$$

In a similar way ψ_5 and ψ_6 may be combined to give

$$\psi(E_2 a) = \psi_5 + \psi_6 = 2\phi_1 - \phi_2 - \phi_3 + 2\phi_4 - \phi_5 - \phi_6$$
$$\psi(E_2 b) = (\psi_5 - \psi_6)/i = -\sqrt{3}\phi_2 + \sqrt{3}\phi_3 - \sqrt{3}\phi_5 + \sqrt{3}\phi_6$$

We now normalize these MO wave functions as described in Section 7.1, neglecting overlap, and obtain the following final expressions:

$$\psi(A) = \frac{1}{\sqrt{6}}(\phi_1 + \phi_2 + \phi_3 + \phi_4 + \phi_5 + \phi_6)$$

$$\psi(B) = \frac{1}{\sqrt{6}}(\phi_1 - \phi_2 + \phi_3 - \phi_4 + \phi_5 - \phi_6)$$

$$\psi(E_1 a) = \frac{1}{\sqrt{12}}(2\phi_1 + \phi_2 - \phi_3 - 2\phi_4 - \phi_5 + \phi_6)$$

$$\psi(E_1 b) = \tfrac{1}{2}(\phi_2 + \phi_3 - \phi_5 - \phi_6)$$

$$\psi(E_2 a) = \frac{1}{\sqrt{12}}(2\phi_1 - \phi_2 - \phi_3 + 2\phi_4 - \phi_5 - \phi_6)$$

$$\psi(E_2 b) = \tfrac{1}{2}(\phi_2 - \phi_3 + \phi_5 - \phi_6)$$

In addition to being normalized the MO's should be mutually orthogonal. That this is true of those given above is easy to verify, although it follows from the procedure used to construct them that they must be.

It is instructive to examine some diagrams showing how the signs of the ψ's vary around the ring. For benzene, using the orbitals constructed above, we can make the following drawings:

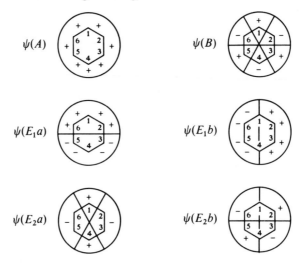

Note that the E_1 orbitals have one nodal plane and the E_2 orbitals two nodal planes.

The energies of these MO's may now be calculated, using the Hückel approximation as discussed in Section 7.1. For $\psi(A)$ we obtain

$$E_A = \tfrac{1}{6}(6\alpha + 12\beta) = \alpha + 2\beta$$

Proceeding in a similar way, we find the energies of the other LCAO-MO's for benzene to be

$$E_B = \alpha - 2\beta$$

$$E_{E_1 a} = E_{E_1 b} = \alpha + \beta$$

$$E_{E_2 a} = E_{E_2 b} = \alpha - \beta$$

Taking α as the zero of energy and β as the unit of energy, we may express these results in the form of an energy level diagram, viz.,

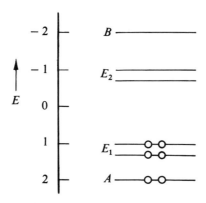

Delocalization (Resonance) Energy

If each of the six π electrons in benzene occupied a single atomic π orbital and there were no interaction, each would have an energy of α. The total energy would then be 6α, which is zero if we assume, as above, that α is the zero of our energy scale. However, when the atomic orbitals interact to produce the MO's, the six electrons will now occupy these MO's according to Hund's rule and the Pauli exclusion principle. The first two will enter the A orbital, and the remaining four occupy the E_1 orbitals. The total energy of the system is then

$$E_T = 2(2\beta) + 4(\beta) = 8\beta$$

Remembering that β is negative, we see that π bonding has stabilized the molecule by 8β. Energies expressed in units of β are not very informative, however, unless we can estimate the value of β. We shall next turn to the calculation of the delocalization energy of benzene in units of β. Since the delocalization energy may be estimated experimentally, we shall then be able to evaluate β. It is not practicable to do this by computation.

Strictly speaking, the concept of delocalization or resonance energy belongs in valence bond theory. It is defined there as the difference between the energy of the most stable canonical structure, which is one of the Kekulé structures, and the actual energy. The actual energy is assumed to be calculable, according to valence bond theory, by taking into account resonance between all possible canonical structures. Of these, usually only five are considered sufficiently low in energy to be significant. These are the two equivalent Kekulé structures and three Dewar structures:

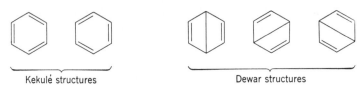

Kekulé structures   Dewar structures

However, it may be assumed that the difference between the energy of one Kekulé structure and the actual energy as calculated by MO theory will also be the resonance energy except that, in keeping with the basic concept of MO theory, we shall call it the delocalization energy.

We have already shown that the energy of the system of six π electrons in benzene is equal to 8β. We must now calculate, in units of β, the energy of a Kekulé structure, that is, the energy of the hypothetical molecule cyclohexatriene. In cyclohexatriene there are three localized π bonds. When two atomic π orbitals, say ϕ_1 and ϕ_2, interact to form a two-center bond, two MO's, ψ_1 and ψ_2, are formed. In order that these be real, normalized, and orthogonal, they must be

$$\psi_1 = \frac{1}{\sqrt{2}}(\phi_1 + \phi_2)$$

$$\psi_2 = \frac{1}{\sqrt{2}}(\phi_1 - \phi_2)$$

Their energies are readily seen to be

$$E_1 = \int \psi_1 \mathcal{H} \psi_1 \, d\tau = \frac{1}{2}\left(\int \phi_1 \mathcal{H} \phi_1 \, d\tau + \int \phi_1 \mathcal{H} \phi_2 \, d\tau \right.$$
$$\left. + \int \phi_2 \mathcal{H} \phi_1 \, d\tau + \int \phi_2 \mathcal{H} \phi_2 \, d\tau \right)$$
$$= \tfrac{1}{2}(2\alpha + 2\beta) = \beta$$
$$E_2 = -\beta$$

Since ψ_1 is the stable MO, the two π electrons will occupy it and their combined energy will be 2β. Thus each of the pairs of localized π electrons in a Kekulé structure contributes 2β to the energy of the molecule, making the total π electron energy of the localized, cyclohexatriene structure 6β. But the actual energy is 8β; hence the resonance or delocalization energy is 2β.

Experimentally, the delocalization energy of benzene is estimated in the following way. The actual enthalpy of formation of benzene can be determined by thermochemical measurements. The energy of the hypothetical molecule cyclohexatriene can be estimated by using the bond energies for $C-C$, $C=C$, and $C-H$ found in other molecules such as ethane and ethylene.

The difference between these energies is the "experimental" value of the delocalization energy. We then evaluate $|\beta|$, since

$$2\,|\beta| = \text{"experimental" delocalization energy}$$

The value of $|\beta|$ obtained for benzene is 18–20 kcal/mole, depending on choices of bond energies.* Practically the same value is obtained when other aromatic molecules, such as naphthalene and anthracene, are treated in the same way, a fact which lends support to the belief that the LCAO method is at least empirically valid.

Other carbocyclic systems of the type $(CH)_n$, belonging to the point groups D_{nh}, which are of interest are those with n equal to 3, 4, 5, 7, and 8. In some cases the actual systems (which may be cations, neutral molecules, or anions) are not necessarily of D_{nh} symmetry, but there is usually some purpose in treating them as though they are, if only as a starting point. The treatment of benzene just developed is applicable, *mutatis mutandis*, to all of them. The results are summarized in Table 7.1. The student can obtain practice in applying the treatment by verifying these results.

The 4n + 2 Rule

From the results we have obtained for the systems C_4H_4, C_6H_6, and C_8H_8 we can infer a rule, first discovered by Hückel and now rather well known, concerning the aromaticity of planar, carbocyclic systems of the type $(CH)_n$.

According to valence bond theory, any such system in which the number of carbon atoms is even would be expected to have resonance stabilization because of the existence of canonical forms of the type illustrated below for the first three members of the homologous series:

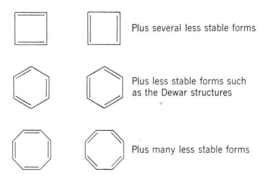

Plus several less stable forms

Plus less stable forms such as the Dewar structures

Plus many less stable forms

* See Appendix IV for some important qualifications concerning the evaluation of β.

Table 7.1 Summary of Hückel MO calculations for $(CH)_n$ Systems

SYSTEM (SYMMETRY)	LCAO-MO EXPRESSIONS	ENERGIES
C_3H_3 (D_{3h}) (Note 1)	$\psi(A) = \dfrac{1}{\sqrt{3}}(\phi_1 + \phi_2 + \phi_3)$	$\alpha + 2\beta$
	$\psi(Ea) = \dfrac{1}{\sqrt{6}}(2\phi_1 - \phi_2 - \phi_3)$	
	$\psi(Eb) = \dfrac{1}{\sqrt{2}}(\phi_2 - \phi_3)$	$\left.\right\}\ \alpha - \beta$
C_4H_4 (D_{4h}) (Note 2)	$\psi(A) = \tfrac{1}{2}(\phi_1 + \phi_2 + \phi_3 + \phi_4)$	$\alpha + 2\beta$
	$\psi(Ea) = \dfrac{1}{\sqrt{2}}(\phi_1 - \phi_3)$	
	$\psi(Ea) = \dfrac{1}{\sqrt{2}}(\phi_2 - \phi_4)$	$\left.\right\}\ \alpha$
	$\psi(B) = \tfrac{1}{2}(\phi_1 - \phi_2 + \phi_3 - \phi_4)$	$\alpha - 2\beta$

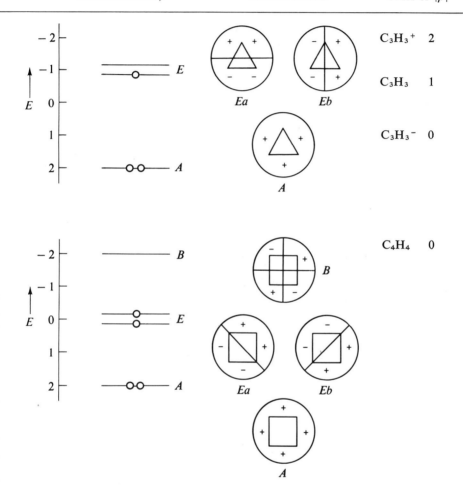

Table 7.1 (continued)

SYSTEM (SYMMETRY)	LCAO-MO EXPRESSIONS	ENERGIES

C_5H_5
(D_{5h})
$\omega = 2\pi/5$
(Note 3)

$$\psi(A) = \frac{1}{\sqrt{5}}(\phi_1 + \phi_2 + \phi_3 + \phi_4 + \phi_5)$$

$\alpha + 2\beta$

$$\psi(E_1 a) = \sqrt{\frac{2}{5}}(\phi_1 + \phi_2 \cos \omega + \phi_3 \cos 2\omega$$
$$+ \phi_4 \cos 2\omega + \phi_5 \cos \omega)$$

$$\psi(E_1 b) = \sqrt{\frac{2}{5}}(\phi_2 \sin \omega + \phi_3 \sin 2\omega$$
$$- \phi_4 \sin 2\omega - \phi_5 \sin \omega)$$

$\alpha + (2 \cos \omega)\beta$

$$\psi(E_2 a) = \sqrt{\frac{2}{5}}(\phi_1 + \phi_2 \cos 2\omega + \phi_3 \cos \omega$$
$$+ \phi_4 \cos \omega + \phi_5 \cos 2\omega)$$

$$\psi(E_2 b) = \sqrt{\frac{2}{5}}(\phi_2 \sin 2\omega - \phi_3 \sin \omega$$
$$+ \phi_4 \sin \omega - \phi_5 \sin 2\omega)$$

$\alpha + (2 \cos 2\omega)\beta$

C_6H_6 See text
(D_{6h})

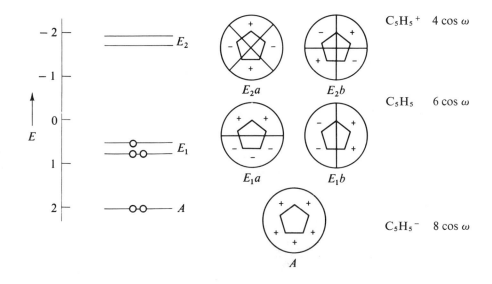

| ENERGY LEVEL DIAGRAM (POPULATED FOR NEUTRAL SPECIES) | ORBITAL SHAPES | DELOCALIZATION ENERGIES, UNITS OF $|\beta|$ |

$C_5H_5{}^+$ $4 \cos \omega$

C_5H_5 $6 \cos \omega$

$C_5H_5{}^-$ $8 \cos \omega$

Table 7.1 (continued)

SYSTEM (SYMMETRY)	LCAO-MO EXPRESSIONS	ENERGIES

C_7H_7
(D_{7h})
$\omega = 2\pi/7$
(Note 3)

$$\psi(A) = \frac{1}{\sqrt{7}}(\phi_1 + \phi_2 + \phi_3 + \phi_4 + \phi_5 + \phi_6 + \phi_7)$$

$\alpha + 2\beta$

$$\psi(E_1 a) = \sqrt{\frac{2}{7}}(\phi_1 + \phi_2 \cos \omega + \phi_3 \cos 2\omega + \phi_4 \cos 3\omega + \phi_5 \cos 3\omega + \phi_6 \cos 2\omega + \phi_7 \cos \omega)$$

$$\psi(E_1 b) = \sqrt{\frac{2}{7}}(\phi_2 \sin \omega + \phi_3 \sin 2\omega + \phi_4 \sin 3\omega - \phi_5 \sin 3\omega - \phi_6 \sin 2\omega - \phi_7 \sin \omega)$$

$\alpha + 2\beta \cos \omega$

$$\psi(E_2 a) = \sqrt{\frac{2}{7}}(\phi_1 + \phi_2 \cos 2\omega + \phi_3 \cos 3\omega + \phi_4 \cos \omega + \phi_5 \cos \omega + \phi_6 \cos 3\omega + \phi_7 \cos 2\omega)$$

$$\psi(E_2 b) = \sqrt{\frac{2}{7}}(\phi_2 \sin 2\omega - \phi_3 \sin 3\omega - \phi_4 \sin \omega + \phi_5 \sin \omega + \phi_6 \sin 3\omega - \phi_7 \sin 2\omega)$$

$\alpha + 2\beta \cos 2\omega$

$$\psi(E_3 a) = \sqrt{\frac{2}{7}}(\phi_1 + \phi_2 \cos 3\omega + \phi_3 \cos \omega + \phi_4 \cos 2\omega + \phi_5 \cos 2\omega + \phi_6 \cos \omega + \phi_7 \cos 3\omega)$$

$$\psi(E_3 b) = \sqrt{\frac{2}{7}}(\phi_2 \sin 3\omega - \phi_3 \sin \omega + \phi_4 \sin 2\omega - \phi_5 \sin 2\omega + \phi_6 \sin \omega - \phi_7 \sin 3\omega)$$

$\alpha + 2\beta \cos 3\omega$

| ENERGY LEVEL DIAGRAM (POPULATED FOR NEUTRAL SPECIES) | ORBITAL SHAPES | DELOCALIZATION ENERGIES, UNITS OF $|\beta|$ |

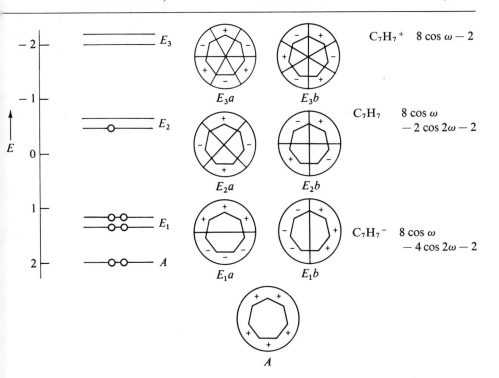

$C_7H_7^+$ $8\cos\omega - 2$

C_7H_7 $8\cos\omega$
$- 2\cos 2\omega - 2$

$C_7H_7^-$ $8\cos\omega$
$- 4\cos 2\omega - 2$

Table 7.1 (continued)

SYSTEM (SYMMETRY)	LCAO-MO EXPRESSIONS	ENERGIES

C_8H_8
(D_{8h})
(Note 4)

$$\psi(A) = \frac{1}{\sqrt{8}}(\phi_1 + \phi_2 + \phi_3 + \phi_4 + \phi_5 + \phi_6 + \phi_7 + \phi_8) \qquad \alpha + 2\beta$$

$$\psi(B) = \frac{1}{\sqrt{8}}(\phi_1 - \phi_2 + \phi_3 - \phi_4 + \phi_5 - \phi_6 + \phi_7 - \phi_8) \qquad \alpha - 2\beta$$

$$\psi(E_1a) = \frac{1}{\sqrt{8}}(\sqrt{2}\phi_1 + \phi_2 - \phi_4 - \sqrt{2}\phi_5 - \phi_6 + \phi_8)$$

$$\psi(E_1b) = \frac{1}{\sqrt{8}}(\phi_2 + \sqrt{2}\phi_3 + \phi_4 - \phi_6 - \sqrt{2}\phi_7 - \phi_8)$$

$$\qquad \alpha + \sqrt{2}\beta$$

$$\psi(E_2a) = \tfrac{1}{2}(\phi_1 - \phi_3 + \phi_5 - \phi_7)$$

$$\psi(E_2b) = \tfrac{1}{2}(\phi_2 - \phi_4 + \phi_6 - \phi_8)$$

$$\qquad \alpha$$

$$\psi(E_3a) = \frac{1}{\sqrt{8}}(\sqrt{2}\phi_1 - \phi_2 + \phi_4 - \sqrt{2}\phi_5 + \phi_6 - \phi_8)$$

$$\psi(E_3b) = \frac{1}{\sqrt{8}}(\phi_2 - \sqrt{2}\phi_3 + \phi_4 - \phi_6 + \sqrt{2}\phi_7 - \phi_8)$$

$$\qquad \alpha - \sqrt{2}\beta$$

1. For a discussion of the C_3H_3 systems, see R. Breslow, *Angew. Chem.*, Intern. Ed. (Engl.), **7**, 565 (1968).

2. The system C_4H_4 is only metastable with minimum energy in a rectangular (D_{2h}) shape; cf. P. Reeves, T. Devon, and R. Pettit, *J. Am. Chem. Soc.*, **91**, 5890 (1969).

3. In obtaining the results for the C_5H_5 and C_7H_7 systems the relation

$$\sum_{k=0}^{n-1} \cos^2\frac{k2\pi}{n} = \sum_{k=0}^{n-1} \sin^2\frac{k2\pi}{n} = \frac{n}{2}$$

has been used, along with other, more familiar trigonometric identities. Although the energies of the doubly degenerate orbitals can be obtained by using the expressions given above for them

ENERGY LEVEL DIAGRAM
(POPULATED FOR
NEUTRAL SPECIES)

ORBITAL SHAPES

DELOCALIZATION
ENERGIES,
UNITS OF $|\beta|$

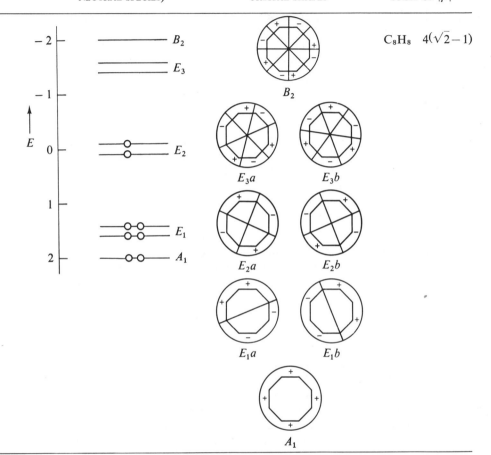

C_8H_8 $4(\sqrt{2}-1)$

and the method described for benzene, some rather messy trigonometric algebra is encountered. A simpler method, which, however, involves some quantum theory not treated in this book, is available. See H. Eyring, J. Walter, and G. E. Kimball, *Quantum Chemistry*, John Wiley, 1944, pp. 254–255.

4. Cyclooctatetraene, C_8H_8, is of course well known to be nonplanar and nonaromatic. It is properly described as a conjugated but nonaromatic tetraolefin. An MO treatment of the *hypothetical* planar $(CH)_8$ is of some interest, however, in respect to the questions of the instability of this configuration for the free molecule, as well as of the possibility of the stabilization of this configuration by formation of the anions $C_8H_8^-$ and $C_8H_8^{2-}$. There is evidence that the dianion is planar with a closed shell configuration. Cf. M. J. S. Dewar, A. Harget, and E. Haselbach, *J. Am. Chem. Soc.*, **91**, 7521 (1969), and references therein.

Since a calculation of the resonance energy of benzene by the valence bond method shows that the greater part of it is due to resonance between the two Kekulé structures shown, we might suppose that its homologs would also have significant resonance stabilization energies. Such conclusions are at variance with experimental fact, however, since cyclobutadiene appears to be too unstable to have any permanent existence, and cyclooctatetraene exists as a nonplanar tetraolefin incapable of having resonance stabilization of the sort considered.

Simple LCAO-MO theory provides a direct and natural explanation for the facts. It may be seen that for C_4H_4, C_6H_6, and C_8H_8 the energy level diagrams have the same general arrangement of levels, namely, a symmetrical distribution of a strongly bonding, nondegenerate A level and a strongly antibonding, nondegenerate B level, with a set of E levels between them. It can be shown that such a pattern will always develop in an even-membered C_nH_n system. Now, in order to attain a closed configuration (i.e., one with all electrons paired) of the general type $(\psi_A)^2(\psi_{E_1})^4 \cdots (\psi_{E_x})^4$, it will be necessary to fill the lowest nondegenerate A level and then to fill completely the first x pairs of degenerate levels above it. For this, $4x + 2$ electrons will be required. It therefore follows that only for systems in which n, which is both the number of π electrons and the ring size of C_nH_n, is a number expressible as $4x + 2$ ($x = 1, 2, 3, \ldots$) can we obtain a closed configuration. The numbers, n, meeting this requirement are 6, 10, 14, ... For the other even integers, namely, 4, 8, 12, ..., we shall always have an electron configuration of the sort $(\psi_A)^2(\psi_{E_1})^4 \cdots (\psi_{E_x})^2$.

The systems with $4n$ electrons (e.g., C_4H_4 and C_8H_8), which as planar systems with D_{nh} symmetry would be diradicals, are more stable with a set of alternating single and double bonds. Thus C_4H_4 is apparently a reactive *rectangular* molecule with two short (double) and two long (single) bonds, while C_8H_8 is definitely known to contain four single and four double bonds. Because of angle strain and H $\cdots$ H repulsions, the planar form of C_8H_8 is something like 17 kcal/mole less stable than the boat form.* Even $C_{16}H_{16}$ has been shown to be a polyolefin rather than an aromatic system, while $C_{18}H_{18}$ and certain related macrocyclic C_nH_n systems (called annulenes) show physical evidence of aromatic character.†

* M. J. S. Dewar, A. Harget, and E. Haselbach, *J. Am. Chem. Soc.*, **91**, 7521 (1969).
† For an extensive review of annulenes, see F. Sondheimer, *Proc. Roy. Soc. (London)*, **297A**, 173 (1967).

7.4   More General Cases of LCAO-MO π Bonding

Tetramethylenecyclobutane

This molecule provides an interesting and entirely genuine illustration of the predictive power of simple Hückel MO theory; it also serves as a good first example of how to apply symmetry methods in more general cases. The molecule was predicted* to have about 30 kcal/mole of delocalization energy in 1952; in 1962 it was synthesized and shown to be stable.†

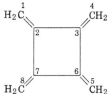

The molecule may be assumed to be planar, and we shall adopt the numbering scheme shown at the right.

The molecule belongs to the point group D_{4h}. By using the set of eight carbon $p\pi$ orbitals as a basis we can obtain a reducible representation which will contain the irreducible representations to which the π MO's must belong:

D_{4h}	E	$2C_4$	C_2	$2C_2'$	$2C_2''$	i	$2S_4$	σ_h	$2\sigma_v$	$2\sigma_d$
Γ_π	8	0	0	0	-4	0	0	-8	0	4

$$\Gamma_\pi = 2A_{2u} + 2B_{1u} + 2E_g$$

There is a very important feature of this situation, which we can turn to advantage. It will be observed that the set of four methylene carbon atoms, numbers 1, 4, 5, and 8, possess D_{4h} symmetry by themselves and that the set of four carbon atoms in the ring, numbers 2, 3, 6, and 7, also by themselves constitute a set having D_{4h} symmetry. Furthermore, the atoms in one set are not equivalent symmetrically to any of those in the other. None of the outer atoms is ever interchanged with any of the inner atoms by any symmetry operation. Thus each of these sets can be used separately as the basis for a representation of the group, and if this is done we obtain from each set a representation, Γ_π', which reduces as follows:

$$\Gamma_\pi' = A_{2u} + B_{1u} + E_g$$

---

* J. D. Roberts, A. Streitwieser, Jr., and C. M. Regan, *J. Am. Chem. Soc.*, **74**, 4579 (1952).
† G. W. Griffin and L. I. Peterson, *J. Am. Chem. Soc.*, **84**, 3398 (1962).

This means that, if we write an expression for an A_{2u} MO as a combination of all eight orbitals, viz.,

$$\psi_{A_{2u}} = N(a_1\phi_1 + a_2\phi_2 + a_3\phi_3 + a_4\phi_4 + a_5\phi_5 + a_6\phi_6 + a_7\phi_7 + a_8\phi_8)$$

we can separate it into two components, one made up only of orbitals of the inner set and one made up only of orbitals of the outer set:

$$\psi_{A_{2u}} = N(\underbrace{a_2\phi_2 + a_3\phi_3 + a_6\phi_6 + a_7\phi_7}_{\text{inner set}}) + N(\underbrace{a_1\phi_1 + a_4\phi_4 + a_5\phi_5 + a_8\phi_8}_{\text{outer set}})$$

Since the symmetry operations cannot interchange orbitals of the two sets, each of the subsets in the expression for the A_{2u} MO must itself have A_{2u} symmetry. Thus, in order to construct an orbital of A_{2u} symmetry for the entire molecule, we can first construct partial orbitals from the atomic orbitals in each of the subsets and then combine them into a complete MO. A similar line of reasoning applies to MO's of any other symmetry.

Our immediate problem, then, is to combine the four outer orbitals into linear combinations having A_{2u}, B_{1u}, and E_g symmetry and also to combine the four inner orbitals into linear combinations having these same symmetries. As in the case of the carbocyclic rings, this process can be simplified by using only the corresponding rotation group C_4 instead of D_{4h}, since the former can discriminate between the orbitals. That is, if we construct an orbital of A symmetry in the group C_4, it will automatically turn out to have A_{2u} symmetry in D_{4h} because of the inherent symmetry of the $p\pi$ orbitals themselves, namely, their antisymmetric character with respect to reflection in the molecular plane.

With these considerations in mind, the process of constructing the correct linear combinations of the subsets proceeds exactly as in the case of the carbocyclic systems. The correct coefficients of the atomic orbitals are simply the characters of the representations. For the E orbitals we will obtain some imaginary coefficients, but these may be eliminated by taking the appropriate linear combinations. We can thus write, almost by direct inspection of the character table of the C_4 group:

$$\psi_{A^i} = \tfrac{1}{2}(\phi_2 + \phi_3 + \phi_6 + \phi_7)$$
$$\psi_{A^o} = \tfrac{1}{2}(\phi_1 + \phi_4 + \phi_5 + \phi_8)$$
$$\psi_{B^i} = \tfrac{1}{2}(\phi_2 - \phi_3 + \phi_6 - \phi_7)$$
$$\psi_{B^o} = \tfrac{1}{2}(\phi_1 - \phi_4 + \phi_5 - \phi_8)$$

$$\psi_{Ea^i} = \frac{1}{\sqrt{2}}(\phi_2 - \phi_6)$$

$$\psi_{Eb^i} = \frac{1}{\sqrt{2}}(\phi_3 - \phi_7)$$

$$\psi_{Ea}^{\,o} = \frac{1}{\sqrt{2}}(\phi_1 - \phi_5)$$

$$\psi_{Eb}^{\,o} = \frac{1}{\sqrt{2}}(\phi_4 - \phi_8)$$

where we use the superscripts i and o to indicate that the combination is made up of inner or outer orbitals.

We may now solve the secular equation, using these symmetry-correct MO's, and obtain the MO energies. Thus, for the A orbitals we have the equation

$$\begin{vmatrix} H_{A^iA^i} - E & H_{A^iA^o} \\ H_{A^oA^i} & H_{A^oA^o} - E \end{vmatrix} = 0$$

The elements of this determinant are easily evaluated by using the Hückel approximation:

$$H_{A^iA^i} = \int \psi_{A^i} \mathcal{H} \psi_{A^i}\, d\tau = \frac{1}{4}\int (\phi_2 + \phi_3 + \phi_6 + \phi_7)\mathcal{H}(\phi_2 + \phi_3 + \phi_6 + \phi_7)\, d\tau$$

$$= \frac{1}{4}\left(\int \phi_2 \mathcal{H} \phi_2\, d\tau + \int \phi_2 \mathcal{H} \phi_3\, d\tau \right.$$

$$\left. + \int \phi_2 \mathcal{H} \phi_6\, d\tau + \cdots + \int \phi_7 \mathcal{H} \phi_7\, d\tau \right)$$

$$= \tfrac{1}{4}(\alpha + \beta + 0 + \cdots + \alpha)$$

$$= \tfrac{1}{4}(4\alpha + 8\beta) = \alpha + 2\beta$$

$$H_{A^oA^o} = \int \psi_{A^o} \mathcal{H} \psi_{A^o}\, d\tau = \frac{1}{4}\int (\phi_1 + \phi_4 + \phi_5 + \phi_8)\mathcal{H}(\phi_1 + \phi_4 + \phi_5 + \phi_8)\, d\tau$$

$$= \tfrac{1}{4}(4\alpha) = \alpha$$

$$H_{A^iA^o} = H_{A^oA^i} = \int \psi_{A^o} \mathcal{H} \psi_{A^i}\, d\tau$$

$$= \frac{1}{4}\int (\phi_1 + \phi_4 + \phi_5 + \phi_8)\mathcal{H}(\phi_2 + \phi_3 + \phi_6 + \phi_7)\, d\tau$$

$$= \tfrac{1}{4}(4\beta) = \beta$$

As before, it is convenient to choose α as the zero of energy and to take β as the unit of energy. The secular equation for the A orbitals then takes the form

$$\begin{vmatrix} 2 - E & 1 \\ 1 & -E \end{vmatrix} = 0$$

which is expanded into a quadratic equation and solved:

$$E^2 - 2E - 1 = 0, \qquad E_A = (1 + \sqrt{2}), (1 - \sqrt{2})$$

Following an analogous procedure for the orbitals of B symmetry, we obtain

$$\begin{vmatrix} H_{B^iB^i} - E & H_{B^iB^o} \\ H_{B^oB^i} & H_{B^oB^o} - E \end{vmatrix} = \begin{vmatrix} -2 - E & 1 \\ 1 & -E \end{vmatrix} = 0$$

The solutions are

$$E_B = (\sqrt{2} - 1) \quad \text{and} \quad (-\sqrt{2} - 1)$$

For the E orbitals we will obtain two two-dimensional determinants, one involving the Ea orbitals and the other involving the Eb orbitals. It is necessary to solve only one of them. Choosing the Ea determinant, we have

$$\begin{vmatrix} H_{E^iE^i} - E & H_{E^iE^o} \\ H_{E^oE^i} & H_{E^oE^o} - E \end{vmatrix} = \begin{vmatrix} -E & 1 \\ 1 & -E \end{vmatrix} = 0$$

which has the roots

$$E_E = \pm 1$$

Remembering that β is intrinsically negative, we may use these results to construct the following energy level diagram, in which the eight electrons have been added to the lower four orbitals:

It can be seen that the order of the levels is such that all of the bonding levels (those with energies <0) are just filled, and all electrons must have their spins paired.

The delocalization energy can be calculated easily. The most stable arrangement of the four electron pairs in localized double bonds would undoubtedly be the one labeled (a) below, all other arrangements, such as (b) or (c), containing fewer than four short, strong double bonds. The energy of this

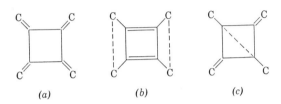

(a)   (b)   (c)

arrangement, taking $\alpha = 0$, can easily be seen, by the argument used in Section 7.3 for benzene in a Kekulé structure, to be 8β. The total energy of the eight electrons occupying the MO's as shown in the energy level diagram above is

$$[2(1 + \sqrt{2}) + 4(1) + 2(\sqrt{2} - 1)]\beta = 9.656\beta$$

Hence the resonance or delocalization energy is 1.66β, which, taking $|\beta| = 20$ kcal/mole, comes to about 33 kcal/mole.

We turn, finally, to the task of finding the actual expressions for the occupied MO's in order that we may compute such properties of the ground-state electron distribution as the π bond orders. For example, $\psi_A^{(1)}$, which has the energy $(1 + \sqrt{2})\beta$, is neither ψ_{A^i}, which has the energy 2β, nor ψ_{A^o}, which has the energy 0. It is a linear combination of both, and the problem is to find the appropriate mixing coefficients. As explained earlier, we do this by returning to the simultaneous equations from which the secular equations arose. For the orbitals of A symmetry, we have

$$c_i(H_{A^iA^i} - E) + c_o H_{A^iA^o} = 0 = c_i(2 - E) + c_o$$

and

$$c_i H_{A^iA^o} + c_o(H_{A^oA^o} - E) = 0 = c_i - c_o E$$

Either of these equations may be used to express the ratio of c_i to c_o, viz.,

$$c_i/c_o = -1/(2 - E)$$

or

$$c_i/c_o = E$$

For the correct values of E, these two equations must give the same ratio.

If we insert the energy of $\psi_A^{(1)}$, $(1 + \sqrt{2})$, we obtain

$$c_i/c_o = -1/(2 - 1 - \sqrt{2}) = -1/(1 - \sqrt{2}) = 1/0.414 = 2.414$$
$$c_i/c_o = 1 + \sqrt{2} = 2.414$$

which are indeed equal as they must be. When this relationship is combined with the normalization condition

$$c_i^2 + c_o^2 = 1$$

the actual values of c_i and c_o can be obtained:

$$c_o = 0.382, \quad \text{and} \quad c_i = 0.924$$

The final expression for the $\psi_A^{(1)}$ MO is then

$$
\begin{aligned}
\psi_A^{(1)} &= c_i \psi_{A^i} + c_o \psi_{A^o} \\
&= (0.924)(\tfrac{1}{2})(\phi_2 + \phi_3 + \phi_6 + \phi_7) \\
&\quad + (0.382)(\tfrac{1}{2})(\phi_1 + \phi_4 + \phi_5 + \phi_8) \\
&= 0.191(\phi_1 + \phi_4 + \phi_5 + \phi_8) + 0.462(\phi_2 + \phi_3 + \phi_6 + \phi_7)
\end{aligned}
$$

Proceeding in the same way, we obtain the following expressions for the other occupied MO's:

$$
\begin{aligned}
\psi_B^{(1)} &= 0.462(\phi_1 - \phi_4 + \phi_5 - \phi_8) \\
&\quad + 0.191(\phi_2 - \phi_3 + \phi_6 - \phi_7) \\
\psi_{Ea}^{(1)} &= 0.500(\phi_1 + \phi_2 - \phi_5 - \phi_6) \\
\psi_{Eb}^{(1)} &= 0.500(\phi_3 + \phi_4 - \phi_7 - \phi_8)
\end{aligned}
$$

As examples of the uses of such LCAO-MO's, let us calculate the bond orders in tetramethylenecyclobutane. The order of the bond between two atoms is defined as the sum of the products of the coefficients of the atomic orbitals of the two atoms in each of the occupied MO's, each product being weighted with the number of electrons occupying the MO. Thus we have for one of the equivalent ring bonds, say the one between C_2 and C_3:

$$
\begin{array}{llll}
\psi_A^{(1)}: & 2 \times (0.462)(0.462) & = & 0.428 \\
\psi_B^{(1)}: & 2 \times (0.191)(-0.191) & = & -0.074 \\
\psi_{Ea}^{(1)}: & 2 \times (0.500)(0) & = & 0.000 \\
\psi_{Eb}^{(1)}: & 2 \times (0.500)(0) & = & 0.000 \\
\hline
& & & 0.354
\end{array}
$$

For one of the exo bonds, say the one between C_1 and C_2 :

$$
\begin{aligned}
\psi_A{}^{(1)}: &\quad 2 \times (0.191)(0.462) = 0.176 \\
\psi_B{}^{(1)}: &\quad 2 \times (0.462)(0.191) = 0.176 \\
\psi_{Ea}{}^{(1)}: &\quad 2 \times (0.500)(0.500) = 0.500 \\
\psi_{Eb}{}^{(1)}: &\quad 2 \times (0)(0) \quad\quad = 0.000 \\
\hline
&\quad\quad\quad\quad\quad\quad\quad\quad\;\; 0.852
\end{aligned}
$$

From these numbers we can see that the π electrons are much more heavily localized in the exo bonds than in the ring bonds.

Bicyclooctatriene

The presumed structure of this compound, which was first reported in 1960,* is shown in Figure 7.2a. It belongs to the point group D_{3h}. An LCAO-MO treatment of it has been described,† but not in as much detail as we shall give here.

Figure 7.2b shows a numbered set of six $p\pi$ atomic orbitals which will be

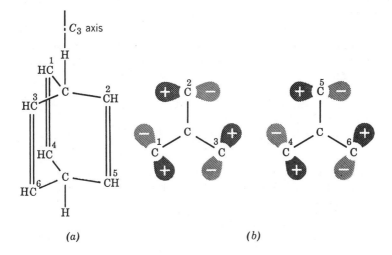

(a)   (b)

Figure 7.2 (a) The molecular structure and the numbering of the carbon atoms for bicyclooctatriene, (b) A sketch showing the orientation of the $p\pi$ orbitals (ϕ_i's) used in the MO treatment.

---

* H. E. Zimmerman and R. M. Paufler, *J. Am. Chem. Soc.*, **82**, 1514 (1960).
† C. F. Wilcox, Jr., S. Winstein, and W. G. McMillan, *J. Am. Chem. Soc.*, **82**, 5450 (1960).

used to construct the π MO's. Using these atomic orbitals as a basis for a representation of the group D_{3h}, we obtain the following results:

D_{3h}	E	$2C_3$	$3C_2$	σ_h	$2S_3$	$3\sigma_v$
Γ_π	6	0	0	0	0	-2

$$\Gamma_\pi = A_2' + A_1'' + E' + E''$$

It is to be noted that in this molecule all of the $p\pi$ orbitals are members of one equivalent set; there is some symmetry operation which will exchange any two of them. Thus we must consider all six in making up MO's of the appropriate symmetries; in fact, it will be entirely impossible to make up orbitals of the correct symmetry using any fewer than the entire six. We now show how this can easily be done.

We first note that all types of A orbitals (in D_{3h}) have the same symmetry properties with respect to the rotations constituting the subgroup C_3; also, both E' and E'' orbitals have the same properties with respect to these rotations. Thus we can use the group C_3 to set up some linear combinations which will be correct to this extent. Since these rotations about the C_3 axis do not interchange any of the orbitals ϕ_1, ϕ_2, ϕ_3 with those of the set ϕ_4, ϕ_5, ϕ_6, we can, *temporarily*, treat the two sets separately. We thus first write down linear combinations corresponding to the A and E representations of C_3. As shown in Section 7.3 for such cyclic systems, the characters are the correct coefficients, and we can thus write, by inspection of the character table for the group C_3:

$$A: \quad \phi_1 + \phi_2 + \phi_3 \qquad \text{and} \qquad \phi_4 + \phi_5 + \phi_6$$

$$E: \begin{cases} \phi_1 + \varepsilon\phi_2 + \varepsilon^*\phi_3 \\ \phi_1 + \varepsilon^*\phi_2 + \varepsilon\phi_3 \end{cases} \quad \text{and} \quad \begin{cases} \phi_4 + \varepsilon\phi_5 + \varepsilon^*\phi_6 \\ \phi_4 + \varepsilon^*\phi_5 + \varepsilon\phi_6 \end{cases}$$

Again using the procedure explained in Section 7.3, we take linear combinations of the above expressions for the E orbitals so as to afford real coefficients, obtaining

$$E: \begin{cases} 2\phi_1 - \phi_2 - \phi_3 \\ \phi_2 - \phi_3 \end{cases} \quad \text{and} \quad \begin{cases} 2\phi_4 - \phi_5 - \phi_6 \\ \phi_5 - \phi_6 \end{cases}$$

We have not bothered to normalize these, since they are not yet actually wave functions.

We now turn back to the character table for D_{3h} and note that an A_2' orbital must go into itself on reflection through the horizontal symmetry plane. The effect of this symmetry operation on the individual atomic orbitals is as follows:

$$\sigma_h(\phi_1) \to \phi_4 \qquad \sigma_h(\phi_4) \to \phi_1$$
$$\sigma_h(\phi_2) \to \phi_5 \qquad \sigma_h(\phi_5) \to \phi_2$$
$$\sigma_h(\phi_3) \to \phi_6 \qquad \sigma_h(\phi_6) \to \phi_3$$

Thus we must combine the two sums which have A symmetry with respect to the threefold rotations into one which goes into itself on reflection through σ_h, making use of the above transformation properties of the individual atomic orbitals. It is obvious that the correct result must be

$$\psi_{A'} = \phi_1 + \phi_2 + \phi_3 + \phi_4 + \phi_5 + \phi_6$$

We can see from the character table that an A_1' orbital would also have the same symmetry properties as those which we have so far consciously built into this LCAO. However, A_1' and A_2' orbitals differ in their behavior upon rotation about a twofold axis or upon reflection in σ_v. The inherent symmetry of the p orbitals is responsible for the fact that the type of orbital we require is A_2' and not A_1' without our having explicitly looked after this. We can easily confirm that an A_2' orbital is needed. If we reflect through the σ_v which passes through carbons 1 and 4, the atomic orbitals transform as follows:

$$\sigma_v(\phi_1) \to -\phi_1 \qquad \sigma_v(\phi_4) \to -\phi_4$$
$$\sigma_v(\phi_2) \to -\phi_3 \qquad \sigma_v(\phi_5) \to -\phi_6$$
$$\sigma_v(\phi_3) \to -\phi_2 \qquad \sigma_v(\phi_6) \to -\phi_5$$

Therefore

$$\sigma_v(\psi_{A'}) = \sigma_v(\phi_1 + \phi_2 + \phi_3 + \phi_4 + \phi_5 + \phi_6)$$
$$= (-\phi_1 - \phi_3 - \phi_2 - \phi_4 - \phi_6 - \phi_5)$$
$$= -(\phi_1 + \phi_2 + \phi_3 + \phi_4 + \phi_5 + \phi_6)$$
$$= -\psi_{A'}$$

Thus, as stated, this orbital is an A_2' orbital. The correct normalization constant in the Hückel approximation is $1/\sqrt{6}$.

The A_1'' orbital, which must change sign on reflection through σ_h and also on reflection through σ_v, obviously has the form

$$\psi_{A_1''} = \frac{1}{\sqrt{6}}(\phi_1 + \phi_2 + \phi_3 - \phi_4 - \phi_5 - \phi_6)$$

Similarly, the characters of the E' and E'' representations under σ_h are 2 and -2, respectively, meaning that each member of an E' pair will go into itself on reflection through σ_h while each member of an E'' set will go into the negative of itself on reflection in σ_h. These requirements are satisfied by combining the above expressions which have E symmetry with respect to the threefold rotations as follows:

$$\psi_{E'a} = 2\phi_1 - \phi_2 - \phi_3 + 2\phi_4 - \phi_5 - \phi_6$$
$$\psi_{E'b} = \phi_2 - \phi_3 + \phi_5 - \phi_6$$
$$\psi_{E''a} = 2\phi_1 - \phi_2 - \phi_3 - 2\phi_4 + \phi_5 + \phi_6$$
$$\psi_{E''b} = \phi_2 - \phi_3 - \phi_5 + \phi_6$$

Now, collecting all of the above results and normalizing each one, we write the following final list of the LCAO-MO's for bicyclooctatriene:

$$\psi_{A_2'} = \frac{1}{\sqrt{6}}(\phi_1 + \phi_2 + \phi_3 + \phi_4 + \phi_5 + \phi_6)$$

$$\psi_{A_1''} = \frac{1}{\sqrt{6}}(\phi_1 + \phi_2 + \phi_3 - \phi_4 - \phi_5 - \phi_6)$$

$$\psi_{E'a} = \frac{1}{\sqrt{12}}(2\phi_1 - \phi_2 - \phi_3 + 2\phi_4 - \phi_5 - \phi_6)$$

$$\psi_{E'b} = \tfrac{1}{2}(\phi_2 - \phi_3 + \phi_5 - \phi_6)$$

$$\psi_{E''a} = \frac{1}{\sqrt{12}}(2\phi_1 - \phi_2 - \phi_3 - 2\phi_4 + \phi_5 + \phi_6)$$

$$\psi_{E''b} = \tfrac{1}{2}(\phi_2 - \phi_3 - \phi_5 + \phi_6)$$

We now consider the energies of these MO's. If we calculate these energies using the Hückel approximation, we set all resonance integrals other than $H_{14} = H_{41}$, $H_{25} = H_{52}$, and $H_{36} = H_{63}$ equal to zero. We then obtain the following results:

ORBITAL	ENERGY	ORBITAL	ENERGY
A_1''	$\alpha - \beta$	A_2'	$\alpha + \beta$
E''	$\alpha - \beta$	E'	$\alpha + \beta$

Thus, in this approximation, the A_1'' orbital is accidentally degenerate with the E'' orbitals and the A_2' orbital is accidentally degenerate with the E' orbitals.

To employ the Hückel approximation in this case, however, is to make the entire process of using an MO treatment pointless, for we then obtain exactly the same answer as we would obtain by assuming the molecule to contain three isolated double bonds. Each double bond can be regarded as resulting from the formation of two two-center MO's, one of energy $\alpha + \beta$ (bonding) and one of energy $\alpha - \beta$ (antibonding). In the Hückel approximation, therefore, we find that bicyclooctatriene has no resonance stabilization energy.

The advantage of the MO treatment is that we can rather easily extend it to take account of interaction between the double bonds. To do this, we recognize that the energy of interaction between two orbitals such as ϕ_1 and ϕ_2, viz., the integral $\int \phi_1 \mathcal{H} \phi_2 \, d\tau$, will not be exactly zero but will have some finite value, say β'. Also, there will be some finite value for the integrals such as $\int \phi_1 \mathcal{H} \phi_5 \, d\tau$, which we may call β''. If we recalculate the energies including these quantities, we will get somewhat different results. Thus for $\psi_{A_1''}$ we obtain

$$
\begin{aligned}
E_{A_1''} &= \frac{1}{6} \int (\phi_1 + \phi_2 + \phi_3 - \phi_4 - \phi_5 - \phi_6) \\
&\quad \times \mathcal{H}(\phi_1 + \phi_2 + \phi_3 - \phi_4 - \phi_5 - \phi_6) \, d\tau \\
&= \frac{1}{6} \left(\int \phi_1 \mathcal{H} \phi_1 \, d\tau + \int \phi_1 \mathcal{H} \phi_2 \, d\tau + \int \phi_1 \mathcal{H} \phi_3 \, d\tau - \int \phi_1 \mathcal{H} \phi_4 \, d\tau \right. \\
&\quad \left. - \int \phi_1 \mathcal{H} \phi_5 \, d\tau - \int \phi_1 \mathcal{H} \phi_6 \, d\tau + \int \phi_2 \mathcal{H} \phi_1 \, d\tau + \cdots \right) \\
&= \tfrac{1}{6}(\alpha + \beta' + \beta' - \beta - \beta'' - \beta'' + \beta' + \cdots) \\
&= \alpha - \beta + 2\beta' - 2\beta''
\end{aligned}
$$

Similarly, for $\psi_{E''b}$, we obtain

$$
\begin{aligned}
E_{E''b} &= \frac{1}{4} \int (\phi_2 - \phi_3 - \phi_5 + \phi_6)\mathcal{H}(\phi_2 - \phi_3 - \phi_5 + \phi_6) \, d\tau \\
&= \frac{1}{4} \left(\int \phi_2 \mathcal{H} \phi_2 \, d\tau - \int \phi_2 \mathcal{H} \phi_3 \, d\tau - \int \phi_2 \mathcal{H} \phi_5 \, d\tau \right. \\
&\quad \left. + \int \phi_2 \mathcal{H} \phi_6 \, d\tau - \int \phi_3 \mathcal{H} \phi_2 \, d\tau + \cdots \right) \\
&= \tfrac{1}{4}(\alpha - \beta' - \beta + \beta'' - \beta' + \cdots) \\
&= \alpha - \beta - \beta' + \beta''
\end{aligned}
$$

Thus, when allowance is made for these additional interactions, we find that the accidental degeneracies are removed. In the same way, we find that the energies of the A_2' and E' orbitals are

$$
E_{A_2'} = \alpha + \beta + 2\beta' + 2\beta''
$$

$$
E_{E'} = \alpha + \beta - \beta' - \beta''
$$

Let us now see what effect this allowance for interaction between the double bonds has on the calculated resonance energy.

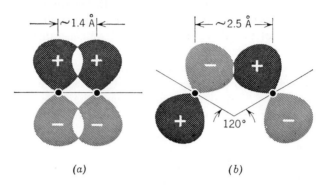

Figure 7.3   (*a*) The relative orientation of $p\pi$ orbitals on adjacent, bonded carbon atoms of bicyclooctatriene. (*b*) The relative orientation of $p\pi$ orbitals on two nonadjacent carbon atoms of bicyclooctatriene.

The quantity β is intrinsically negative, since it is a measure of the interaction between adjacent $p\pi$ orbitals so oriented (Figure 7.3*a*) as to give a bonding interaction. The integral β' measures the interaction between p orbitals oriented as in Figure 7.3*b*. It can be seen that this interaction will be antibonding, so that β' is positive and also that it should be smaller in absolute magnitude than β, since the orbitals concerned are much farther apart and overlap less. The ratio $-\beta'/\beta$ can be roughly estimated by using overlap integrals: it is ~ 0.1. Also by means of overlap integrals, it may be shown that β'' is still smaller and for the present we shall neglect it entirely. With these considerations in mind we can draw the following energy level diagram, in which, as usual, we take α as the zero of energy:

$$A_1'' \quad \underline{\hspace{3cm}} \quad -\beta + 2\beta' \approx -1.2\beta$$

$$E'' \left\{ \underline{\overline{\hspace{2cm}}} \quad -\beta - \beta' \approx -0.9\beta \right.$$

$$A_2' \quad \underline{\hspace{1cm}\circ\hspace{0.3cm}\circ\hspace{1cm}} \quad \beta + 2\beta' \approx 0.8\beta$$

$$E' \left\{ \underline{\overline{\hspace{1cm}\circ\hspace{0.3cm}\circ\hspace{1cm}}} \quad \beta - \beta' \approx 1.1\beta \right.$$

The energy of the six electrons occupying these orbitals as shown is given by

$$4(\beta - \beta') + 2(\beta + 2\beta') = 6\beta$$

The same answer is obtained even when the complete expressions using β'' are employed. Thus the conclusion is that, although interaction between the double bonds causes certain displacements of the energy levels, it does not result in any increased stabilization of the molecule. The delocalization energy remains zero.

7.5 A Worked Example: Naphthalene

In order to summarize and illustrate the methods of MO theory as they have been developed so far, an example which is elaborate enough to be general and yet is within the scope of noncomputerized numerical solution may be useful. The naphthalene molecule is suitable for this purpose, and an outline of an MO calculation in the Hückel approximation is presented here. In addition, the naphthalene molecule affords an excellent vehicle for introducing some basic ideas concerning the electronic spectra of unsaturated organic molecules, including the concept of configuration interaction. These matters will be considered in the next section. The reader who has mastered the material in Sections 7.1–7.4 should have no difficulty in verifying all of the results given in this section, which in turn provide a basis for Section 7.6.

The naphthalene molecule, as mentioned previously, belongs to the point group D_{2h}. A set of coordinate axes and a numbering scheme for the atoms have already been shown in Figure 7.1. The $p\pi$ orbitals $\phi_1, \phi_2, \ldots, \phi_{10}$, form three subsets; the members of each are symmetry-equivalent to each other but not to those in other sets. These sets and the irreducible representations for which they form bases are as follows:

Set 1: $\phi_1, \phi_4, \phi_5, \phi_8.$ $A_u, B_{1u}, B_{2g}, B_{3g}$

Set 2: $\phi_2, \phi_3, \phi_6, \phi_7.$ $A_u, B_{1u}, B_{2g}, B_{3g}$

Set 3: $\phi_9, \phi_{10}.$ B_{1u}, B_{3g}

Thus, there are two A_u MO's, two B_{2g} MO's, three B_{1u} MO's, and three B_{3g} MO's. By constructing SALC's corresponding to these representations, the well-nigh hopeless problem of solving a 10×10 determinantal equation is reduced to the tractable task of solving two quadratic and two cubic equations. This has already been illustrated in Section 7.2.

Projection operators may be used to obtain the SALC's. The procedure is straightforward, since only one-dimensional representations are involved.

For example:

$$\hat{P}^{B_{2g}}\phi_2 \approx (1)\hat{E}\phi_2 + (-1)\hat{C}_2^z\phi_2 + (1)\hat{C}_2^y\phi_2 + (-1)\hat{C}_2^x\phi_2 + (1)\hat{\imath}\phi_2$$
$$+ (-1)\hat{\sigma}^{xy}\phi_2 + (1)\hat{\sigma}^{xz}\phi_2 + (-1)\hat{\sigma}^{yz}\phi_2$$
$$= \phi_2 - \phi_6 - \phi_7 + \phi_3 - \phi_6 + \phi_2 + \phi_3 - \phi_7$$
$$\approx \phi_2 + \phi_3 - \phi_6 - \phi_7$$
$$\approx \tfrac{1}{2}(\phi_2 + \phi_3 - \phi_6 - \phi_7)$$

The complete set of SALC's (which constitute the solution to Exercise 6.1) is as follows:

$$A_u: \quad \psi_1 = \tfrac{1}{2}(\phi_1 - \phi_4 + \phi_5 - \phi_8)$$
$$\psi_2 = \tfrac{1}{2}(\phi_2 - \phi_3 + \phi_6 - \phi_7)$$
$$B_{1u}: \quad \psi_3 = \tfrac{1}{2}(\phi_1 + \phi_4 + \phi_5 + \phi_8)$$
$$\psi_4 = \tfrac{1}{2}(\phi_2 + \phi_3 + \phi_6 + \phi_7)$$
$$\psi_5 = \frac{1}{\sqrt{2}}(\phi_9 + \phi_{10})$$
$$B_{2g}: \quad \psi_6 = \tfrac{1}{2}(\phi_1 + \phi_4 - \phi_5 - \phi_8)$$
$$\psi_7 = \tfrac{1}{2}(\phi_2 + \phi_3 - \phi_6 - \phi_7)$$
$$B_{3g}: \quad \psi_8 = \tfrac{1}{2}(\phi_1 - \phi_4 - \phi_5 + \phi_8)$$
$$\psi_9 = \tfrac{1}{2}(\phi_2 - \phi_3 - \phi_6 + \phi_7)$$
$$\psi_{10} = \frac{1}{\sqrt{2}}(\phi_9 - \phi_{10})$$

The following secular equations may then be set up:

$$A_u: \quad \begin{vmatrix} \alpha - E & \beta \\ \beta & \alpha - \beta - E \end{vmatrix} = 0$$

$$B_{1u}: \quad \begin{vmatrix} \alpha - E & \beta & \sqrt{2}\beta \\ \beta & \alpha + \beta - E & 0 \\ \sqrt{2}\beta & 0 & \alpha + \beta - E \end{vmatrix} = 0$$

$$B_{2g}: \quad \begin{vmatrix} \alpha - E & \beta \\ \beta & \alpha - \beta - E \end{vmatrix} = 0$$

$$B_{3g}: \quad \begin{vmatrix} \alpha - E & \beta & \sqrt{2}\beta \\ \beta & \alpha - \beta - E & 0 \\ \sqrt{2}\beta & 0 & \alpha - \beta - E \end{vmatrix} = 0$$

These determinants can be expanded into polynomial equations (with energies measured in units of β and referred to α as the zero of energy):

$$A_u: \quad E^2 + E - 1 = 0; \quad E = \frac{-1 \pm \sqrt{5}}{2} = -1.618, \ +0.618$$

$$B_{1u}: \quad (E - 1)(E^2 - E - 3) = 0; \quad E = 1$$

$$E = \frac{1 \pm \sqrt{13}}{2} = 2.303, \ -1.303$$

$$B_{2g}: \quad E^2 - E - 1 = 0; \quad E = \frac{1 \pm \sqrt{5}}{2} = 1.618, \ -0.618$$

$$B_{3g}: \quad (E + 1)(E^2 + E - 3) = 0; \quad E = -1$$

$$E = \frac{-1 \pm \sqrt{13}}{2} = -2.303, \ 1.303$$

These results lead to the energy level diagram shown in Figure 7.4. The delocalization energy in units of β is given by

$$2(2.303 + 1.618 + 1.303 + 1.000 + 0.618) - 10 = 3.684$$

The SALC's ψ_1 through ψ_{10} may then be combined into MO's. For the A_u MO's, the simultaneous equations from which the secular equation originates are as follows:

$$c_1(\alpha - E) + c_2 \beta = 0$$
$$c_1 \beta + c_2(\alpha - \beta - E) = 0$$

Taking α as the zero of energy and β as the unit of energy, we obtain an expression for the ratio c_1/c_2:

$$c_1/c_2 = 1/E = 1 + E$$

For the A_u orbital of energy 0.618, $\psi_{A_u}^{(1)}$, we have $c_1/c_2 = 1.618$. Normalization requires that $c_1^2 + c_2^2 = 1$. Solving for c_1 and c_2, we obtain $c_1 = 0.850$ and $c_2 = 0.526$; the bonding A_u MO may thus be written

$$\psi_{A_u}^{(1)} = 0.850\psi_1 + 0.526\psi_2$$
$$= 0.425(\phi_1 - \phi_4 + \phi_5 - \phi_8) + 0.263(\phi_2 - \phi_3 + \phi_6 - \phi_7)$$

Proceeding in the same way for all the MO's of naphthalene, we obtain the results shown in Table 7.2. The student may test his understanding of the procedure by verifying some of these results.

There are four nonequivalent types of C-C bonds in naphthalene, these being represented by C_1-C_2, C_2-C_3, C_4-C_{10}, and C_9-C_{10}. By using the MO expressions in Table 7.2, the π bond order of each type may be

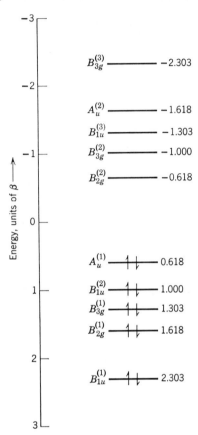

Figure 7.4   Energy level diagram for π orbitals of naphthalene.

computed. As before, in treating tetramethylenecyclobutane, the bond order, p_{mn}, of the bond between atoms m and n is defined as the sum of contributions from each occupied MO, each contribution being given by twice (for two electrons) the product of the coefficients of ϕ_m and ϕ_n in that MO. For p_{12} in naphthalene we have

$$\psi_{B_{1u}}^{(1)}: \ 2 \times (0.301 \times 0.231) = 0.139$$
$$\psi_{B_{2g}}^{(1)}: \ 2 \times (0.263 \times 0.425) = 0.225$$
$$\psi_{B_{3g}}^{(1)}: \ 2 \times (0.400 \times 0.174) = 0.139$$
$$\psi_{B_{1u}}^{(2)}: \ 2 \times (0.000 \times 0.408) = 0.000$$
$$\psi_{A_u}^{(1)}: \ 2 \times (0.425 \times 0.263) = \underline{0.225}$$
$$0.728$$

The orders of the other bonds may be figured similarly, to give the results expressed by the following diagram:

Table 7.2   Molecular Orbitals of Naphthalene

MO	ENERGY, UNITS OF β ($\alpha = 0$)	LCAO EXPRESSION
$\psi_{B_{1u}}^{(1)}$	2.303	$0.301(\phi_1 + \phi_4 + \phi_5 + \phi_8) + 0.231(\phi_2 + \phi_3 + \phi_6 + \phi_7)$ $+ 0.461(\phi_9 + \phi_{10})$
$\psi_{B_{2g}}^{(1)}$	1.618	$0.263(\phi_1 + \phi_4 - \phi_5 - \phi_8) + 0.425(\phi_2 + \phi_3 - \phi_6 - \phi_7)$
$\psi_{B_{3g}}^{(1)}$	1.303	$0.400(\phi_1 - \phi_4 - \phi_5 + \phi_8) + 0.174(\phi_2 - \phi_3 - \phi_6 + \phi_7)$ $+ 0.347(\phi_9 - \phi_{10})$
$\psi_{B_{1u}}^{(2)}$	1.000	$0.408(\phi_2 + \phi_3 + \phi_6 + \phi_7) - 0.408(\phi_9 + \phi_{10})$
$\psi_{A_u}^{(1)}$	0.618	$0.425(\phi_1 - \phi_4 + \phi_5 - \phi_8) + 0.263(\phi_2 - \phi_3 + \phi_6 - \phi_7)$
$\psi_{B_{2g}}^{(2)}$	-0.618	$0.425(\phi_1 + \phi_4 - \phi_5 - \phi_8) - 0.263(\phi_2 + \phi_3 - \phi_6 - \phi_7)$
$\psi_{B_{3g}}^{(2)}$	-1.000	$0.408(\phi_2 - \phi_3 - \phi_6 + \phi_7) - 0.408(\phi_9 - \phi_{10})$
$\psi_{B_{1u}}^{(3)}$	-1.303	$0.400(\phi_1 + \phi_4 + \phi_5 + \phi_8) - 0.174(\phi_2 + \phi_3 + \phi_6 + \phi_7)$ $- 0.347(\phi_9 + \phi_{10})$
$\psi_{A_u}^{(2)}$	-1.618	$0.263(\phi_1 - \phi_4 + \phi_5 - \phi_8) - 0.425(\phi_2 - \phi_3 + \phi_6 - \phi_7)$
$\psi_{B_{3g}}^{(3)}$	-2.303	$0.301(\phi_1 - \phi_4 - \phi_5 + \phi_8) - 0.231(\phi_2 - \phi_3 - \phi_6 + \phi_7)$ $- 0.461(\phi_9 - \phi_{10})$

7.6   Electronic Excitations of Naphthalene: Selection Rules and Configuration Interaction

From the energy level diagram, Figure 7.4, it can be seen that the lowest energy transition for naphthalene might be expected to involve excitation of an electron from the $A_u^{(1)}$ orbital to the $B_{2g}^{(2)}$ orbital. The energy should be equal to the energy difference between the two orbitals, which is $0.618 - (-0.618) = 1.236\beta$.* The next two transitions, involving $A_u^{(1)}$ to $B_{3g}^{(2)}$ and $B_{1u}^{(2)}$ to $B_{2g}^{(2)}$ excitations, would be expected to have identical energies, viz., 1.618β, but this is not actually the case. In this section we shall look

---

* At this point the comments on the value of β made in Appendix IV should be noted.

more closely at the three lowest electronic excitations of naphthalene and shall discuss selection rules, polarizations, and the effect of what is called configuration interaction on two of these transitions.

In order to understand the electronic spectra of molecules we must first recognize that, although we have so far found it convenient to think exclusively in terms of the *configurations* of the electrons (i.e., the way in which the one-electron orbitals are populated), this does not *directly* provide a satisfactory basis for describing electronic transitions. One-electron orbitals and the electron configurations are, in themselves, fictional; it is the *states* arising out of electron configurations which are real. Wave functions, to be genuine, must describe *states*, not individual orbitals (unless only one electron is present) or configurations. Therefore, as we apply symmetry arguments to the analysis of electronic transitions in naphthalene or any other molecule, it is the symmetries of the states which must always be considered, not directly the symmetries of the orbitals to which we assign individual electrons. We can, however, determine the symmetries of the states from the symmetries of the occupied orbitals. In this elementary discussion we wish to emphasize spatial symmetry properties. We shall therefore omit explicit consideration of electron spin by dealing only with excited states which, like the ground state, have a spin quantum number of zero. The designation of state will then refer only to the spatial or orbital distribution of the electron density.

Finally, it is convenient to introduce a form of notation which allows us to specify electron configurations as simply as possible. An electron occupying a B_{1u} orbital will be represented by b_{1u}, the lower-case letter denoting that this is the symmetry of the orbital of *one* electron, not of a true *state* wave function. When two electrons occupy the same orbital we shall write b_{1u}^2. In this way, the electron configuration of the naphthalene molecule in its ground state is written

$$b_{1u}^2 b_{2g}^2 b_{3g}^2 b_{1u}'^2 a_u^2$$

Note that the electrons are listed from left to right in increasing order of the energies of the orbitals they occupy and that different b_{1u} orbitals are distinguished by using a prime instead of the more cumbersome superscript numbers, which might now be confused with the superscripts indicating the presence of two electrons.

It is important to realize that any configuration of electrons in which there are only completely filled orbitals, such as that just considered, gives use to only one state and that state is totally symmetric.* Thus, the ground state of the naphthalene molecule has A_{1g} symmetry.

---

* This can easily be demonstrated, using direct product representations for orbitals alone, when only nondegenerate orbitals are involved. The same is true when degenerate orbitals are involved, but more sophisticated methods of proof are required.

The Lower, Singly Excited Configurations

In an excited configuration one or more electrons occupy orbitals other than the lowest ones available. In a *singly* excited configuration only one electron is promoted from the orbital it occupies in the ground configuration to an orbital of higher energy. Most observed electronic transitions are from the ground state to one arising from a singly excited configuration. In addition, the transitions most likely to be in the conveniently observable spectral regions (visible, near ultraviolet) are to states arising from the excited configurations with the lowest energies. These energies may be estimated in an approximate manner by considering the energies of the orbitals whose populations are changed in going from ground to excited configurations.

For naphthalene, the three lowest-energy, singly excited configurations and their energies, relative to the ground-state energy as zero, are as follows:

$$B_{2u}: \; b_{1u}{}^2 b_{1g}{}^2 b_{3g}{}^2 b_{1u}{}'^2 \, a_u b_{2g} \quad E = 0.618 - (-0.618) = 1.236$$

$$B_{3u}: \; b_{1u}{}^2 b_{1g}{}^2 b_{3g}{}^2 b_{1u}{}'^2 \, a_u b_{3g} \quad E = 0.618 - (-1.000) = 1.618$$

$$B_{3u}: \; b_{1u}{}^2 b_{1g}{}^2 b_{3g}{}^2 b_{1u}' a_u{}^2 b_{2g} \quad E = 1.000 - (-0.618) = 1.618$$

Each of these configurations gives rise to a state with the symmetry which is specified at the left. The state symmetries are determined in the following way. First, we neglect all of the electrons which are in fully occupied orbitals, since that part of the entire configuration is in each case totally symmetric. Second, we invoke the fact that for the remaining electrons a product wave function can be written. These product wave functions are $a_u b_{2g}$, $a_u b_{3g}$, and $b'_{1u} b_{2g}$ in the three cases above. The symmetry of each is found by forming the direct product representation, as explained in Section 5.2.

Selection Rules and Polarizations

Transitions from the ground state to each of the excited states just discussed may or may not be allowed, in the sense discussed in Section 5.3. We can employ the criterion presented there to find out which transitions are allowed and to ascertain their polarizations.

The criterion for a transition being electric dipole allowed is that the direct product representation for the ground and excited states be or contain an irreducible representation to which one or more of the Cartesian coordinates belongs. Since the ground state belongs to the totally symmetric representation, the direct product representations will in each case be the same as the representations to which the excited state belongs; these are B_{2u}, B_{3u}, and B_{3u}. As the D_{2h} character table shows, the y coordinate belongs to the B_{2u} representation and the x coordinate belongs to the B_{3u} representation.

Thus, the $A_{1g} \to B_{2u}$ and both of the $A_{1g} \to B_{3u}$ transitions are electric dipole allowed. In all cases (see Figure 7.1) the transitions are "in-plane" polarized, the first one occurring by absorption of radiation with its electric vector vibrating along the y or short axis of the molecule, the other two having x or "long-axis" polarization.

Experimentally, three transitions are observed in the near ultraviolet, and polarization measurements indicate that one is "short-axis" polarized while the other two are "long-axis" polarized. The results are given in Table 7.3.

Table 7.3   Electronic Transitions in Naphthalene

ENERGY, CM^{-1}	POLARIZATION	ASSIGNMENT
31,800	Long axis	$A_{1g} \to B_{3u}$
34,700	Short axis	$A_{1g} \to B_{2u}$
45,200	Long axis	$A_{1g} \to B_{3u}$

Configuration Interaction

Although the results in Table 7.3 seem to be in generally good agreement with theory, there is one notable discrepancy. From orbital energies we would have estimated that the two $A_{1g} \to B_{3u}$ transitions have the same energy and that this energy is higher than that for the $A_{1g} \to B_{2u}$ transition. Experiment shows, however, that the energy of the $A_{1g} \to B_{2u}$ transition lies between the energies of the two $A_{1g} \to B_{3u}$ transitions. The cause of this is configuration interaction.

Configuration interaction is simply a manifestation of the rule (Section 5.3) that an integral of the form

$$\int \psi_1 \mathscr{H} \psi_2 \, d\tau$$

may have a nonzero value only if ψ_1 and ψ_2 belong to the same representation. Previously, we have emphasized the converse, namely, that the integral *must* equal zero when ψ_1 and ψ_2 belong to different representations. Now we are concerned with the fact that only in special cases—by accident, as it were— will such an integral be zero if the two wave functions *do* have the same symmetry. In the present case this means that the two excited configurations of B_{3u} symmetry will have a net interaction with each other. The result is that neither of the actual B_{3u} *states* will be derived purely from *one* of the excited B_{3u} *configurations*, as we have thus far tacitly assumed. Instead, there will be a mixing and splitting, described by a second-order secular equation,

$$\begin{vmatrix} E^0 - E & H_{12} \\ H_{12} & E^0 - E \end{vmatrix} = 0$$

with

$$H_{12} = \int \psi_{B_{3u}} \mathscr{H} \psi'_{B_{3u}} \, d\tau$$

Obviously, this leads to two different energy values, $E^0 \pm H_{12}$, instead of the same one, E^0, for both states. The entire situation is formally analogous to the occurrence of a 2×2 determinantal equation for the energies of two individual orbitals of the same symmetry. In practice, however, it is a little different, since there is no simple way to estimate the magnitude of H_{12}, the interaction energy. This is due to interelectronic repulsion and is difficult to compute accurately.

The way in which the two excited configurations mix and split apart in energy is called configuration interaction. It is depicted in the energy level diagram of Figure 7.5 for the case at hand.

Actually, the example of configuration interaction that we have just examined is special in the sense that the two interacting configurations have the same energy before interaction. In its most general form, configuration interaction involves *any* two configurations with the same symmetry. The

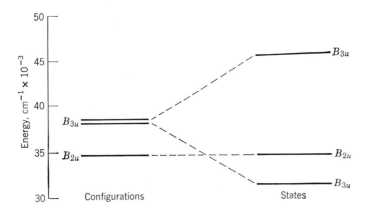

Figure 7.5   Energy level diagram showing how configuration interaction in the B_{3u} excited configurations of naphthalene leads to two widely separated B_{3u} states. The energies are measured from the ground-state energy as zero.

result of the interaction is always to produce an energy difference between the two states which is greater than that between the two configurations.

When this kind of interaction occurs between vibrational states instead of electronic states it is called Fermi resonance; we shall discuss this later (page 330). In fact, the whole qualitative concept of resonance stabilization as used in the valence bond theory is just the same principle in still another guise.

7.7   Three-Center Bonding

It is now recognized that there are many molecules in which bonding must be treated with three-atom units as the smallest ones considered. In other words, the three atoms must be regarded as one indivisible entity instead of as a pair of two-center systems with an atom in common. We will thus assume that electrons may be delocalized over the framework of three atoms instead of localized between only two of them. If our three-atom framework constitutes a complete molecule or ion in itself, then our treatment will be an MO one in the strict sense of the word. If the three-atom entity is instead only a portion of a larger molecule, then our analysis will not be an MO one in the strict sense of considering the possibility of delocalization of electrons over the entire molecule. Nevertheless, its principles and results as they apply to the selected group of three atoms will not, of course, differ in any essential way from those obtained when the three atoms are the entire molecule.

As examples of three-center bonding, we will take the following:

(i) Open three-center bonding as found, for example, in the π system of the allyl ion, $[H_2CCHCH_2]^-$, and in the bridge bonding in diborane.

(ii) Closed three-center bonding as found, for example, in one of the types of B-B-B bonding occurring in certain boron hydrides.

Open Three-Center Bonding

The nuclear framework, $p\pi$ atomic orbitals, and a set of reference axes for the allyl ion are shown below:

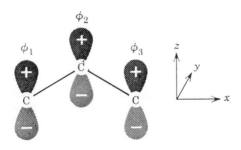

The ion belongs to the C_{2v} point group, and the set of three $p\pi$ orbitals forms a basis for the following representation:

C_{2v}	E	C_2	σ_{yz}	σ_{xy}
Γ_π	3	-1	1	-3

$$\Gamma_\pi = A_2 + 2B_1$$

We must now write SALC's which belong to these representations. By applying projection operators to ϕ_1 and then normalizing, we readily obtain

$$\psi_{A_2} = \frac{1}{\sqrt{2}}(\phi_1 - \phi_3)$$

$$\psi_{B_1} = \frac{1}{\sqrt{2}}(\phi_1 + \phi_3)$$

Since ϕ_2 is unique, it must, by itself, form a basis for the B_1 representation. It is easy to confirm this by examining its behavior under each symmetry operation. Thus, the other SALC of B_1 symmetry is

$$\psi'_{B_1} = \phi_2$$

The secular equation to be solved for the B_1 orbitals is then

$$\begin{vmatrix} H_{BB} - E & H_{BB'} \\ H_{BB'} & H_{B'B'} - E \end{vmatrix} = 0$$

The elements of the determinant are evaluated as follows:

$$H_{BB} = \int \phi_2 \mathcal{H} \phi_2 \, d\tau = \alpha$$

$$H_{BB'} = \frac{1}{\sqrt{2}} \int (\phi_2) \mathcal{H} (\phi_1 + \phi_3) \, d\tau$$

$$= \frac{1}{\sqrt{2}} \left(\int \phi_2 \mathcal{H} \phi_1 \, d\tau + \int \phi_2 \mathcal{H} \phi_3 \, d\tau \right)$$

$$= \frac{1}{\sqrt{2}} (\beta + \beta) = \sqrt{2}\beta$$

$$H_{B'B'} = \frac{1}{2} \int (\phi_1 + \phi_3) \mathcal{H} (\phi_1 + \phi_3) \, d\tau$$

$$= \frac{1}{2} \left(\int \phi_1 \mathcal{H} \phi_1 \, d\tau + \int \phi_1 \mathcal{H} \phi_3 \, d\tau \right.$$

$$\left. + \int \phi_3 \mathcal{H} \phi_1 \, d\tau + \int \phi_3 \mathcal{H} \phi_3 \, d\tau \right)$$

$$= \tfrac{1}{2}(\alpha + 0 + 0 + \alpha) = \alpha$$

Taking $\alpha = 0$, as usual, and inserting into the determinant, we obtain

$$\begin{vmatrix} -E & \sqrt{2}\beta \\ \sqrt{2}\beta & -E \end{vmatrix} = 0$$

which has the roots $\pm\sqrt{2}\beta$.

The single A_2 orbital, which contains only the nonadjacent pair of orbitals ϕ_1 and ϕ_3, can be seen on inspection to have the energy α ($=0$).

It is now possible to draw the following energy level diagram for the allyl anion:

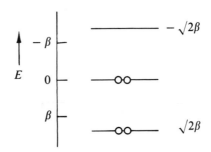

The energy of the allyl ion in the localized form $H_2\overset{..}{C}-CH=CH_2$ would be $2\alpha + (2\alpha + 2\beta)$ or, with $\alpha = 0$, 2β. The energy of the electrons distributed as shown above is $2\sqrt{2}\beta$. Hence the delocalization energy is calculated to be $(2\sqrt{2} - 2)\beta = 0.828\beta \approx 16$ kcal/mole.

To find the actual form of the occupied B_1 orbital, we first write the simultaneous equations from which the B_1 secular equation is obtained:

$$c_1(H_{BB} - E) + c_2 H_{BB'} = 0$$

$$c_1 H_{BB'} + c_2(H_{B'B'} - E) = 0$$

The appropriate value of E ($\sqrt{2}$ in units of β) and the values of the H's may then be inserted into either one of these equations, leading to an equation for the ratio c_1/c_2. For example, with the first of the two equations we obtain

$$c_1(0 - \sqrt{2}) + c_2\sqrt{2} = 0$$

$$-\sqrt{2}c_1 + \sqrt{2}c_2 = 0$$

$$c_1 = c_2$$

Normalization then requires that

$$c_1 = c_2 = \frac{1}{\sqrt{2}}$$

We thus obtain

$$\psi_{B_1}{}^{(1)} = \frac{1}{\sqrt{2}} \left[\frac{1}{\sqrt{2}} (\phi_1 + \phi_3) + \phi_2 \right]$$

$$= \tfrac{1}{2}(\phi_1 + \sqrt{2}\phi_2 + \phi_3)$$

The antibonding B_1 orbital, $\psi_{B_1}{}^{(2)}$ can similarly be shown to be

$$\psi_{B_1}{}^{(2)} = \tfrac{1}{2}(\phi_1 - \sqrt{2}\,\phi_2 + \phi_3)$$

Generalization of the Results

We have so far treated only the specific case of π bonding in a three-center system of identical atoms using identical π orbitals. It is easy to make a semiquantitative generalization of these results. A few examples should suffice to show how this is done.

Suppose that we have a system of three atoms each with a π orbital but with the center atom different from the end atoms, as in NO_2 or NO_2^-. The symmetry is still C_{2v}, and so we still expect MO's belonging to the representations A_2 and $2B_1$ of C_{2v}. The expression for the A_2 orbital will still be, for reasons of symmetry alone,

$$\psi_{A_2} = \frac{1}{\sqrt{2}} (\phi_1 - \phi_3)$$

and its energy will be α_O, where the subscript shows that this is the energy of an oxygen p orbital. We can again set up the same two linear combinations of orbitals having B_1 symmetry. The elements of the secular determinant will have the following values:

$$\begin{vmatrix} \alpha_N - E & \sqrt{2}\beta \\ \sqrt{2}\beta & \alpha_O - E \end{vmatrix} = 0$$

Because we now have two different α's, we cannot obtain the extremely simple result that we previously did, but qualitatively the results will be very similar, as indicated in Figure 7.6. The two B_1 orbitals are no longer

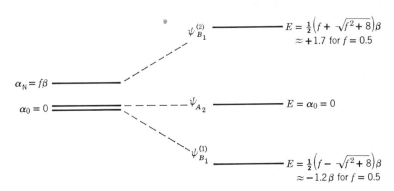

Figure 7.6   Energy level diagram for π bonding in NO_2.

symmetrically placed with respect to the A_2 orbital. The degree to which the diagram becomes unsymmetrical depends on how much α_N and α_O differ. In Figure 7.6 the difference has been expressed in units of β as $f\beta$. Experimental data show that the value of f is probably in the range $0 < f < 1$. If it is assumed that $f = 0.5$, the relative placement of levels turns out as shown in Figure 7.6.

Let us now look at some cases of σ bonding instead of π bonding. A three-center approach to σ bonding finds application in the boron hydrides,

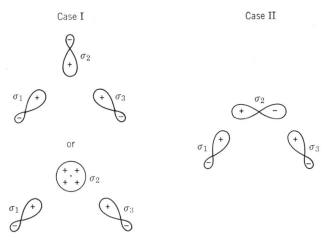

Figure 7.7   The two general cases of three-center bonding. Case I: center atom employs a symmetric orbital. Case II: center atom employs an antisymmetric orbital.

in certain cases of very strong hydrogen bonding, and in many compounds of the heavier posttransition elements, such as XeF_2 and SF_4. Basically, two cases arise, as shown in Figure 7.7. In case I the center atom uses an orbital which is symmetric to reflection in a plane perpendicular to the $B \cdots B$ line; in case II, the center atom uses an antisymmetric orbital. It should especially be noted that, even if the system happens to be linear, the results of interest here are unchanged. The only orbital symmetry property of importance to us is the behavior upon reflection in a plane passing through the center atom and perpendicular to a line connecting the terminal atoms. Whether that line happens to go through the center atom is irrelevant. In effect, then, we employ the group C_s, which is a subgroup of both C_{2v} and $D_{\infty h}$, one of which will be the full symmetry group of the system. Although C_s symmetry allows only the distinction between symmetric (A) and antisymmetric (B) orbitals, that is as much symmetry information as we need. For cases I and II the following results are easily obtained, assuming that all three orbitals have the same α, which we set equal to zero:

<table>
<tr><td></td><td align="center">*Case I*</td><td align="center">*Case II*</td></tr>
</table>

1. *SALC's*

   Case I

   A: $\dfrac{1}{\sqrt{2}}(\sigma_1 + \sigma_3)$, σ_2

   B: $\dfrac{1}{\sqrt{2}}(\sigma_1 - \sigma_3)$

   Case II

   A: $\dfrac{1}{\sqrt{2}}(\sigma_1 + \sigma_3)$

   B: $\dfrac{1}{\sqrt{2}}(\sigma_1 - \sigma_3)$, σ_2

2. *Secular Equations and Energies*

   Case I

   A: $\begin{vmatrix} -E & \sqrt{2}\beta_{12} \\ \sqrt{2}\beta_{12} & \beta_{13} - E \end{vmatrix} = 0$

   $E = \pm\sqrt{2}\beta_{12}\,(\beta_{13} = 0)$

   B: $E = 0\,(\beta_{13} = 0)$

   Case II

   A: $E = 0\,(\beta_{13} = 0)$

   B: $\begin{vmatrix} -E & \sqrt{2}\beta_{12} \\ \sqrt{2}\beta_{12} & \beta_{13} - E \end{vmatrix} = 0$

   $E = \pm\sqrt{2}\beta_{12}\,(\beta_{13} = 0)$

3. *Expressions for Orbitals*

   Case I

   $\psi_A{}^{(1)} = \tfrac{1}{2}(\sigma_1 + \sqrt{2}\sigma_2 + \sigma_3)$

   $\psi_A{}^{(2)} = \tfrac{1}{2}(\sigma_1 - \sqrt{2}\sigma_2 + \sigma_3)$

   $\psi_B = \dfrac{1}{\sqrt{2}}(\sigma_1 - \sigma_3)$

   Case II

   $\psi_A = \dfrac{1}{\sqrt{2}}(\sigma_1 + \sigma_3)$

   $\psi_B{}^{(1)} = \tfrac{1}{2}(\sigma_1 + \sqrt{2}\sigma_2 - \sigma_3)$

   $\psi_B{}^{(2)} = \tfrac{1}{2}(\sigma_1 - \sqrt{2}\sigma_2 - \sigma_3)$

speak of "ionic character" in the bonds. For general and fruitful applications of these ideas, the reader may consult various articles in the research and review literature.*

Closed Three-Center Bonding

In the cases treated so far, we have neglected any direct interaction between the end atoms. Open three-center bonding can be defined as the situation in which such neglect is justified. When the effect of β_{13} on the results becomes too great to neglect, we have closed three-center bonding. In the secular equations written earlier for cases I and II of open three-center bonding, β_{13} was included but then set equal to zero in solving for the energies. Suppose that we now go to the other extreme for case I and assume that $\beta_{13} = \beta_{12}$. The secular equation for the A orbitals takes the form

$$\begin{vmatrix} -E & \sqrt{2}\beta \\ \sqrt{2}\beta & \beta - E \end{vmatrix} = 0$$

which has the roots $-\beta$ and 2β. Meanwhile the energy of the B orbital is also changed, becoming $-\beta$ instead of 0 (for $\alpha = 0$). Since there must be a continuous change in the energies of the MO's as the magnitude of β_{13} changes, we can draw the following correlation diagram:

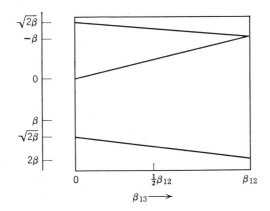

It will be seen that, when we set $\beta_{13} = \beta_{12}$ and all atoms are assigned the same α value, we are imposing at least C_3 symmetry on the system. Thus, the results in this limit could have been obtained directly by treating a

* An especially good place to start is the following: R. E. Rundle, *Record Chem. Progr.*, **23**, 195 (1962).

system of three equivalent σ orbitals on three atoms related by a threefold axis. In that case we should regard the two orbitals with the same energy $(-\beta)$ as components of a doubly degenerate pair. An entirely analogous situation, differing only in unimportant details, arises if we rework the treatment of the allyl radical, including $\beta_{13} = \beta_{12}$. The limiting case here is equivalent to the π system of the cyclopropenyl radical.

7.8 Symmetry-Based "Selection Rules" for Cyclization Reactions

It has been shown in Section 5.1 that acceptable molecular wave functions must form bases for irreducible representations of the point symmetry group to which the molecule in question belongs. In earlier sections of this chapter this most basic of symmetry restrictions has been employed to find satisfactory wave functions for a variety of molecules in their ground or excited states. The science of chemistry deals not only with what molecules *are* but also with what they *do*—that is, with chemical reactions. In principle a reacting system must conform at all stages to the requirements of the quantum theory, including symmetry restrictions, but generally the explicit analysis of a reacting system is forbiddingly complex. There are certain important cases, however, in which symmetry considerations enter relatively straightforwardly, and by proper analysis very powerful and general rules may be discovered. Although rules of this kind have been discussed by several workers, the most comprehensive studies have been done by R. B. Woodward and R. Hoffmann, and such rules are often called "Woodward-Hoffmann rules." An extended account of the subject has been published by these workers.*

Symmetry-based "selection rules" of the Woodward-Hoffmann type are relevant when two conditions are fulfilled. (1) The reaction must have as its rate-determining step a *concerted process*. By this is meant a process in which the reacting entities come together and are transformed into products in one, continuous, progressive encounter, without any intermediates or any intervention by nonreacting species, such as catalysts. (2) During the *entire* course of the concerted process one or more symmetry elements of the *entire* reacting system must persist. Wave functions for the system must then *continuously* conform to the requirements imposed by these persisting symmetry elements.

As a practical matter, the second condition need not be fulfilled rigorously. If the reacting skeleton of atoms by itself has certain symmetry elements,

---

* R. B. Woodward and R. Hoffmann, *Angew. Chem.*, Intern. Ed. (Engl.), **8**, 781 (1969).

rules based on these symmetry elements may often be expected to hold even if substituents on some of the skeletal atoms formally destroy the skeletal symmetry. This will be true when the substituents do not differ much from each other in an electronic sense. For instance, the diene 7.8-I has C_{2v} symmetry. If one methyl group is replaced by an ethyl group, to give 7.8-II, the symmetry is reduced to no more than C_s. However, the electronic

similarity of CH_3 and C_2H_5 means that the electronic structure of the skeleton of 7.8-II will differ very little from that of 7.8-I. Symmetry-based rules pertaining rigorously to 7.8-I will hold in practice also for 7.8-II. Clearly, if one methyl group in 7.8-I were replaced by a group *very* different electronically from CH_3, say COOH or F, then the transferability of rules based on C_{2v} symmetry to the substituted compound might be questionable. The chemist must use his judgment and be cautious in dealing with symmetry-destroying substituents, but experience to date suggests that a good deal of deviation from the idealized symmetry is tolerable.

Among the many types of reactions which may be treated by Woodward-Hoffmann rules, cyclizations in which open-chain olefins are converted, either thermally or photochemically, to cyclic species are especially important and can serve well to illustrate the principles involved in this type of analysis. We shall therefore discuss bimolecular and unimolecular cyclizations, beginning with the dimerization of ethylene and proceeding to the important and celebrated Diels-Alder reaction.

Dimerization of Ethylene

This is the simplest practical example of an olefin cyclization. The reaction is represented schematically as follows:

$$\| + \| \rightarrow \square \tag{7.8-1}$$

If the reactant molecules approach each other with their molecular planes parallel and then pass into the product, cyclobutane, with a planar C_4 skeleton, there are a number of persisting symmetry elements. Until the product molecule is actually reached, the symmetry is D_{2h}; for the final product with a planar ring the symmetry is D_{4h}. Since D_{2h} is a subgroup of D_{4h}, the symmetry elements giving rise to D_{2h} are the persisting ones.

The group D_{2h} can be generated by only three symmetry elements, namely, the three planes of symmetry. Each pair of reflections in the planes generates a C_2 rotation about the axis formed by the intersection of the planes; thus the symmetry properties of a wave function with respect to the planes will automatically determine its symmetry with respect to the rotations, and the latter need not be explicitly considered.

The reaction we are considering transforms two π bonds into two σ bonds—or, more precisely, transfers two electron pairs from π bonding orbitals to σ bonding orbitals. All other bonds in the molecules remain more or less unaltered and can be ignored. We focus attention on the electrons and orbitals which undergo major change. The same strategy will apply in all similar analyses. Moreover, since all the atomic orbitals to be considered lie in the plane containing the four carbon atoms, all atomic and molecular orbitals, σ or π, involved will be symmetric to reflection in this plane. Hence, symmetry with respect to this plane is invariant and may be ignored. Only the symmetry properties relative to the other two planes, perpendicular to the plane of the carbon atoms, need be considered. This means that the symmetry group C_{2v}, a subgroup of both D_{2h} and D_{4h}, is sufficient, and the arguments to follow will be framed using only the two mutually perpendicular planes which define the group C_{2v}.

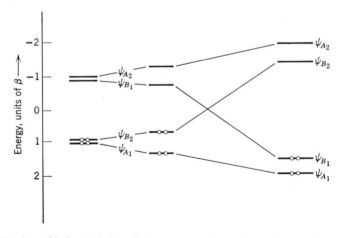

Figure 7.8   An orbital correlation diagram for ethylene dimerization. *Left*: two widely separated ethylene molecules. *Center*: two ethylene molecules close enough for significant interactions to occur. *Right*: cyclobutane; electron configurations correspond to the ground state for each stage.

Each C-C π bond is formed by overlap of $p\pi$ orbitals on the carbon atoms; positive overlap leads to the bonding orbital, stabilized by the energy $|\beta|$ and containing two electrons. There is also the empty antibonding orbital in which there is negative overlap; this is destabilized by the energy $|\beta|$. When the two ethylene molecules are so far apart that interaction between them is negligible, the energy level diagram shown at the left in Figure 7.8 is appropriate. The two bonding and the two antibonding π orbitals are degenerate. As the ethylene molecules approach, however, and interaction begins, the diagram will change in the way shown in the center of Figure 7.8.

The reason for the changes in energy and the basis for the symmetry designations of the orbitals can be easily understood by referring to the accompanying sketches. It is clear from comparison of these sketches with the C_{2v}

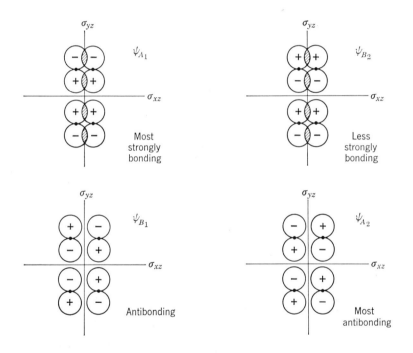

character table that the functions shown have the symmetry properties required by the representations to which they have been assigned. It will also be clear that, as the ethylene molecules approach closely enough for some intermolecular overlap to come into play, the positive overlap involved in ψ_{A_1} and the negative overlap in ψ_{B_2} will cause their energies to diverge as shown in Figure 7.8.

We now turn to the σ orbitals, which must arise in order to form the cyclic product. In the sketches below these are drawn schematically. The right

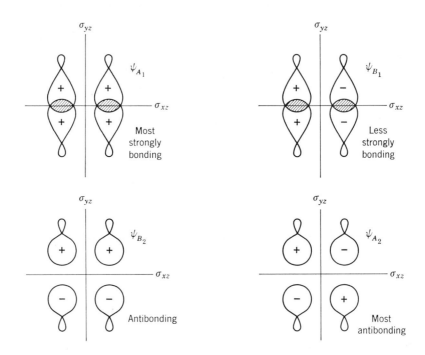

side of Figure 7.8 shows the relative energies of the σ orbitals in a semi-quantitative way. Their energies relative to one another follow from the relative degrees of overlap, which are obvious. Their energies relative to the energies of the π orbitals have been adjusted to take qualitative account of the fact that the σ overlaps are greater than the π overlaps and, hence, σ bonds are stronger than π bonds.

Inspection of Figure 7.8 as it is now drawn reveals a highly significant inconsistency. In its present form the figure shows each stage of the system with the electron configuration of lowest energy, that is, in its ground state. Since the two symmetry planes σ_{xz} and σ_{yz} exist continuously, as the relative importance of the different interactions changes, leading from the left to the right side of Figure 7.8 (or vice versa), there should be a continuous correlation of the orbitals of each symmetry type. Lines representing these correlations have been drawn. We see, however, that the B_2 orbital changes its nature from bonding at the left to antibonding at the right. Therefore, the two electrons occupying an orbital of B_2 symmetry would end up in an antibonding orbital, according to the orbital correlations shown, and the cyclobutane molecule would be formed in a high-energy excited state instead of in the ground state. It could then, of course, lose energy, radiatively or otherwise, and reach the ground state, but the point is that sufficient energy to reach the excited state in the first place would have to have been acquired in the course of the concerted transformation. The amount of energy required is so great as to be unavailable from thermal excitation, and thus the reaction should be "thermally forbidden." This is in agreement with the experimental fact that ethylene dimerization to cyclobutane and other, essentially similar, intermolecular reactions of monoolefins do not proceed at a useful rate under purely thermal activation, even though they are thermodynamically favorable.

Although the preceding argument is correct as far as it goes, and leads to a satisfying insight into the source of the inhibition of ethylene dimerization, it is possible and desirable to carry it a step further and thus obtain an even better understanding of this kind of problem. As stressed already in Section 7.6, orbital descriptions leading to the specification of electron configurations are a limited and not entirely adequate basis for understanding the electronic structures of molecules. It is frequently necessary to look explicitly at the states which arise from the configurations. When degeneracies are not involved, the symmetries of the states can easily be found by forming the direct product representations of the various occupied orbitals. Let us now do this for the lowest energy states of both reactants and products in the ethylene dimerization reaction. In the process we shall employ the shorthand notation previously (Section 7.6) introduced, whereby an electron in, say, an A_2 orbital is represented by a_2 and two electrons occupying this orbital are represented by a_2^2.

The lowest energy configuration of the π electrons in two weakly interacting ethylene molecules is, as shown in Figure 7.8, $a_1^2 b_2^2$. The representation of the direct product $A_1 \times A_1 \times B_2 \times B_2$ is easily seen to be A_1. The next most stable configuration must be $a_1^2 b_2 b_1$, which uniquely defines a state of A_2 symmetry. The only other configuration of interest to us here is the one which would correlate directly with the ground configuration of the product, namely, the $a_1^2 b_1^2$ configuration, which gives a state of symmetry A_1. In a similar fashion, we find that the lowest and next to lowest configurations for the product, as well as the configuration of the product which correlates directly with the ground configuration of the reactants, and the states they give are as follows:

$$a_1^2 b_1^2 \rightarrow A_1$$

$$a_1^2 b_1 b_2 \rightarrow A_2$$

$$a_1^2 b_2^2 \rightarrow A_1$$

It is now possible to draw a correlation diagram between the states, as shown in Figure 7.9. The crucial feature to note here is that the A_1 to A_1 correlations which would seem to follow from direct orbital correlations cannot and do not actually occur, because of what is called the *noncrossing rule*. Two states of the same symmetry cannot cross, in the manner indicated by the dotted lines, because of electron repulsion. Instead, as they approach they turn away from each other so that the lowest A_1 states on each side are correlated with each other as shown by the full lines. The repulsive interaction is similar in essence to that involved in configuration interaction in naphthalene, as discussed in Section 7.6. Indeed, the noncrossing rule is no more than a special but straightforward instance of configuration interaction.

From Figure 7.9 it is clear that the thermally activated dimerization, that is, a ground-state to ground-state process, is inhibited by a substantial energy barrier. We reach the same conclusion as before, namely, that the thermal reaction is "forbidden." But Figure 7.9 tells us something more. We see that, if one of the reactant molecules is photoexcited, so that the

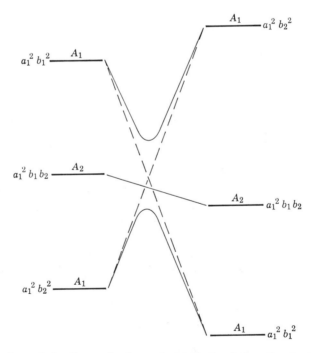

Figure 7.9   A correlation diagram for electronic states in the ethylene dimerization reaction. *Left*: for the π electrons in the two ethylene molecules. *Right*: for the new σ electrons in the cyclobutane molecule.

system comprised of the two reactant molecules is formed in its first excited state, of A_2 symmetry, it may cross directly, without any further electronic excitation, to the first excited state, also of A_2 symmetry, of the product. Thus, the reaction is photochemically "allowed." Although this conclusion could have been drawn from Figure 7.8, it is shown in a more obvious and certain way in Figure 7.9. It is an experimental fact that intermolecular olefin dimerizations of the type in question proceed at a useful rate under irradiation.

The Diels-Alder Reaction

In its simplest form this may be shown as follows:

$$\text{(diene)} + \text{(olefin)} \longrightarrow \text{(product)} \tag{7.8-2}$$

Experimental data indicate that this reaction proceeds thermally under mild conditions (often at temperatures below 0°C), apparently in a concerted manner, and that it is not, in general, accelerated by irradiation with visible or ultraviolet light.

In order for a concerted reaction to occur, the diene and the olefin must approach each other in the way shown in Figure 7.10. In this configuration the only persisting symmetry element is a plane of symmetry which is perpendicular to the single bond in the diene and to the double bonds in the

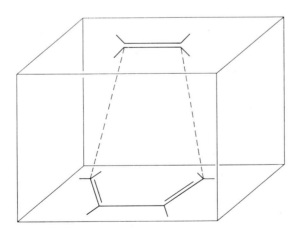

Figure 7.10 The approach configuration required for a concerted Diels-Alder reaction.

reactant and product monoolefins. All orbitals which change in the course of the reaction must now be classified with respect to this symmetry element. The analysis here will obviously follow along lines similar to those used for the ethylene dimerization reaction and can be presented in a less discursive fashion.

As the student will be asked to show in Exercise 7.2, the π MO's for butadiene have the forms and relative energies shown below. Symmetry designations are relative to the perpendicular plane, *A* signifying symmetric and *B* anti-symmetric.

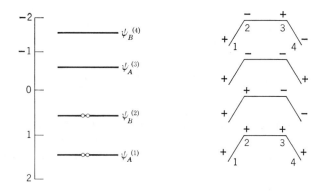

For the π bonds in both the reactant and the product monoolefins, the bonding orbitals are of *A* symmetry and the antibonding orbitals are of *B* symmetry. Finally, the symmetries of the new σ orbitals, between atom pairs 1, 6 and 4, 5, and their relative energies are as follows:

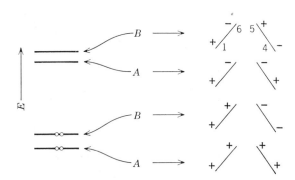

All of the foregoing discussion of orbital symmetries and relative energies is summarized in Figure 7.11. From this orbital correlation diagram it is seen that all filled bonding orbitals in the reactants correlate with filled

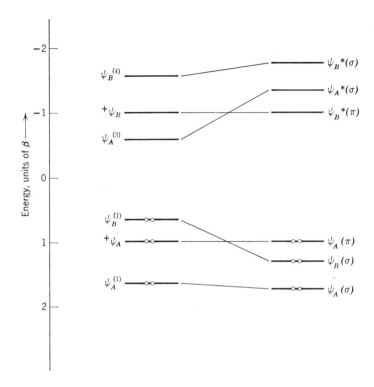

Figure 7.11   An orbital correlation diagram for the Diels-Alder reaction. The $^+\psi_A$ and $^+\psi_B$ orbitals at the left are for ethylene, while the others at the left are for butadiene. The orbitals on the right are for the product.

bonding orbitals in the ground state of the product, and thus the reaction would be expected to be thermally allowed, in agreement with experiment.

It is possible to show very generally that for two olefins, having m_1 and m_2 π electrons, coming together to form a cyclic olefin with $(m_1 + m_2 - 4)/2$ π bonds, as shown below, the reaction will be thermally allowed when $m_1 + m_2 = 4n + 2$ (e.g., $4 + 2 = 6$ in the case of the Diels-Alder reaction). On the contrary, when $m_1 + m_2 = 4n$ (e.g., $2 + 2 = 4$ for the ethylene dimerization) the reaction is thermally forbidden but photochemically allowed. For a discussion of this generalization the article of Woodward and Hoffmann should be consulted.

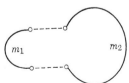

Intramolecular Cyclization: the Butadiene-Cyclobutene Interconversion

One further example of selection rules for reactions is provided by the intramolecular conversion of an open-chain, conjugated polyene to a cyclic olefin with one less pair of π electrons. The simplest example is the butadiene-cyclobutene interconversion:

$$\text{(7.8-3)}$$

while in general we have the following reaction:

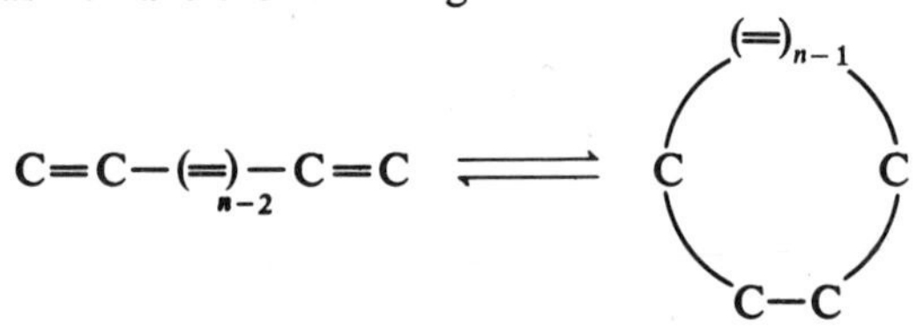

$$C=C-(=)_{n-2}-C=C \rightleftharpoons \quad\text{(7.8-4)}$$

When a reaction like 7.8-3 is carried out thermally with a 1,4-disubstituted butadiene, the stereochemistry is very specific. Thus we have:

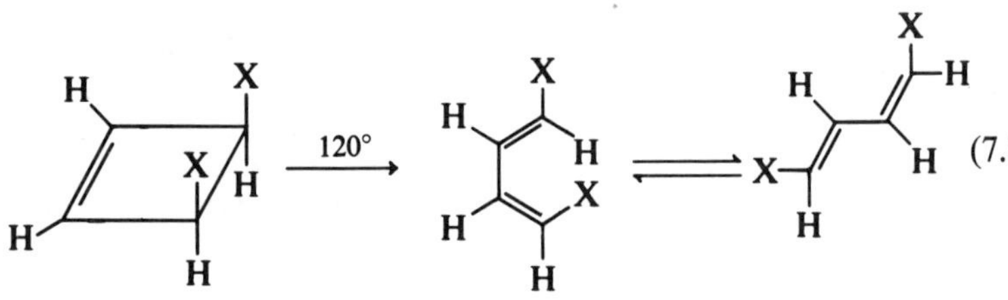

$$\text{(7.8-5)}$$

where $X = CH_3$ or $COOR$, but no

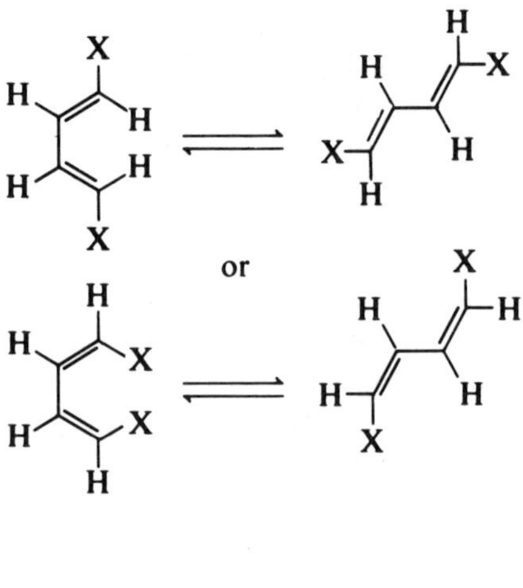

or

A steric explanation is not adequate, since the second isomer is sterically as good as the one obtained or even better. Also, the $-COOR$ and $-CH_3$ groups have quite different inductive properties, so that the course of the reaction seems to depend on some inherent characteristics of the olefin systems.

The product obtained in reaction 7.8-5 arises by a concerted bond-breaking/rotatory rearrangement, as shown in Figure 7.12a. In this process the two

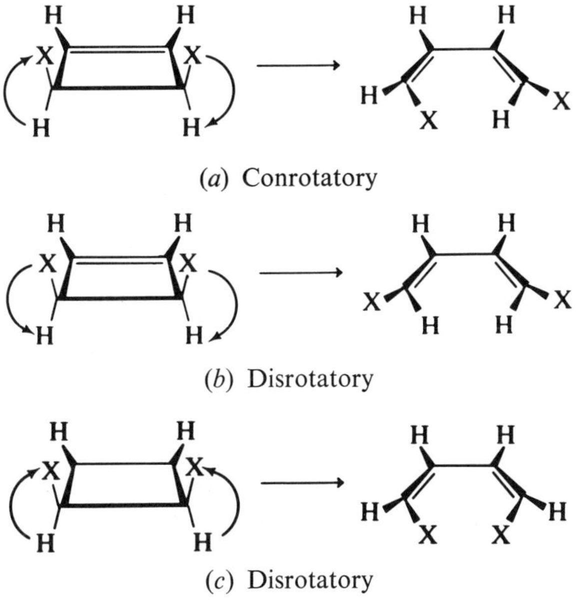

(a) Conrotatory

(b) Disrotatory

(c) Disrotatory

Figure 7.12   Conrotatory and disrotatory ring openings.

rotations are in the same direction; the process is therefore called *con*rotatory. The processes shown in Figures 7.12b and 7.12c would lead to the other geometric isomers, which do not form to any significant extent in the thermal reaction. The latter processes involve rotations in opposite directions and hence are called *dis*rotatory. We could sum up the experimental results neatly by saying that the ring opening is thermally allowed by a conrotatory process and thermally forbidden by a disrotatory process. We now seek an explanation for these simple and striking observations.

In the disrotatory processes, the system has one persisting symmetry element; a plane perpendicular to the skeleton of carbon atoms. In the conrotatory process the one persisting element of symmetry is a C_2 axis, which

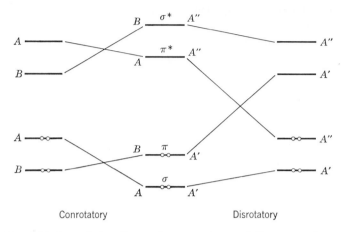

Figure 7.13   Orbital correlation diagram for conrotatory and disrotatory ring openings of cyclobutenes.

is a bisector of the original double bond. The symmetry groups concerned are C_2 and C_s, respectively, and the wave functions must in each case be either symmetric (A or A') or antisymmetric (B or A'') to the relevant symmetry operation. Figure 7.13 shows the orbital symmetry correlations for the two cases. The symmetries and relative energies of the orbitals should be evident without further explanation. The energies and symmetries of the π orbitals for butadiene have been given on page 186.

Figure 7.13 shows that in conrotatory ring opening bonding orbitals correlate only with bonding orbitals; thus the reaction should be thermally allowed. In disrotatory ring opening, on the contrary, the product cannot be produced in the ground state because of correlations of bonding with antibonding orbitals and such a process should be thermally forbidden, again in agreement with experiment.

Finally, let us look at the corresponding state correlation diagram, as shown in Figure 7.14. In constructing this, we use the direct product rules:

$$\left.\begin{array}{l} A \times A = A \\ B \times B = A \\ A \times B = B \end{array}\right\} \text{ for } C_2$$

$$\left.\begin{array}{l} A' \times A' = A' \\ A'' \times A'' = A' \\ A' \times A'' = A'' \end{array}\right\} \text{ for } C_s$$

This diagram confirms our earlier conclusion, namely, that conrotatory ring opening is thermally allowed and disrotatory ring opening is thermally forbidden. It shows further, however, that a singly excited cyclobutene

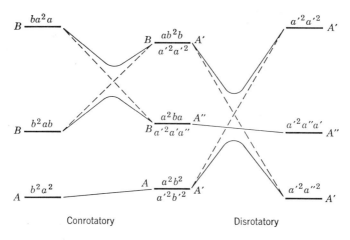

Figure 7.14   State correlation diagram for conrotatory and disrotatory ring openings of cyclobutenes.

molecule can undergo an allowed disrotatory ring opening but is unlikely to undergo a conrotatory one.

It is possible to extend the results just obtained to the general case of a cyclic conjugated olefin with m π electrons (say cyclohexadiene, $m = 4$) opening to a straight-chain polyolefin with $m + 2$ π electrons (hexatriene), whence the following rules are found. For $m = 4n + 2$ ($n = 0, 1, 2, \ldots$) the ring opening (or closing) is thermally conrotatory and photochemically disrotatory (as we have just seen for $n = 0$), whereas for $m = 4n$ ($n = 1, 2, \ldots$) the reaction is thermally disrotatory and photochemically conrotatory. For further discussion, see the article of Woodward and Hoffmann.

Exercises

7.1   Derive the results given in Table 7.1 for any of the systems other than C_3H_3 and C_6H_6.

7.2   Show that for butadiene [which may be treated either in its *trans*-planar (C_{2h}) or *cis*-planar (C_{2v}) form], the four MO's have energies of 1.62β, 0.62β, -0.62β, and -1.62β, the delocalization energy is $0.48\,|\beta|$, and the π bond orders are 0.90 and 0.45.

7.3   The following molecules will afford additional practice in using the Hückel approximation for hydrocarbons. In each case the π bond orders, the delocalization energy (in units of $|\beta|$), and an indication (T) of whether the molecule has a triplet ground state are given.

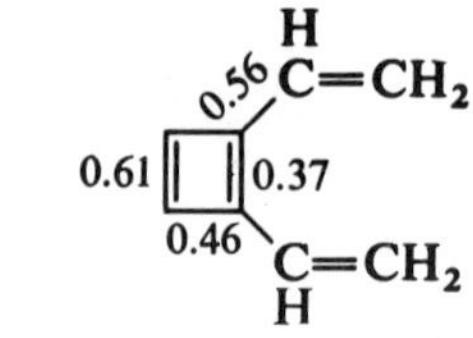

H
0.56 C=CH₂
0.61 ⬚ 0.37
0.46 C=CH₂
H

1.21(*T*)

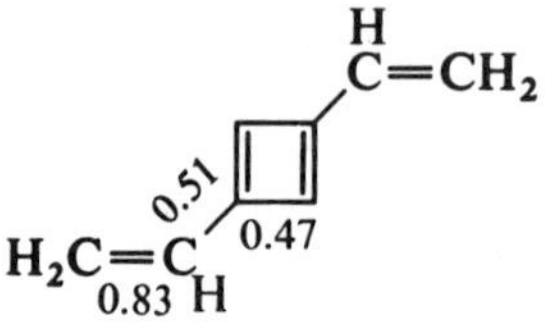

0.65
0.17 0.50
0.41

2.25

H
C=CH₂
0.51
H₂C=C 0.47
0.83 H

1.15(*T*)

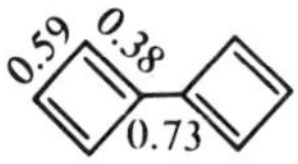

0.71
0.00 → 0.50

1.66

0.59 0.38
0.73

1.25(*T*)

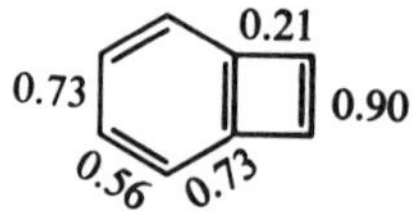

0.45 0.72
0.72 ↘
0.28

2.47

0.21
0.73 0.90
0.56 0.73

2.38

CH₂
0.83 0.39
H₂C CH₂

1.30

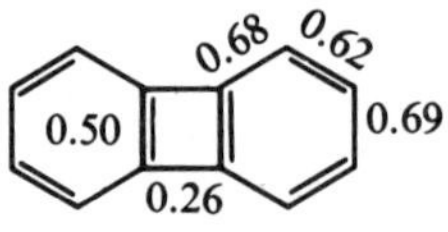

0.68 0.62
0.50 0.69
0.26

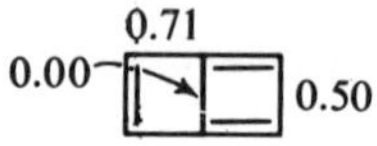

4.51

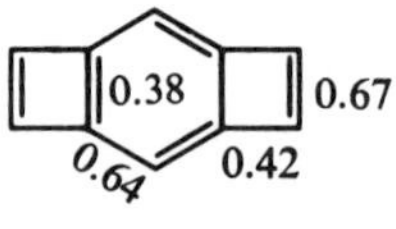

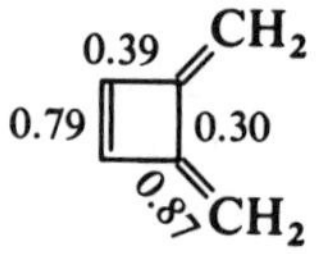

0.39 CH₂
0.79 0.30
0.87 CH₂

1.21

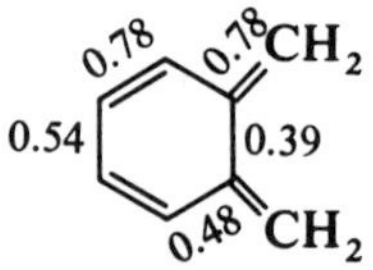

0.38 0.67
0.64 0.42

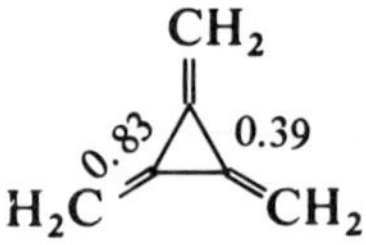

2.68(*T*)

0.78 0.78 CH₂
0.54 0.39
0.48 CH₂

1.95

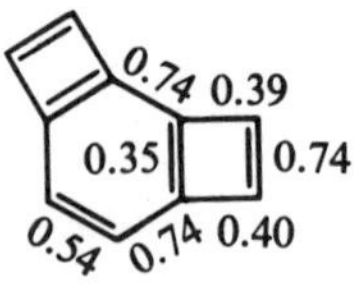

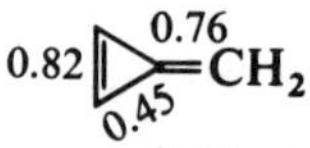

0.74 0.39
0.35 0.74
0.54 0.74 0.40

3.07

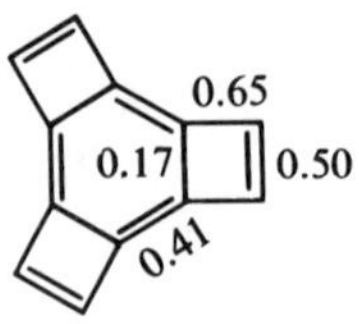

0.76
0.82 CH₂
0.45

0.96

0.29
0.79 0.89

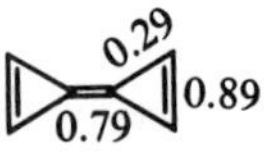

1.46

7.4   Tetramethylenethane

$$\begin{array}{ccc} H_2C & & CH_2 \\ & \diagdown\;C\;\diagup & \\ & | & \\ & C & \\ \diagup & & \diagdown \\ H_2C & & CH_2 \end{array}$$

has recently been detected (P. Dowd, *J. Am. Chem. Soc.*, **92**, 1067(1970)). What does a simple Hückel calculation predict about its delocalization energy, bond orders, spin multiplicity and distribution of unpaired electrons (if any)?

7.5   Investigate the symmetry restrictions on the reaction of the allyl anion with the ethylene molecule to form the cyclopentadienyl anion. If the reaction is not thermally allowed, which reactant should be singly excited for an allowed photochemical reaction?

7.6   The Cope rearrangement is as follows:

Carry out the appropriate symmetry analysis to show that it is thermally allowed.

Hybrid Orbitals and Molecular Orbitals

for AB_n-Type Molecules

8.1 Introduction

In Chapter 7 MO theory was introduced and discussed with emphasis on its application to organic systems in which π systems extending over planar skeletons constitute the major part, if not the entirety, of the problem.

In another very broad class of molecular systems a central atom, A, is surrounded by a set of n other atoms, B_n, B_bC_{n-b}, $B_bC_cD_{n-b-c}$, etc, all of which are bonded to the central atom but not to each other. Included in this class are all of the mononuclear coordination complexes, polynuclear complexes in which direct metal-to-metal interactions are negligible, all of the oxo anions such as NO_3^- and SO_4^{2-}, and all of the molecular halides, oxides, sulfides, etc., such as BF_3, PF_5, SF_4, XeF_4, SO_3, OsO_4, and PCl_3S.

In this chapter we shall discuss the symmetry considerations basic to MO theory as it is specialized for such molecules and also the symmetry concepts underlying the hybrid orbital formulation of the bonding in such molecules. It is convenient to begin with a discussion of hybrid orbitals for atom A and then turn to the question of MO's for this class of molecules.

8.2 Transformation Properties of Atomic Orbitals

The symmetry group to which an $A(B, C, \ldots)_n$ molecule belongs is determined by the arrangement of the pendent atoms. The A atom, being unique, must lie on all planes and axes of symmetry. The orbitals which atom A uses in forming the A-(B,C, ...) bonds must therefore be discussed and classified in terms of the set of symmetry operations generated by these axes and planes—that is, in terms of the overall symmetry of the molecule. Thus, our first order of business is to examine the wave functions for atomic orbitals and consider their transformation (symmetry) properties under the various operations which constitute the point group of the $A(B,C,\ldots)_n$ molecule.

The wave functions for the hydrogen atom are known exactly. They are functions of the three spatial coordinates of the electron and take their most

4.   *Energy Level Diagrams*

$-\beta_{12}$ ——— | ——————— $\psi_A{}^{(2)}$ $-\beta_{12}$ ——— | ——————— $\psi_B{}^{(2)}$

0 ——— | ——————— ψ_B 0 ——— | ——————— ψ_A

β_{12} ——— | ——————— $\psi_A{}^{(1)}$ β_{12} ——— | ——————— $\psi_B{}^{(1)}$

It will be perfectly obvious on inspecting the results that the two seemingly different cases are essentially similar in their ultimate results; only the labeling differs. Moreover, bearing in mind that the effect of changing α for the center atom to a different value from that for the end atoms has only a small qualitative effect on the end results (unless the difference in α's is made very great), we see that in all open three-center bonding situations we shall be dealing with essentially the same set of energy levels; one bonding level, one approximately nonbonding level, one antibonding level. It may be observed that the question of what happens if the α values differ a great deal is not really very important because a large difference would lead to poor bonding and instability.

There are two general types of three-center bonding, based on the number of electron pairs involved. If only one pair of electrons is available, as in the B-H-B bonds for B_2H_6, the electrons will occupy the bonding three-center MO. The expression for the bonding MO, $\frac{1}{2}(\sigma_1 + \sqrt{2}\sigma_2 + \sigma_3)$, shows that the distribution of electronic charge is roughly even for this case. We thus have an essentially nonpolar system with two bonding electrons serving to unite two pairs of atoms.

In the second important type of three-center bonding a total of four electrons must occupy the MO's. One pair will enter the bonding orbital, tending to give a fairly even distribution of charge and a strongly bonding contribution. The second pair will enter the nonbonding MO, where the electrons have little effect on the bond strength but have a marked effect on the bond polarity. The nonbonding orbital, described by the expression

$$\frac{1}{\sqrt{2}}(\sigma_1 \pm \sigma_3)$$

concentrates its electrons entirely on the end atoms and thus makes them more negative than the center atom. The presence of electrons in the nonbonding as well as the bonding MO leads to a situation where one might simple form when we choose these coordinates to be the polar coordinates shown in Figure 8.1 in relation to a set of Cartesian axes. The point at x, y, and z in Cartesian coordinates is fixed by r, the radial distance, OP, from the origin of the coordinate system (always considered positive); θ, the angle be

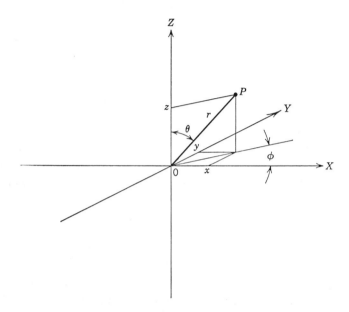

Figure 8.1 Diagram showing the relation of polar coordinates, r, θ, ϕ, to Cartesian coordinates for the point P.

tween the z axis and the line OP; and ϕ, the angle between the x axis and the projection of OP on the xy plane.

The wave functions for the electron in the hydrogen atom are all products of two functions. First there is the radial function, $R(n, r)$ which depends on the principal quantum number, n, and the coordinate, r. Then there is the angular part, $A(\theta, \phi)$, which is independent of both n and r but is a function of θ and ϕ. Both $R(n, r)$ and $A(\theta, \phi)$ are assumed to be separately normalized* to unity, that is,

$$\int_0^\infty [R(n, r)]^2 r^2 \, dr = 1$$

$$\int_0^{2\pi} \int_0^\pi [A(\theta, \phi)]^2 \sin \theta \, d\theta \, d\phi = 1$$

* It should be recalled that the differential element of volume in polar coordinates is $r^2 \sin \theta \, dr \, d\theta \, d\phi$.

Their product, the complete orbital wave function, is then also normalized to unity.

Because no symmetry operation can alter the value of $R(n, r)$, we need not consider the radial wave functions any further. Symmetry operations do alter the angular wave functions, however, and so we shall now examine them in more detail. It should be noted that, since $A(\theta, \phi)$ does not depend on n, the angular wave functions for all s, all p, all d, etc., orbitals of a given type are the same regardless of the principal quantum number of the shell to which they belong. Table 8.1 lists the angular wave functions for s, p, d, and f orbitals.

In an example worked out at the end of Section 5.2 it was noted in passing that the p orbital with an angular dependence on $\sin \theta \cos \phi$ was called a p_x orbital because the function $\sin \theta \cos \phi$ has the same transformation properties as does the Cartesian coordinate x. At this point we shall discuss the transformation properties and hence the notation for the various orbitals more fully. To do so we should recognize that the x, y, and z coordinates of a point (see Figure 8.1) are related to its polar coordinates in the following way:

$$x = r \sin \theta \cos \phi$$
$$y = r \sin \theta \sin \phi \qquad (8.2\text{-}1)$$
$$z = r \cos \theta$$

These relations mean that, since x is equal to $\sin \theta \cos \phi$ times a constant, which is of course unaltered by any transformation of the kind occurring in a point group, $\sin \theta \cos \phi$ must transform in the same way as does x. On this basis the assignment of the subscripts x, y, and z to the p orbitals is clear.

The notation of the d and f orbitals also may be deduced by using relations 8.2-1, as shown in the following examples:

(1) $\sin^2 \theta \sin 2\phi = 2 \sin^2 \theta \sin \phi \cos \phi$
$\qquad = 2(\sin \theta \cos \phi)(\sin \theta \sin \phi)$
$\qquad = 2(x/r)(y/r) = (2/r^2)xy$
$\qquad = \text{constant} \cdot xy$

(2) $3 \cos^2 \theta - 1 = 3 \cos^2 \theta - \cos^2\theta - \sin^2\theta$
$\qquad = 2 \cos^2 \theta - \sin^2\theta$

Now

$$(x/r)^2 = \sin^2 \theta \cos^2 \theta$$
$$(y/r)^2 = \sin^2 \theta \sin^2 \phi$$

hence

$$(1/r^2)(x^2 + y^2) = \sin^2 \theta (\sin^2 \phi + \cos^2 \phi)$$
$$= \sin^2 \theta$$

Table 8.1   Angular Wave Functions, $A(\theta, \phi)$, of s, p, d and f Orbitals (Normalized to Unity)

	ORBITAL			$A(\theta, \phi)$
LETTER TYPE	FULL POLYNOMIAL[a]	SIMPLIFIED POLYNOMIAL	NORMALIZING FACTOR	ANGULAR FUNCTION
s			$\dfrac{1/\sqrt{\pi}}{2}$	
p	z		$\dfrac{\sqrt{3/\pi}}{2}$	$\cos\theta$
	x		$\dfrac{\sqrt{3/\pi}}{2}$	$\sin\theta\cos\phi$
	y		$\dfrac{\sqrt{3/\pi}}{2}$	$\sin\theta\sin\phi$
d	$2z^2 - x^2 - y^2$	z^2	$\dfrac{\sqrt{5/\pi}}{4}$	$(3\cos^2\theta - 1)$
	xz		$\dfrac{\sqrt{15/\pi}}{2}$	$\sin\theta\cos\theta\cos\phi$
	yz		$\dfrac{\sqrt{15/\pi}}{2}$	$\sin\theta\cos\theta\sin\phi$
	$x^2 - y^2$		$\dfrac{\sqrt{15/\pi}}{4}$	$\sin^2\theta\cos 2\phi$
	xy		$\dfrac{\sqrt{15/\pi}}{4}$	$\sin^2\theta\sin 2\phi$
f[b]	xyz		$\dfrac{\sqrt{105/\pi}}{4}$	$\sin^2\theta\cos\theta\sin 2\phi$
	$x(z^2 - y^2)$		$\dfrac{\sqrt{105/\pi}}{4}$	$\sin\theta\cos\phi(\cos^2\theta - \sin^2\theta\sin^2\phi)$
	$y(z^2 - x^2)$		$\dfrac{\sqrt{105/\pi}}{4}$	$\sin\theta\sin\phi(\cos^2\theta - \sin^2\theta\cos^2\phi)$
	$z(x^2 - y^2)$		$\dfrac{\sqrt{105/\pi}}{4}$	$\sin^2\theta\cos\theta\cos 2\phi$
	$x(5x^2 - 3r^2)$	x^3	$\dfrac{\sqrt{7/\pi}}{4}$	$\sin\theta\cos\phi(5\sin^2\theta\cos^2\phi - 3)$
	$y(5y^2 - 3r^2)$	y^3	$\dfrac{\sqrt{7/\pi}}{4}$	$\sin\theta\sin\phi(5\sin^2\theta\sin^2\phi - 3)$
	$z(5z^2 - 3r^2)$	z^3	$\dfrac{\sqrt{7/\pi}}{4}$	$5\cos^3\theta - 3\cos\theta$

[a] $r^2 = x^2 + y^2 + z^2$
[b] See Appendix V for further discussion of f orbitals.

Therefore we can write

$$3 \cos^2 \theta - 1 = 2(z^2/r^2) - (1/r^2)(x^2 + y^2)$$
$$= \text{constant} \cdot (2z^2 - x^2 - y^2)$$

Thus the d orbital whose angular wave function is a constant times $3 \cos^2 \theta - 1$ should be written $d_{2z^2-x^2-y^2}$. Since in most groups z^2 and $x^2 + y^2$ transform in the same way, $2z^2 - x^2 - y^2$ will transform in the same way as z^2 and the shorter notation d_{z^2} is used.

(3) $\sin^2 \theta \cos \theta \sin 2\phi = \sin^2 \theta \cos \theta (2 \sin \phi \cos \phi)$
$$= 2(\sin \theta \cos \phi)(\sin \theta \sin \phi) \cos \theta$$
$$= 2 \quad . \quad x \quad . \quad y \quad . \quad z$$

Hence the f orbital with the above angular functions is called the f_{xyz} orbital.

As a result of the fact that the polynomial subscript to an orbital symbol tells us that the orbital transforms in the same way as the subscript, we can immediately determine the transformation properties of any orbital on an atom lying at the center of the coordinate system by looking up its subscript in the appropriate column on the right of a character table, if it is a p or d orbital. An s orbital always transforms according to the totally symmetric representation, since it has no angular dependence. For f orbitals, the usual character tables do not give the desired information directly because f orbitals are not used sufficiently often. However, the assignment of each f orbital to the appropriate irreducible representation can be worked out for any given group when needed. Consider, for example, the f_{xyz} orbital in the group D_{2h}. The function xyz is transformed into $+1$ times itself by the operations E, $C_2(z)$, $C_2(x)$, and $C_2(y)$ but into -1 times itself by the operations i, $\sigma(xy)$, $\sigma(xz)$, and $\sigma(yz)$. Thus it is a basis for the A_u representation. In the following sections we shall not explicitly discuss f orbitals, but it is to be emphasized that their inclusion would not require any new principles.

As an example of using the character tables directly for p and d orbitals, we will consider the phosphorus atom in PCl_3. By looking at the character table for the group C_{3v} we immediately learn that the phosphorus orbitals belong to the following representations:

$$A_1: s, p_z, d_{z^2}$$
$$A_2: \text{none}$$
$$E: (d_{xy}, d_{x^2-y^2})(d_{xz}, d_{yz}), (p_x, p_y)$$

It should be recalled that, when we say a certain orbital (or group of orbitals) "belongs" to a certain irreducible representation, we mean that it is a basis for that irreducible representation.

The preceding discussion of the symmetry properties of atomic orbitals has referred explicitly to the one-electron orbitals of the hydrogen atom. However, the principles can be carried over to the treatment of many-electron atoms. The wave functions for these atoms may be written as products of one-electron wave functions. For each electron in a many-electron atom we write a wave function consisting of an angular function which is the same as the angular function of an analogous electron in the hydrogen atom, and a radial function which differs from the radial function the electron would have in a hydrogen atom because of mutual shielding and repulsion effects among the electrons. The important point is that, since the angular properties of an electron in a many-electron atom can be taken as being the same as those of a corresponding hydrogen electron, the symmetry properties of the one-electron wave functions used to build up the total wave functions for atoms have the same transformation properties as the simple and exact one-electron wave functions obtained by solving the wave equation for the hydrogen atom.

8.3   Hybridization Schemes for σ Orbitals

It is perhaps easiest to explain this subject by going directly to an example. Let us take, as a typical but relatively simple one, the case of tetrahedral hybridization. We wish to know what atomic orbitals on atom A in the tetrahedral molecule AB$_4$ are required to construct a set of four σ orbitals on atom A which have their lobes directed to the B atoms, that is, toward the vertices of a tetrahedron. This set of four orbitals will form a basis for a representation of the symmetry group of the molecule, in this case the group T_d. We may represent each hybrid by a vector pointing in the appropriate direction and number these vectors r_1, r_2, r_3, r_4, as shown in Figure 8.2. Let us now determine the characters of the representation for which this set of vectors forms a basis.

Applying the identity operation, we obtain

$$r_1 \rightarrow r_1 \quad + 0r_2 + 0r_3 + 0r_4$$
$$r_2 \rightarrow 0r_1 + r_2 \quad + 0r_3 + 0r_4$$
$$r_3 \rightarrow 0r_1 + 0r_2 + r_3 \quad + 0r_4$$
$$r_4 \rightarrow 0r_1 + 0r_2 + 0r_3 + r_4$$

The matrix of the coefficients on the right is a unit matrix of dimension 4 and hence $\chi(E) = 4$.

If we rotate the set of vectors by $2\pi/3$ about the C_3 axis which is coincident with r_1, we get

$$r_1 \rightarrow r_1 \quad + 0r_2 + 0r_3 + 0r_4$$
$$r_2 \rightarrow 0r_1 + 0r_2 + r_3 \quad + 0r_4$$
$$r_3 \rightarrow 0r_1 + 0r_2 + 0r_3 + r_4$$
$$r_4 \rightarrow 0r_1 + r_2 \quad + 0r_3 + 0r_4$$

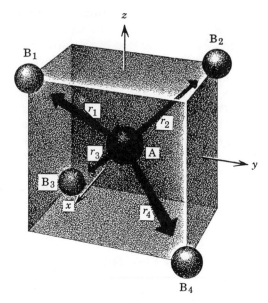

Figure 8.2   A set of vectors, $r_1, r_2, r_3,$ and r_4, representing the four hybrid σ orbitals used by atom A to bond the four B atoms in a tetrahedral AB_4 molecule.

The character of the matrix of the coefficients here, $\chi(C_3)$, is equal to 1.

Proceeding in the same way with a C_2, an S_4, and a σ_d, we obtain the following set of characters for the representation generated:

	E	$8C_3$	$3C_2$	$6S_4$	$6\sigma_d$
Γ_{tetra}	4	1	0	0	2

Reference to the T_d character table shows that this representation is not among the irreducible ones but that it can be reduced in the following way:

$$\Gamma_{tetra} = A_1 + T_2$$

This means that the four atomic orbitals which are combined to make the set of four hybrid orbitals must be chosen so as to include one orbital of A_1 symmetry and a set of three orbitals belonging to the T_2 representation. The character table also tells us that atomic orbitals of atom A falling into these categories are as follows:

A_1 ORBITALS	T_2 ORBITALS
s	(p_x, p_y, p_z)
	(d_{xy}, d_{xz}, d_{yz})

Hence our set of hybrid orbitals may be sp^3 or sd^3 (where, of course, the d^3 means specifically d_{xy}, d_{xz}, d_{yz}, and no others). From the point of view of symmetry there is no difference between sp^3 and sd^3 hybrids.

Because the sp^3 and sd^3 hybrids have exactly the same symmetry properties it is impossible for any atom A in a tetrahedral molecule AB_4 to use purely one set or the other. It must always use a mixture of both. In many cases, however, other reasoning based on knowledge or estimates of the energies of the various orbitals may lead us to believe that the contribution of one set is of minor or perhaps totally negligible magnitude. For example, carbon can form a set of sp^3 hybrid orbitals using its $2s$ and $2p$ orbitals. The most stable, that is, lowest energy, d orbitals available to it are its $3d$ orbitals, so the most stable sd^3 hybrids it could form would be constructed from $2s$, $3d_{xy}$, $3d_{xz}$, and $3d_{yz}$. However, in carbon, the $3d$ orbitals lie some 230 Kcal/mole higher than the $2p$ orbitals. Therefore, in order for the bonds formed using an sd^3 set of hybrids to be more stable than a set using sp^3 hybrids, each sd^3 bond would have to be some $3 \times \frac{230}{4} \sim 170$ kcal/mole stronger than each sp^3 bond This is quite impossible, and we can be sure that carbon will not use an sd^3. set. In fact the $3d$ orbitals are so high in energy relative to the $2p$ orbitals that even partial usage of the d orbitals is unlikely to be important. Thus we can correctly say that carbon (and any other element in the first short period, lithium through fluorine) will form a set of four tetrahedrally directed bonds using $2s2p^3$ hybrid orbitals. But it should be borne in mind that this is so for reasons of energy and not for reasons of symmetry.

In contrast, it is quite likely that the sd^3 set is important if not dominant in the formation of four σ bonds by manganese and chromium in the ions MnO_4^-, MnO_4^{2-}, and CrO_4^{2-}. Here the lowest usable d orbitals are $3d$, and the lowest usable p orbitals are $4p$; the $3d$ orbitals are of somewhat lower energy than the $4p$ orbitals.

Before proceeding to some other illustrative examples we shall develop a rule that will simplify the process of working out the characters of the reducible representation for which a set of hydrid orbitals is a basis. If, on carrying out a symmetry operation on a set of hybrid orbitals or the set of vectors representing them, a certain vector remains unshifted, there appears in the matrix a diagonal element equal to 1. If, however, that vector and some other one are interchanged by the operation, two corresponding diagonal elements are equal to 0. Hence, to determine the character of the matrix corresponding to a given operation we can use the following simple rule:

The character is equal to the number of vectors which are unshifted by the operation.

Suppose that we wish to find what atomic orbitals phosphorus may use to form the σ bonds to five fluorine atoms in PF_5, assuming PF_5 to have a trigonal bipyramid structure. The symmetry group is then D_{3h}. The set of five σ orbitals on phosphorus forms a basis for a representation with the following characters:

D_{3h}	E	$2C_3$	$3C_2$	σ_h	$2S_3$	$3\sigma_v$
Γ_σ	5	2	1	3	0	3

These numbers were obtained as follows, where Figure 8.3 shows the coordinate system and numbering of the bonds. All bonds are unshifted upon

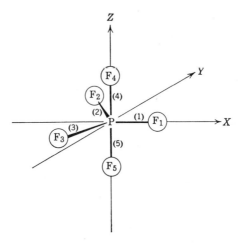

Figure 8.3   Coordinate axes and numbering system for PF_5.

performing E; hence $\chi(E) = 5$. Only (4) and (5) remain unshifted on performing a C_3; hence $\chi(C_3) = 2$. On performing a C_2, only one bond remains unshifted; if the C_2 is $C_2(x)$ only (1) remains fixed; hence $\chi(C_2) = 1$. A σ_h leaves (1), (2), and (3) unshifted; hence $\chi(\sigma_h) = 3$. S_3 leaves no bonds unshifted; the rotation part of it shifts (1), (2), and (3), and the reflection part then interchanges (4) and (5); hence $\chi(S_3) = 0$. A σ_v leaves three bonds unshifted; for example $\sigma(xz)$ interchanges (2) and (3) but does not shift (1), (4), or (5); hence $\chi(\sigma_v) = 3$.

Now it is easily shown that

$$\Gamma_\sigma = 2A'_1 + A''_2 + E'$$

Orbitals falling into these symmetry classes are

A'_1	A''_2	E'
s	p_z	(p_x, p_y)
d_{z^2}		$(d_{xy}, d_{x^2-y^2})$

Thus any of the following combinations could be used:

$$(1) \ ns, (n+1)s, p_z, \begin{cases} p_x, p_y & (a) \\ d_{xy}, d_{x^2-y^2} & (b) \end{cases}$$

$$(2) \ nd_{z^2}, (n+1)d_{z^2}, p_z, \begin{cases} p_x, p_y & (a) \\ d_{xy}, d_{x^2-y^2} & (b) \end{cases}$$

$$(3) \ s, d_{z^2}, p_z, \begin{cases} p_x, p_y & (a) \\ d_{xy}, d_{x^2-y^2} & (b) \end{cases}$$

In molecules which are known to have or might have a trigonal bipyramid structure, it is rather unlikely, for energetic reasons, that combination (1) or (2) would be used. In PF_5 for example, combination (3a) seems likely to be a good description of the composition of the hybrid orbitals used. On the other hand, in gaseous $MoCl_5$ it is probable that some mixture of schemes (3a) and (3b) is required to represent the situation faithfully, since the Mo $4d$ orbitals are comparable in energy to the $5p$ orbitals. In short notation, schemes (3a) and (3b) can be written dsp^3 and d^3sp, but it should be remembered when appropriate that specific d and p orbitals are required.

To emphasize this, let us determine what combinations of atomic orbitals can hybridize to give the σ orbitals required in the other commonly occurring shape for an AB_5 molecule, viz., a tetragonal pyramid. Such a molecule belongs to the group C_{4v}, and the characters of the representation, Γ_σ, for which the five σ orbitals of atom A form a basis are as follows:

C_{4v}	E	$2C_4$	C_2	$2\sigma_v$	$2\sigma_d$
Γ_σ	5	1	1	3	1

This Γ_σ reduces as

$$\Gamma_\sigma = 2A_1 + B_1 + E$$

There are the following orbitals in the required symmetry classes:

A_1	B_1	E
s	$d_{x^2-y^2}$	(p_x, p_y)
p_z		(d_{xz}, d_{yz})
d_{z^2}		

One way in which we can obtain two A_1 orbitals, one B_1 orbital, and one pair of E orbitals is to select $s, p_z, d_{x^2-y^2}, p_x, p_y$, which set may be abbreviated dsp^3. In this C_{4v} case it must be understood that the d refers specifically and only to $d_{x^2-y^2}$, whereas in the dsp^3 set discussed above for the D_{3h} case d must be understood to mean specifically and only d_{z^2}. In a situation where the atom A in AB_5 is likely for energetic reasons to use hybrids composed of one s, one d, and three p orbitals, the geometry of the molecule will depend on whether a $d_{x^2-y^2}$ or a d_{z^2} orbital is used. Of course, for the C_{4v} case, other possible hybridization schemes, written in abbreviated notation, are

$$sd^4, \ spd^3, \ sd^2p^2, \ pd^4, \ p^3d^2$$

The type of analysis discussed rather fully above for several important molecules will now be presented in summary form for some other important cases.

AB_3 (planar): symmetry D_{3h}. Examples: BF_3, AlR_3, NO_3^-, SO_3.

D_{3h}	E	$2C_3$	$3C_2$	σ_h	$2S_3$	$3\sigma_v$
Γ_σ	3	0	1	3	0	1

$\Gamma_\sigma = A_1' + E'$

Possible combinations: (s, p_x, p_y), $(s, d_{xy}, d_{x^2-y^2})$, (d_{z^2}, p_x, p_y), $(d_{z^2}, d_{xy}, d_{x^2-y^2})$, which are, in brief notation, sp^2, sd^2, dp^2, d^3.

AB_4 (planar): symmetry D_{4h}. Examples: $AuCl_4^-$, XeF_4, $Ni(CN)_4^{-2}$.

D_{4h}	E	$2C_4$	C_2	$2C_2'$	$2C_2''$	i	$2S_4$	σ_h	$2\sigma_v$	$2\sigma_d$
Γ_σ	4	0	0	2	0	0	0	4	2	0

$\Gamma_\sigma = A_{1g} + B_{1g} + E_u$

Possible combinations: $(s, d_{x^2-y^2}, p_x, p_y)$, $(d_{z^2}, d_{x^2-y^2}, p_x, p_y)$, which are, in brief notation, dsp^2 and d^2p^2.

AB_6 (octahedral): symmetry O_h. Examples: SF_6, PF_6^-, $Fe(CN)_6^{-3}$.

O_h	E	$8C_3$	$6C_2$	$6C_4$	$3C_2$	i	$6S_4$	$8S_6$	$3\sigma_h$	$6\sigma_d$
Γ_σ	6	0	0	2	2	0	0	0	4	2

$\Gamma_\sigma = A_{1g} + E_g + T_{1u}$

Possible combination: *only* $s, p_x, p_y, p_z, d_{z^2}, d_{x^2-y^2}$, which is, in brief notation, d^2sp^3.

8.4   Hybridization Schemes for π Bonding

We begin by discussing in detail the characteristics of π orbitals as compared to σ orbitals. The difference can be stated in terms of the number of nodal planes possessed by each. A nodal plane—or surface, more generally—is the locus of all points at which the wave function has zero amplitude as a result of its changing sign on passing from one side of the surface to the other. A σ orbital or bond is defined as one having *no* nodal surface which contains the bond axis. A π orbital or bond is defined as one which has *one* nodal surface or plane containing the bond axis. This system of classification extends further to include, for example, δ orbitals and δ bonds which have *two* nodal surfaces intersecting along the bond axis, but we shall not go beyond the π cases here.

To illustrate these definitions, Figure 8.4 shows some cross sections of orbitals or bonds. The closed curves are lines of constant value of the wave function, those nearest the center generally representing higher amplitudes, though this is not necessary, since there may also be nodal surfaces concentric around the bond axis. Figure 8.4a shows a section through a σ orbital which is circularly symmetric. This might be a pure s orbital, a pure p_q orbital viewed along the q axis, or the main lobe of a d_{z^2} orbital viewed along

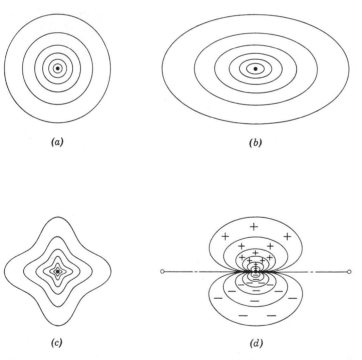

(a)   (b)

(c)   (d)

Figure 8.4   (*a*), (*b*), (*c*) Sections through σ bonds or σ orbitals perpendicular to the internuclear axis. Note absence of nodal planes. (*d*) Section through a π bond or π orbital. o—————o is the trace of a nodal plane. The curves are loci of all points with the same value of the electronic wave function.

the z axis. Figure 8.4*b* shows a cross section of a σ orbital which has elliptical symmetry, and 8.4*c* represents a section which is even more complex in its shape. All, however, have the defining property of a σ orbital: if we follow a circle concentric about the bond axis, there is never any change in the sign of the wave function. Note also that for a σ orbital or bond there is a finite value of the wave function along the bond axis.

Figure 8.4*d* shows a cross section through a π orbital. The nodal plane passing through the bond axis is easily seen. If such a bond is to be formed by overlap of two atomic orbitals (AO's) one on each of the two bonded atoms, it is obviously necessary that each AO have π character with respect to the internuclear axis (i.e., have a nodal plane containing the axis), *and* that these two nodal planes coincide. These considerations are illustrated in Figure 8.5.

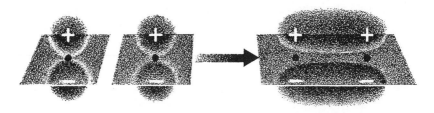

Figure 8.5   Schematic diagram showing formation of a π bond from two atomic or hybrid orbitals with a common nodal plane.

Finally, it may also be easily seen that it is possible to have two, and only two, orthogonal π bonds between the same two atoms. These two π bonds will have their nodal planes mutually perpendicular, as shown in Figure 8.6. It can be seen that the positive overlap of the positive lobe of the first π bond with the positive lobe of the second will be exactly canceled by the overlap of the positive lobe of the first bond with the negative lobe of the second one. An analogous cancellation occurs in the overlaps of the negative lobe of the first orbital with the lobes of the second. However, no third π orbital having net zero overlap with *both* of the first two is possible.

In consequence of the properties of π bonds discussed above, the question of what orbitals are required on a central atom in order to form π bonds to each of a set of atoms surrounding it can be approached in the following way. We want to have altogether $2n$ π-type hybrid orbitals on atom A in the molecule AB_n, two for each B atom. These $2n$ hybrids on A must include an orbital to match (in the sense discussed above and shown in Figure 8.5) each of the π orbitals on the atoms B. Thus the set of $2n$ hybrids on A and the $2n$ π AO's on the B atoms will both form bases for the same representation of the

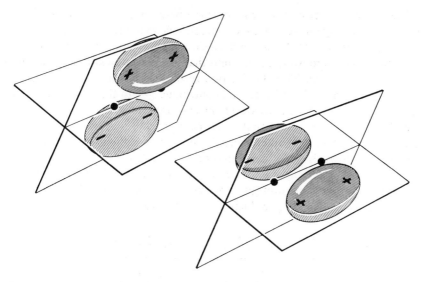

Figure 8.6. Schematic diagram showing two π bonds between the same pair of atoms with the nodal planes of the two orbitals mutually perpendicular.

symmetry group of the molecule. Moreover, each π AO on one of the B atoms can be represented by a vector perpendicular to the nodal plane and pointing in the direction of the positive values of the wave function. There may be two such vectors at right angles on each B atom. Rather than attempting to carry the discussion further on a general basis, let us turn to a specific example.

We will consider a planar, symmetrical AB$_3$ molecule, such as BF$_3$ or NO$_3{}^-$, belonging to the group D_{3h}. The six A-B π bonds which are, in principle, possible will transform, as noted above, in the same way as a set of six vectors attached to the B atoms. The two vectors on a B atom need only be perpendicular to one another and be in a plane perpendicular to the A—B axis. Their orientation within that plane can, in principle, be chosen arbitrarily. However, in cases such as the present one where there is a molecular plane, it is usually advantageous, both in working out the results and in grasping their significance, if we orient one vector on each B atom perpendicular to the molecular plane, and then, necessarily, have the other one lying in the molecular plane, as shown in Figure 8.7.

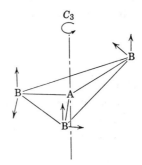

Figure 8.7 Six vectors representing π orbitals on the B atoms in an AB$_3$ molecule belonging to the group D_{3h}.

This set of six vectors is now used as the basis for a representation, Γ_π, of the group D_{3h}. Again we can use the simple rule that any vectors shifted by a symmetry operation contribute zero to the character and any left unchanged contribute $+1$. Here we shall also find that a vector may not be shifted to a new position but will have its direction reversed. Thus it is transformed into the negative of itself and therefore contributes -1 to the character. In this way Γ_π is easily obtained. However, in carrying out the various symmetry operations, it will be found that in no case are any of the vectors perpendicular to the plane ever interchanged with those in the plane. This means that each set gives rise to a representation independently of the other.

Let us call the representation given by the vector perpendicular to the plane $\Gamma_\pi(\perp)$ and the representation given by the set in the plane $\Gamma_\pi(\|)$. The entire body of results, including reduction of the several representations, is as follows:

D_{3h}	E	$2C_3$	$3C_2$	σ_h	$2S_3$	$3\sigma_v$
Γ_π	6	0	-2	0	0	0
$\Gamma_\pi(\perp)$	3	0	-1	-3	0	1
$\Gamma_\pi(\|)$	3	0	-1	3	0	-1

$$\Gamma_\pi = \Gamma_\pi(\perp) + \Gamma_\pi(\|)$$
$$\Gamma_\pi(\perp) = A_2'' + E''$$
$$\Gamma_\pi(\|) = A_2' + E'$$

Thus, in order for atom A to form a π bond perpendicular to the molecular plane [a $\pi(\perp)$ bond] to each of the B atoms, it must use three hybrid orbitals built of one AO transforming as A_2'' and a degenerate pair of AO's transforming as E''. Reference to the D_{3h} character table shows that s, p, or d orbitals meeting these requirements are:

$$A_2'' : p_z$$
$$E'' : (d_{xz}, d_{yz})$$

Thus a set of three equivalent hybrid orbitals constructed from these is the only one possible for forming the $\pi(\perp)$ bonds.

Turning now to the $\pi(\|)$ bonds, we find that AO's of the necessary types are as follows:

$$A_2' : \text{none}$$
$$E' : (p_x, p_y) \quad \text{and} \quad (d_{x^2-y^2}, d_{xy})$$

Since there are no AO's of A_2' symmetry, it is impossible to form a set of *three* equivalent $\pi(\|)$ A-B bonds. This general situation, that is, lack of a complete set of AO's to form a complete set of π bonds, arises very frequently, as we shall see in several subsequent examples. The conclusion stated here will therefore be of general value.

The unavailability of the A_2' orbital does not mean that no $\pi(\|)$ bonds can be formed, nor does it mean that only two of the B atoms can be $\pi(\|)$ bonded. It means rather that *there can be only two $\pi(\|)$ bonds shared equally among the three B atoms*. The treatment of the octahedral AB$_6$ molecule, which follows immediately, will provide an opportunity to discuss this concept more fully.

We shall now consider several more of the important cases, starting with the octahedral AB$_6$ molecule. To determine the representation of O_h for which the twelve possible A-B π bonds form a basis, we attach two vectors to each B atom as shown in Figure 8.8, and consider the effects of the group operations

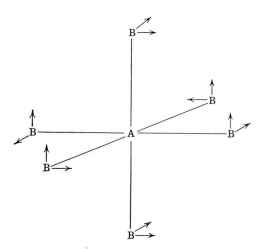

Figure 8.8   An octahedral AB$_6$ molecule with a set of twelve vectors representing the π orbitals of the B atoms.

upon them. It should be noted that each vector is exchanged with each of the other eleven by one symmetry operation or another. This means that all twelve π bonds and hence all twelve orbitals required on atom A fall in the same set. The results are as follows:

O_h	E	$8C_3$	$6C_2$	$6C_4$	$3C_2 (= C_4{}^2)$	i	$6S_4$	$8S_6$	$3\sigma_h$	$6\sigma_d$
Γ_π	12	0	0	0	-4	0	0	0	0	0

$$\Gamma_\pi = T_{1g} + T_{2g} + T_{1u} + T_{2u}$$

Referring again to the O_h character table, we find that atom A will possess, among its s, p, and d orbitals, the following members of each of the above symmetry classes:

$$T_{1g}: \text{none}$$
$$T_{2g}: (d_{xy}, d_{xz}, d_{yz})$$
$$T_{1u}: (p_x, p_y, p_z)$$
$$T_{2u}: \text{none}$$

The first conclusion to be drawn from these results is that it is impossible for the entire set of twelve A-B π bonds to exist, since the A atom does not have all of the necessary orbitals. Moreover, the only T_{1u} orbitals are the p orbitals. If we assume that a full set of A-B σ bonds has been formed, then these p orbitals are already fully used for that purpose (see page 204) and cannot be of any use for π bonding. The correctness of this assumption is not of *a priori* certainty, but in most cases at least it would seem a reasonable one. Thus only three orbitals on atom A, the T_{2g} d orbitals, are available for π bonding, and, as before, the correct interpretation is to conceive of three π bonds shared equally among the six A—B pairs.

In this octahedral case, it is easy to see geometrically the correctness of this interpretation. If we look at the d_{xy} orbital, we see (Figure 8.9b) that it can form π bonds with B atoms 1, 2, 3, and 4 equally well. There is no reason why it should form a π bond exclusively to one and not to the others. Similarly, the d_{xz} orbital can form π bonds with B atoms 1, 2, 5, and 6 equally well (Figure 8.9c), and the d_{yz} can form π bonds equally well with B atoms 3, 4, 5, and 6 (Figure 8.9d). Looking at the situation from the opposite direction, so to speak, we see that each B atom has the same prospects for forming π bonds with each of two d orbitals on atom A. Thus we must regard the π bonding as being shared equally among all six A-B bonds, giving, in effect, one-half of a π bond per A—B pair.

We shall next consider the square planar AB_4 molecule, which belongs to the point group D_{4h}. Here, as in the planar AB_3, we may divide the eight possible π bonds into two subsets, four perpendicular to the molecular plane and four lying in the molecular plane. When the representations for which these two sets form bases are worked out and reduced, the following results are obtained:

D_{4h}	E	$2C_4$	C_2	$2C_2'$	$2C_2''$	i	$2S_4$	σ_h	$2\sigma_v$	$2\sigma_d$
$\Gamma_\pi(\perp)$	4	0	0	-2	0	0	0	-4	2	0
$\Gamma_\pi(\|)$	4	0	0	-2	0	0	0	4	-2	0

$$\Gamma_\pi(\perp) = A_{2u} + B_{2u} + E_g$$
$$\Gamma_\pi(\|) = A_{2g} + B_{2g} + E_u$$

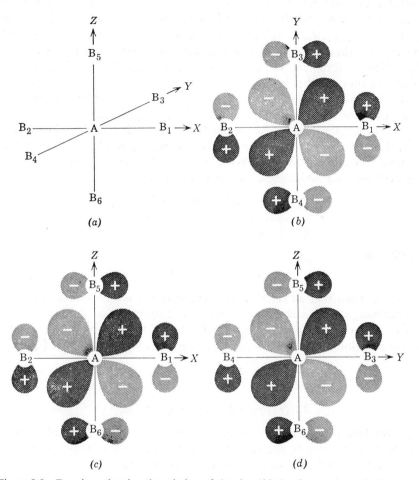

Figure 8.9 Drawings showing the relation of the $d\pi$ orbitals of atom A to the B atoms in an octahedral AB$_6$ molecule.

The D_{4h} character table also tells us that for each of the required representations there are the following s, p, or d orbitals on atom A:

$$A_{2u}: p_z \qquad A_{2g}: \text{none}$$
$$B_{2u}: \text{none} \qquad B_{2g}: d_{xy}$$
$$E_g: (d_{xz}, d_{yz}) \qquad E_u: (p_x, p_y)$$

Thus neither of the sets of π bonds can possibly be complete, since the required B_{2u} and A_{2g} orbitals on atom A simply do not exist (among s, p, and d orbitals). There may, however, be as many as three perpendicular π bonds shared among the four A—B sets, for the p_z, d_{xz}, and d_{yz} orbitals cannot be used at all in A-B σ bonding (see Section 8.2). Since the A-B σ bonds require the use of s, p_x, p_y, $d_{x^2-y^2}$ hybrid orbitals on atom A, however, and we usually assume that the formation of σ bonds takes precedence, there is only the d_{xy} orbital to form one in-plane π bond shared equally (see Figure 8.9*b*) among all four A—B pairs.

We shall conclude this section by working out the π bonding possibilities in a tetrahedral AB_4 molecule. Following the usual practice, we assign a pair of vectors to each B atom, so oriented that they will transform identically with the A-B π bonds which may be formed and hence identically with the required orbitals on atom A. We obtain the following characters of the representation for which these vectors form a basis:

T_d	E	$8C_3$	$3C_2$	$6S_4$	$6\sigma_d$
Γ_π	8	-1	0	0	0

which reduces to

$$\Gamma_\pi = E + T_1 + T_2$$

It will be noted that all eight vectors belong to the same set, since any one is interchanged with each of the others by one operation or another.

Again referring to the T_d character table, we find that the following s, p, and d orbitals belong to the irreducible representations which constitute Γ_π:

$$E: (d_{x^2-y^2}, d_{z^2})$$
$$T_1: \text{none}$$
$$T_2: (p_x, p_y, p_z) \quad \text{and} \quad (d_{xy}, d_{xz}, d_{yz})$$

Thus the first conclusion is that, of the eight physically conceivable A-B π bonds, only five may be formed when atom A has only s, p, and d orbitals at its disposal. Only one set of T_2 orbitals may be used. This set may be the three pure p orbitals, the three pure d orbitals, *or* a set of three p-d hybrids constructed from the two limiting sets.

In this case we have a particularly important example of a phenomenon already encountered, namely, the requirement of orbitals belonging to the same representation for both σ bonding and π bonding. In the situation in which there is only one orbital or one set of orbitals of the required type, we usually assume, as noted already, that the formation of σ bonds takes precedence. The present case, however, is a little more complex. As shown in Section 8.3, a set of tetrahedrally directed σ orbitals on atom A requires AO's belonging to the representations A_1 and T_2. Thus both the σ and the π sets require a group of T_2 orbitals. Since two sets of T_2 orbitals are available, each may satisfy its requirement, but it is impossible, purely on the basis of symmetry, to make any definite allocation of one set of orbitals to σ bonding and another set to π bonding. Three cases, however, may be considered. If the metal uses pure sp^3 hybrids for σ bonding, then it can use a pure d^5 set for π bonding. In a second limiting case atom A will use a pure sd^3 set for σ bonding and will then use a p^3d^2 set for π bonding. Finally, there is a whole range of inter-mediate cases in which σ orbitals are a mixture of the sp^3 and sd^3 limiting cases, and the π orbitals are a complementary mixture of the d^5 and p^3d^2 limiting cases. Symmetry considerations alone can tell us only that these are the possibilities.

8.5   Hybrid Orbitals as Linear Combinations of Atomic Orbitals

We have now seen how to answer the qualitative questions as to what atomic orbitals on the central atom A are required in order to construct the hybrid orbitals for σ and/or π bonding to the set of pendent atoms. The question which naturally arises next concerns how to write the explicit expressions for each of the hybrid orbitals, so as to state how much each AO contributes to each hybrid orbital. In this section we shall explain how to derive such expressions in a systematic way.

As an expository illustration, we shall consider a set of three σ orbitals in a planar AB$_3$ molecule (D_{3h} symmetry). We have already seen (page 204) that AO's of symmetry types A_1' and E' are required, and that the set s, p_x, p_y meets this requirement if we assume the molecule to lie in the xy plane (Figure 8.10a). We now seek the coefficients, c_{ij}, in the following LCAO expressions for the hybrid orbitals Φ_1, Φ_2, and Φ_3:

$$\Phi_1 = c_{11}s + c_{12}p_x + c_{13}p_y$$
$$\Phi_2 = c_{21}s + c_{22}p_x + c_{23}p_y$$
$$\Phi_3 = c_{31}s + c_{32}p_x + c_{33}p_y$$

The procedure to be presented for evaluating these coefficients is based on several considerations which will be explained first. The set of coefficients we seek forms a matrix, and the above set of equations can be written in matrix form:

$$\begin{bmatrix} \Phi_1 \\ \Phi_2 \\ \Phi_3 \end{bmatrix} = \begin{bmatrix} c_{11} & c_{12} & c_{13} \\ c_{21} & c_{22} & c_{23} \\ c_{31} & c_{32} & c_{33} \end{bmatrix} \begin{bmatrix} s \\ p_x \\ p_y \end{bmatrix}$$

This matrix tells us how to take a set of atomic wave functions, each belonging to a particular irreducible representation and listed in a specified order, and

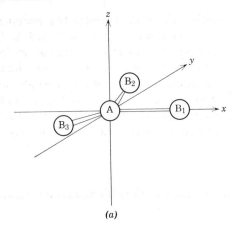

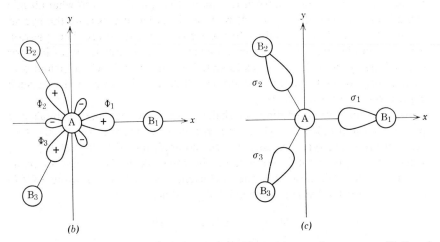

Figure 8.10   (*a*) Orientation of AB₃ molecule in Cartesian coordinate system. (*b*) Set of equivalent hybrid orbitals, Φ_1, Φ_2, Φ_3. (*c*) Set of equivalent σ orbitals on pendent atoms, σ_1, σ_2, σ_3.

combine them into a set of three equivalent functions. It should evidently be possible to carry out the inverse transformation and to express it in matrix form, viz.,

$$
\begin{bmatrix} s \\ p_x \\ p_y \end{bmatrix} = \begin{bmatrix} d_{11} & d_{12} & d_{13} \\ d_{21} & d_{22} & d_{23} \\ d_{31} & d_{32} & d_{33} \end{bmatrix} \begin{bmatrix} \Phi_1 \\ \Phi_2 \\ \Phi_3 \end{bmatrix}
$$

where the $\mathscr{D}$ matrix is the inverse of the $\mathscr{C}$ matrix. Thus, one way to determine the elements of the $\mathscr{C}$ matrix would be to have the $\mathscr{D}$ matrix and take its inverse. Taking the inverse is, in itself, a very simple process, since $\mathscr{C}$ and $\mathscr{D}$ are orthogonal matrices and thus the inverse is simply the transpose.

Now the point of the preceding discussion is that we already have a procedure for writing the $\mathscr{D}$ matrix. The $\mathscr{D}$ matrix describes the transformation of a set of three equivalent basis functions into a set of linear combinations having the symmetry of the AO's, which, in turn, have symmetry corresponding to certain irreducible representations of the molecular symmetry group. As we know, projection operators generate such symmetry-adapted linear combinations, the coefficients of which are the elements of the desired matrix. Thus, if we use the projection operator technique to transform a set of equivalent σ orbitals—either the hybrid orbitals on the central atom or the σ orbitals on the pendent atoms—into SALC's we obtain a set of coefficients which constitute the $\mathscr{D}$ matrix. We can now state and illustrate the three steps in a systematic procedure for forming hybrid orbitals.

1. *Form SALC's from the set of equivalent orbitals on the pendent atoms.* As noted above and emphasized in Figure 8.10, we could use either the hybrid orbitals on atom A or the σ orbitals on the B atoms, since their symmetry properties are the same. We choose the orbitals on the pendent atoms for two reasons. First, the application of projection operators to these is exactly as previously explained in Chapter 6. Second, we shall have further use for the SALC's later in this chapter. The results obtained are:

$$\psi_1(A_1') = \frac{1}{\sqrt{3}}(\sigma_1 + \sigma_2 + \sigma_3)$$

$$\psi_2(E_a') = \frac{1}{\sqrt{6}}(2\sigma_1 - \sigma_2 - \sigma_3)$$

$$\psi_3(E_b') = \frac{1}{\sqrt{2}}(\sigma_2 - \sigma_3)$$

2. *The matrix of the coefficients is written and its inverse taken.* The matrix is

$$\begin{bmatrix} 1/\sqrt{3} & 1/\sqrt{3} & 1/\sqrt{3} \\ 2/\sqrt{6} & -1/\sqrt{6} & -1/\sqrt{6} \\ 0 & 1/\sqrt{2} & -1/\sqrt{2} \end{bmatrix}$$

and the inverse of this matrix (its transpose) is

$$\begin{bmatrix} 1/\sqrt{3} & 2/\sqrt{6} & 0 \\ 1/\sqrt{3} & -1/\sqrt{6} & 1/\sqrt{2} \\ 1/\sqrt{3} & -1/\sqrt{6} & -1/\sqrt{2} \end{bmatrix}$$

3. *The matrix so obtained is applied to a column vector of the atomic orbitals (in the correct order of the representations to which they belong) to generate the hybrids.* We therefore write

$$
\begin{bmatrix}
1/\sqrt{3} & 2/\sqrt{6} & 0 \\
1/\sqrt{3} & -1/\sqrt{6} & 1/\sqrt{2} \\
1/\sqrt{3} & -1/\sqrt{6} & -1/\sqrt{2}
\end{bmatrix}
\begin{bmatrix}
s \\
p_x \\
p_y
\end{bmatrix}
$$

$$
=
\begin{bmatrix}
(1/\sqrt{3})\,s + (2/\sqrt{6})\,p_x \\
(1/\sqrt{3})\,s - (1/\sqrt{6})\,p_x + (1/\sqrt{2})\,p_y \\
(1/\sqrt{3})\,s - (1/\sqrt{6})\,p_x - (1/\sqrt{2})\,p_y
\end{bmatrix}
\begin{matrix}
= \Phi_1 \\
= \Phi_2 \\
= \Phi_3
\end{matrix}
$$

As a second example, which will be presented in outline form, we take the A-B σ bonds in a planar AB_4 molecule with D_{4h} symmetry. The coordinate axes and numbering are shown below.

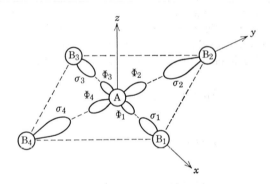

It has already been found (page 204) that the hybrid orbitals give a representation

$$\Gamma_\sigma = A_{1g} + B_{1g} + E_u$$

and that a suitable set of AO's on atom A will be as follows:

$$A_{1g}: s$$
$$B_{1g}: d_{x^2-y^2}$$
$$E_u: p_x, p_y$$

Construction of the correct linear combinations of these may now proceed by the three-step procedure.

1. *Form SALC's.* The pure rotational symmetry C_4 is sufficient. Using this, we obtain

$$\psi_A = \tfrac{1}{2}(\sigma_1 + \sigma_2 + \sigma_3 + \sigma_4)$$

$$\psi_B = \tfrac{1}{2}(\sigma_1 - \sigma_2 + \sigma_3 - \sigma_4)$$

$$\psi_{E_a} = \frac{1}{\sqrt{2}}(\sigma_1 - \sigma_3)$$

$$\psi_{E_b} = \frac{1}{\sqrt{2}}(\sigma_2 - \sigma_4)$$

2. *Form and invert the matrix of the coefficients.*

$$
\begin{bmatrix}
\tfrac{1}{2} & \tfrac{1}{2} & \tfrac{1}{2} & \tfrac{1}{2} \\
\tfrac{1}{2} & -\tfrac{1}{2} & \tfrac{1}{2} & -\tfrac{1}{2} \\
1/\sqrt{2} & 0 & -1/\sqrt{2} & 0 \\
0 & 1/\sqrt{2} & 0 & -1/\sqrt{2}
\end{bmatrix}^{-1}
=
\begin{bmatrix}
\tfrac{1}{2} & \tfrac{1}{2} & 1/\sqrt{2} & 0 \\
\tfrac{1}{2} & -\tfrac{1}{2} & 0 & 1/\sqrt{2} \\
\tfrac{1}{2} & \tfrac{1}{2} & -1/\sqrt{2} & 0 \\
\tfrac{1}{2} & -\tfrac{1}{2} & 0 & -1/\sqrt{2}
\end{bmatrix}
$$

3. *Apply the inverse matrix to a column vector of the atomic orbitals.*

$$
\begin{bmatrix}
\tfrac{1}{2} & \tfrac{1}{2} & 1/\sqrt{2} & 0 \\
\tfrac{1}{2} & -\tfrac{1}{2} & 0 & 1/\sqrt{2} \\
\tfrac{1}{2} & \tfrac{1}{2} & -1/\sqrt{2} & 0 \\
\tfrac{1}{2} & -\tfrac{1}{2} & 0 & -1/\sqrt{2}
\end{bmatrix}
\begin{bmatrix}
s \\
d_{x^2 - y^2} \\
p_x \\
p_y
\end{bmatrix}
=
$$

$$\tfrac{1}{2}(s + d_{x^2-y^2} + \sqrt{2}\,p_x) = \Phi_1$$

$$\tfrac{1}{2}(s - d_{x^2-y^2} + \sqrt{2}p_y) = \Phi_2$$

$$\tfrac{1}{2}(s + d_{x^2-y^2} - \sqrt{2}p_x) = \Phi_3$$

$$\tfrac{1}{2}(s - d_{x^2-y^2} - \sqrt{2}p_y) = \Phi_4$$

8.6   Molecular Orbital Theory for AB$_n$-Type Molecules

As in any LCAO-MO treatment, we begin with a *basis set* of atomic orbitals. For an AB$_n$-type molecule this set is usually taken to include all the valence shell orbitals on each of the atoms, since the influence of filled, low-lying inner orbitals or empty, high-lying outer orbitals is relatively small. For example, to treat BF$_3$ we would normally consider only the $2s$ and $2p$ orbitals of each of the four atoms, thus giving a basis set of sixteen orbitals.

The secular equation for the BF_3 problem must involve a determinant of order 16; symmetry factorization is therefore required to make the problem tractable, as well as to allow a clearer understanding of the bonding. In order to achieve maximum symmetry factorization, a five-step procedure is followed. First, the basis set is separated into two parts: one is the set of valence shell orbitals on the unique central atom A; the other part consists of the orbitals on the pendent atoms. Second, the orbitals on the set of pendent atoms are grouped into sets of equivalent orbitals, that is, orbitals which are interchanged with each other by symmetry operations. Third, each of the sets found in step 2 is used as a basis for a representation, and this representation is resolved into its component irreducible representations. Fourth, the basis-set orbitals are combined into SALC's corresponding to the irreducible representations found in step 3. Fifth, the valence shell orbitals of the central atom are classified according to the irreducible representations for which they form a basis. The secular determinant will then be factored into blocks, each involving AO's of A and SALC's made up of B atom orbitals, all of which belong to one irreducible representation. This procedure for setting up the secular determinant will now be illustrated for BF_3.

Step 1. For the central boron atom we have four basis orbitals, s^B, $p_x{}^B$, $p_y{}^B$, $p_z{}^B$. For each fluorine atom there are also four orbitals: s^{F1}, $p_x{}^{F1}$, $p_y{}^{F1}$, $p_z{}^{F1}$, s^{F2}, $p_x{}^{F2}$, ..., $p_y{}^{F3}$, $p_z{}^{F3}$. The arrangement of these basis orbitals is specified by the coordinate axes in the following diagram.

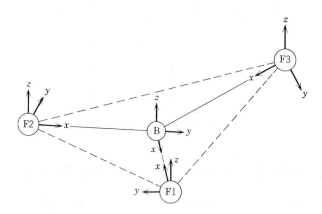

Step 2. The symmetry of the BF_3 molecule is D_{3h}. Inspection shows that the operations in this group interchange basis orbitals only within the following four sets and never interchange orbitals of different sets:

(1) s^{F1}, s^{F2}, s^{F3} ($s\sigma$ orbitals)
(2) p_x^{F1}, p_x^{F2}, p_x^{F3} ($p\sigma$ orbitals)
(3) p_y^{F1}, p_y^{F2}, p_y^{F3} ($\| p\pi$ orbitals)
(4) p_z^{F1}, p_z^{F2}, p_z^{F3} ($\perp p\pi$ orbitals)

Step 3. Each of the above sets gives a reducible representation which reduces as indicated:

$$\Gamma_{s\sigma} = A_1' + E'$$
$$\Gamma_{p\sigma} = A_1' + E'$$
$$\Gamma_{\| p\pi} = A_2' + E'$$
$$\Gamma_{\perp p\pi} = A_2'' + E''$$

Step 4. In applying projection operators to form SALC's, it is expeditious to use the uniaxial pure rotational subgroup C_3, as explained in Section 6.3, where the SALC's for a set of orbitals identical to the three $\perp p\pi$ orbitals (p_z's) involved here were constructed. The SALC's for the other orbital sets can be constructed in the same way. Actually, a little thought will show that they must have the same form in every case. Thus, we have

$$\psi_A = \frac{1}{\sqrt{3}}(\phi_1 + \phi_2 + \phi_3)$$

$$\psi_{Ea} = \frac{1}{\sqrt{6}}(2\phi_1 - \phi_2 - \phi_3)$$

$$\psi_{Eb} = \frac{1}{\sqrt{2}}(\phi_2 - \phi_3)$$

where A represents A_1', A_2' or A_2'', E represents E' or E'', and ϕ_1, ϕ_2, ϕ_3 stand for the appropriate set of AO's on the fluorine atoms.

Step 5. By examination of the D_{3h} character table we find that the valence shell orbitals of the central boron atom transform as follows:

$$A_1': s^B$$
$$A_2'': p_z^B$$
$$E': p_x^B, p_y^B$$

It can now be seen that the secular equation will be factored into the following blocks:

A_1' block (3 × 3): $\psi_A^{s\sigma}$, $\psi_A^{p\sigma}$, s^B
A_2' block (1 × 1): $\psi_A^{\| p\pi}$
two E' blocks (4 × 4): $\psi_{Ea}^{s\sigma}$, $\psi_{Ea}^{p\sigma}$, $\psi_{Ea}^{\| p\pi}$, p_x^B
                          $\psi_{Eb}^{s\sigma}$, $\psi_{Eb}^{p\sigma}$, $\psi_{Eb}^{\| p\pi}$, p_y^B
A_2'' block (2 × 2): $\psi_A^{\perp p\pi}$, p_z^B
two E'' blocks (1 × 1): $\psi_{Ea}^{\perp p\pi}$
                          $\psi_{Eb}^{\perp p\pi}$

This is as far as symmetry considerations alone can take us. We now have the following set of secular equations to solve for the energies of the MO's:

$$A_1': \begin{vmatrix} H_{s^B, s^B} - E & H_{s^B, s\sigma} & H_{s^B, p\sigma} \\ H_{s^B, s\sigma} & H_{s\sigma, s\sigma} - E & H_{s\sigma, p\sigma} \\ H_{s^B, p\sigma} & H_{s\sigma, p\sigma} & H_{p\sigma, p\sigma} - E \end{vmatrix} = 0$$

$$A_2': E = H_{\|p\pi, \|p\pi}$$

$$E': \begin{vmatrix} H_{p_x{}^B, p_x{}^B} - E & H_{p_x{}^B, s\sigma} & H_{p_x{}^B, p\sigma} & H_{p_x{}^B, \|p\pi} \\ H_{p_x{}^B, s\sigma} & H_{s\sigma, s\sigma} - E & H_{s\sigma, p\sigma} & H_{s\sigma, \|p\pi} \\ H_{p_x{}^B, p\sigma} & H_{s\sigma, p\sigma} & H_{p\sigma, p\sigma} - E & H_{p\sigma, \|p\pi} \\ H_{p_x{}^B, \|p\pi} & H_{s\sigma, \|p\pi} & H_{p\sigma, \|p\pi} & H_{\|p\pi, \|p\pi} - E \end{vmatrix} = 0$$

$$A_2'': \begin{vmatrix} H_{p_z{}^B, p_z{}^B} - E & H_{p_z{}^B, \perp p\pi} \\ H_{p_z{}^B, \perp p\pi} & H_{\perp p\pi, \perp p\pi} - E \end{vmatrix} = 0$$

$$E'': E = H_{\perp p\pi, \perp p\pi}$$

There is a total of only 21 distinct matrix elements to evaluate instead of $(16^2 + 16)/2 = 136$ if there had been no symmetry factorization. This is clearly going to save much time and labor, but evaluation of even 21 matrix elements can be laborious, depending on the accuracy desired.

Various approximation schemes are employed to evaluate the matrix elements. In the crudest ones, overlap of orbitals on nonbonded atoms is set equal to zero. In the present case this would eliminate four more distinct matrix elements altogether, viz.,

$$A_1' \text{ block:} \int \psi_A{}^{s\sigma} \mathscr{H} \psi_A{}^{p\sigma} \, d\tau$$

$$E' \text{ block:} \int \psi_E{}^{s\sigma} \mathscr{H} \psi_E{}^{p\sigma} \, d\tau$$

$$\int \psi_E{}^{s\sigma} \mathscr{H} \psi_E{}^{\|p\pi} \, d\tau$$

$$\int \psi_E{}^{p\sigma} \mathscr{H} \psi_E{}^{\|p\pi} \, d\tau$$

Even when no matrix elements are arbitrarily set equal to zero, semi-empirical rules are used for their evaluation. The commonest such rules, which may be varied in minor ways, or subjected to various small "improvements," are:

1. Each diagonal matrix element is directly related, though the expressions for the SALC, with or without considering overlap, to the "spectroscopic" energies of the atomic orbitals constituting the SALC in question.

2. The off-diagonal elements are made proportional to overlap, in some plain or fancy way.

Figure 8.11 shows an orbital energy level diagram for BF$_3$ based on a calculation using the two empirical rules just mentioned. Some of the details of this calculation are given in Appendix VI.

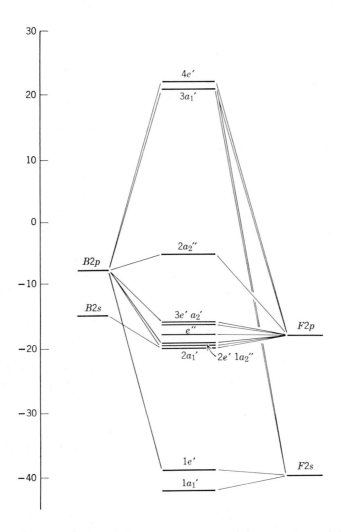

Figure 8.11   An orbital energy level diagram for BF$_3$ based on a simple "extended Hückel" calculation, described in Appendix VI. The 24 electrons present (neglecting 1s electrons) just fill the MO's up to the 3e' orbital.

A final word of warning must be given here. The type of calculation we are considering deals with one-electron orbitals. It takes no *explicit* account of electron exchange or interelectronic repulsions, both of which make *major* contributions to the true total energy of the electron configuration. In the empirical schemes, *implicit* allowance is made for these effects mainly in the way in which the numerical values of the off-diagonal matrix elements are chosen. This whole approach is certainly a " ruthless approximation " and hence the results cannot be taken literally as to fine details.

8.7   The Relationship of the Molecular Orbital and the Hybridization Treatments

The thoughtful student of the earlier sections of this chapter will doubtless have realized that the hybridization and MO treatments are intimately related because of their common reliance on the symmetry of the molecule. The purpose of this section is to comment directly on the relationship between them. The BF_3 molecule provides a suitable illustration.

Let us consider first the σ bonds. The hydridization approach *emphasizes* the equivalence of the three B-F bonds and makes an *explicit* statement as to the central atom s and p character in them; it does *not*, however, provide any direct way to compute the energy of these bonds. The MO treatment only *implicitly* describes the three B—F σ interactions as equivalent and assigns central atom s and p orbitals participation in correct proportions; it *does*, however, afford a straightforward approach to computing the energy of interaction.

In the hybridization approach we think of three electron pairs forming the set of equivalent, localized σ bonds. Although not mentioned as yet, there are also three more electron pairs occupying σ orbitals on the fluorine atoms which are directed away from the central boron atom. An intuitively appealing, and common, way of specifying more precisely how this total of six electron pairs is distributed is to suppose that each fluorine atom forms sp hybrid orbitals, one directed toward the boron atom and the other away from it. Each of the former hybrids overlaps one of the sp^2 hybrid orbitals of the boron atom, and a pair of electrons is then localized in the region of overlap to form a σ bond. Each of the sp hybrids directed away from the boron atom then contains a " lone pair " of electrons.

As Figure 8.11 shows, the s^B, $p_x^{\ B}$, $p_y^{\ B}$, s^{F1}, s^{F2}, s^{F3}, $p_x^{\ F1}$, $p_x^{\ F2}$, and $p_x^{\ F3}$ orbitals, that is, the same set of orbitals discussed in the preceding paragraph as responsible for localized σ bonds and localized σ "lone pairs," gives rise to the A_1', and E' MO's. It is seen that one A_1' and one E' MO are strongly bonding and are occupied by a total of three electron pairs. These correspond to the three pairs of localized σ bonding electrons in the hybridization picture. There are, then, three more or less nonbonding electron pairs occupying the next A'_1 and E' orbitals. These correspond to the lone pairs in the hybridization picture.

The correspondences just mentioned show that the hybridization and MO treatments, despite their diverse starting points and contradictory ways of conceptualizing the bonding, end up saying qualitatively the same thing about it. They just convey the information in different ways, each of which may have advantages for certain purposes. The hybridation approach is conceptually simple and emphasizes directly the equivalence of the bonds. The MO treatment is more flexible and affords a mechanism for making calculations.

Turning now to the π bonding in BF$_3$, we note first that the MO approach shows that in-plane π bonding is not a practical possibility. The set of in-plane π orbitals on the fluorine atoms, the p_y^{F1}, p_y^{F2}, and p_y^{F3} orbitals, form SALC's belonging to the A'_2 and E' representations. The boron atom has no valence shell orbital of A'_2 symmetry, and its p_x and p_y orbitals, which do have E' symmetry, are able to overlap with the fluorine s and p_x orbitals much better than with the p_y orbitals. The kind of calculations which lead to Figure 8.11 support this conclusion. The hybridization treatment leads of course, to the same result, since it is *assumed* that s^B, p_x^B, and p_y^B are fully employed in forming the σ hybrids, leaving no in-plane orbital to form π bonds. It should be stressed, however, that neither approach necessarily excludes in-plane π interaction rigorously; rather, in each case, the pre-eminence of σ bonding is assumed or indicated.

Finally, we come to the out-of-plane or $\perp$ π bonding. In the valence bond approach, with sp^2 hybridization assumed as the first step, only one central atom orbital, p_z^B, remains. This has an equal possibility of overlapping with any one of the p_z^F orbitals, and the notion of a resonance hybrid has to be invoked. Resonance among the three equivalent structures, as shown below, then leads to an average of one-third of a π bond per B $\cdots$ F pair:

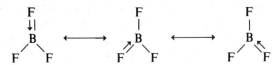

The MO approach arrived at the same conclusion by virtue of the fact that the set of $\perp p\pi$ orbitals, p_z^{F1}, p_z^{F2}, and p_z^{F3}, forms SALC's of symmetries A''_2 and E''. Since there are no orbitals of the latter symmetry on the boron atom, the E'' SALC's are, in themselves, a degenerate pair of nonbonding MO's. The A''_2 SALC interacts with the p_z^B orbital, which has A''_2 symmetry, to form one bonding and one antibonding $\perp$ π MO, to which three fluorine p_z orbitals contribute equally.

In closing, one more point of comparison must be mentioned. An MO description of the electronic structure of the *ground state* of a molecule is fundamentally equivalent to the description provided by the valence bond treatment, employing the concept of hybridization, and is not *qualitatively* superior. Wherever spectra or other phenomena involving excited states are concerned, however, there is generally a distinct qualitative advantage to the MO approach because the valence bond theory does not explicitly consider any empty orbitals, thus excluding all the antibonding ones and often some nonbonding ones as well. The MO theory on the other hand, explicitly deals with all the orbitals which can be made from the basis set, whether they are bonding, nonbonding, or antibonding, full or empty.

8.8 Molecular Orbitals for Regular Octahedral and Tetrahedral Molecules

Regular octahedral, AB_6, and regular tetrahedral, AB_4, molecules or complex ions have special importance because they occur often, especially in transition metal chemistry, but even more because they are prototypes to which many molecules and complexes of lower symmetry can be referred. We present here in summary form the symmetry analysis leading to an MO calculation for each of these cases. The steps followed are the same five explained in Section 8.6 and illustrated there for BF_3. Verification of the results at each stage will form a very useful exercise for the student.

Octahedral AB_6

Step 1. We shall assume that the central atom, A has s, p, and d orbitals in its valence shell and that each B atom has s and p orbitals in its valence shell. There are then thirty-three orbitals in the basis set: nine for the central atom and twenty-four for the set of six equivalent B atoms.

Step 2. The symmetry is O_h. The operations of this group sort the ligand atom orbitals into three sets, and we employ Figure 8.12 to define atom numbers and local coordinates.

Set 1: the six s orbitals

Set 2: the six p_z orbitals

Set 3: the twelve p_x and p_y orbitals

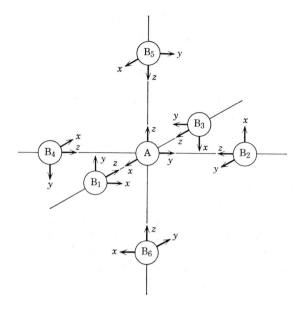

Figure 8.12 Coordinate system for an octahedral AB$_6$ molecule or complex ion.

Step 3. Each of the above sets gives a reducible representation which reduces as indicated.

$$\text{Set 1: } \Gamma_{s\sigma} = A_{1g} + E_g + T_{1u}$$

$$\text{Set 2: } \Gamma_{p\sigma} = A_{1g} + E_g + T_{1u}$$

$$\text{Set 3: } \Gamma_{p\pi} = T_{1g} + T_{2g} + T_{1u} + T_{2u}$$

Step 4. The derivation of the complete set of SALC's is not trivial in this case because of the degeneracies and because of the fact that no one pure rotational subgroup exists which breaks all of these down into distinct sets of complex, one-dimensional representations. However, a little consideration shows that it would be foolish to plunge ahead, with grim determination and an ample supply of paper, pencils and patience, to form all of the possible SALC's. First, we shall look ahead to step 5 and see what representations the valence shell orbitals span, since it is only for these that A-B interactions are possible and hence only these that we need to know for most purposes.

Step 5. Inspection of the O_h character table reveals directly that the s, p and d orbitals span representations as follows:

$$s: A_{1g}$$
$$d_{x^2-y^2}, d_{z^2}: E_g$$
$$p_x, p_y, p_z: T_{1u}$$
$$d_{xy}, d_{xz}, d_{yz}: T_{2g}$$

Step 4 Revisited. We now form only the useful SALC's. The A_{1g} SALC's can be formed systematically by projection operators and are

$$\psi_{A_{1g}} = \frac{1}{\sqrt{6}}(\phi_1 + \phi_2 + \phi_3 + \phi_4 + \phi_5 + \phi_6) \quad \phi_i = s^i \text{ or } p_z^{\,i}$$

The easiest procedure now is to combine the ligand basis orbitals, taking those of each set separately, into the required E_g, T_{1u}, and T_{2g} orbitals by inspection, so as to make them match, on a one-to-one basis, the central atom orbitals with which they are to interact. For example, the $s\sigma$ and $p\sigma$ T_{1u} sets must consist of $\phi_1 - \phi_3$, $\phi_2 - \phi_4$, and $\phi_5 - \phi_6$ to match the central atom p_x, p_y, and p_z orbitals, repectively. Proceeding in this way and normalizing the SALC's we obtain:

$$\psi_{E_g} = \begin{cases} \dfrac{1}{\sqrt{12}}(2\phi_5 + 2\phi_6 - \phi_1 - \phi_2 - \phi_3 - \phi_4) \\[2mm] \tfrac{1}{2}(\phi_1 - \phi_2 + \phi_3 - \phi_4) \end{cases} \text{matching} \begin{cases} d_{z^2} \\ d_{x^2-y^2} \end{cases}$$

$$\psi_{T_{1u}} = \cdot \begin{cases} \dfrac{1}{\sqrt{2}}(\phi_1 - \phi_3) \\[2mm] \dfrac{1}{\sqrt{2}}(\phi_2 - \phi_4) \\[2mm] \dfrac{1}{\sqrt{2}}(\phi_5 - \phi_6) \end{cases} \cdot \text{matching} \begin{cases} p_x \\ p_y \\ p_z \end{cases}$$

$$\psi_{T_{2g}} = \begin{cases} \tfrac{1}{2}(p_y^{\,1} + p_x^{\,5} + p_x^{\,3} + p_y^{\,6}) \\[2mm] \tfrac{1}{2}(p_x^{\,2} + p_y^{\,5} + p_y^{\,4} + p_x^{\,6}) \\[2mm] \tfrac{1}{2}(p_x^{\,1} + p_y^{\,2} + p_y^{\,3} + p_x^{\,4}) \end{cases} \text{matching} \begin{cases} d_{xz} \\ d_{yz} \\ d_{xy} \end{cases}$$

The T_{1u} SALC's consisting of $p\pi$ orbitals have been omitted, since calculations generally show that the central atom p orbitals interact with them only weakly, as compared to their interactions with T_{1u} SALC's made of σ-type ligand orbitals.

An approximate MO diagram of the usual type is shown in Figure 8.13. Obviously, the details depend on the relative energies of the various AO's;

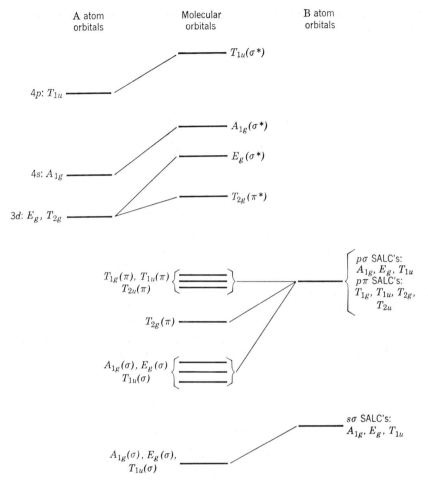

Figure 8.13.   An approximate MO diagram for an octahedral AB$_6$, molecule or ion, where A is a $+2$ or $+3$ ion from the first transition series and the B's are F, O, or Cl atoms.

the diagram shown assumes relationships typical for a central atom in the first transition series in a normal oxidation state (i.e., $+2$ or $+3$) and light ligand atoms such as fluorine or oxygen.

Tetrahedral AB$_4$

Step 1. Again the central atom will use s, p, and d orbitals, and for the B atoms we shall use s and p σ-type orbitals and two $p\pi$ orbitals on each, giving a basis set of twenty-five orbitals. These are defined as shown in Figure 8.14.

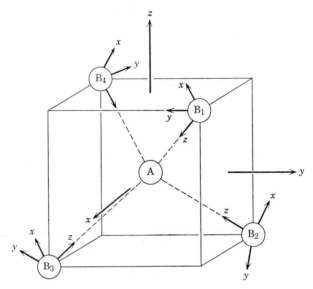

Figure 8.14   Coordinate system for a tetrahedral AB_4 molecule or complex ion. The x vector on each B atom lies in a plane which also contains the z axis of atom A.

Step 2. The operations of the group T_d sort the ligand atom orbitals into the following sets:

$$\text{Set 1: the four } s\sigma \text{ orbitals}$$
$$\text{Set 2: the four } p\sigma \text{ orbitals}$$
$$\text{Set 3: the eight } \pi \text{ orbitals}$$

Step 3. The representations given by each of these sets are

$$\text{Sets 1, 2: } \Gamma_\sigma = A_1 + T_2$$
$$\text{Set 3: } \Gamma_\pi = E + T_1 + T_2$$

Step 4. The SALC's are:

$$\psi_{A_1} = \tfrac{1}{2}(\phi_1 + \phi_2 + \phi_3 + \phi_4) \quad \text{matching } s^A, \text{ where } \phi_i = p_z{}^i \text{ or } s^i$$

$$\psi_E = \begin{cases} \tfrac{1}{2}(p_x{}^1 - p_x{}^2 - p_x{}^3 + p_x{}^4) \\ \tfrac{1}{2}(p_y{}^1 - p_y{}^2 - p_y{}^3 + p_y{}^4) \end{cases} \quad \text{matching } \begin{cases} d_{z^2} \\ d_{x^2 - y^2} \end{cases}$$

$$\psi_{T_2(\sigma)} = \begin{cases} \tfrac{1}{2}(\phi^1 - \phi^2 + \phi^3 - \phi^4) \\ \tfrac{1}{2}(\phi^1 + \phi^2 - \phi^3 - \phi^4) \\ \tfrac{1}{2}(\phi^1 - \phi^2 - \phi^3 + \phi^4) \end{cases} \quad \text{matching } \begin{pmatrix} p_x \\ p_y \\ p_z \end{pmatrix}$$

$$\psi_{T_2(\pi)} = \begin{cases} \frac{1}{4}(p_x^{\ 1} + p_x^{\ 2} - p_x^{\ 3} - p_x^{\ 4}) \\[4pt] \quad + \frac{\sqrt{3}}{4}(-p_y^{\ 1} - p_y^{\ 2} + p_y^{\ 3} + p_y^{\ 4}) \\[4pt] \frac{1}{4}(p_x^{\ 1} - p_x^{\ 2} + p_x^{\ 3} - p_x^{\ 4}) \\[4pt] \quad + \frac{\sqrt{3}}{4}(p_y^{\ 1} - p_y^{\ 2} + p_y^{\ 3} - p_y^{\ 4}) \\[4pt] -\frac{1}{2}(p_x^{\ 1} + p_x^{\ 2} + p_x^{\ 3} + p_x^{\ 4}) \end{cases} \quad \text{matching} \quad \begin{cases} d_{yz} \\ d_{xz} \\ d_{xy} \end{cases}$$

Step 5. The valence shell orbitals of the central atom belong to the following representations:

$$s: A_1$$
$$d_{z^2}, d_{x^2-y^2}: E$$
$$p_x, p_y, p_z: T_2$$
$$d_{xy}, d_{xz}, d_{yz}: T_2$$

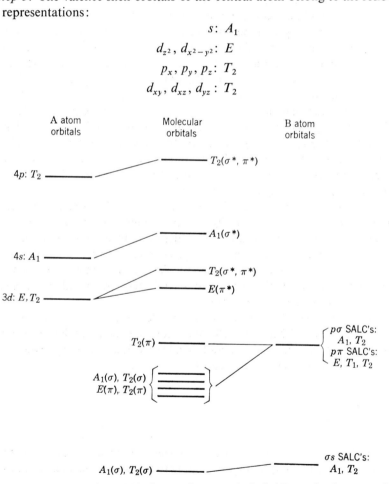

Figure 8.15   An approximate MO diagram for a tetrahedral, AB$_4$, molecule or complex, where A is a $+2$ ion from the first transition series and the B's are O or Cl atoms.

An energy level diagram based on parameters typical for a first transition series M^{2+} ion and Cl^- ions is given in Figure 8.15.

8.9   Molecular Orbitals for Metal Sandwich Compounds

The term "metal sandwich compounds" applies strictly to compounds of the type $(C_nH_n)_2M$, such as $(C_5H_5)_2Fe$ and $(C_6H_6)_2Cr$, in which a metal atom is "sandwiched" symmetrically between two parallel carbocyclic ring systems, but is commonly used in a broader sense to include, in addition, all compounds, in which at least one carbocyclic ring, C_nH_n, such as C_4H_4, C_5H_5, C_6H_6, C_7H_7, is bound to a metal atom in such a way that the metal atom lies along the n-fold symmetry axis of the ring and is thus equivalently bonded to all of the carbon atoms in the ring. Thus, in addition to the highly symmetrical molecules mentioned above, the term refers also to such mono-ring compounds as C_5H_5NiNO, $C_6H_6Cr(CO)_3$, $[C_7H_7Mo(CO)_3]^+$, and $C_5H_5Fe(CO)_2C_2H_5$, and such mixed-ring systems as $(C_5H_5)(C_7H_7)V$ and $(C_5H_5)(C_6H_6)Mn$. It should be noted that there are also compounds, such as $(C_5H_5)_2MoH_2$, in which the rings are not exactly parallel, although it is believed that the metal-ring bonding is still symmetrical about the symmetry axis of each ring. The bonding in such cases will have essentially the same features as in the more symmetrical molecules but cannot of course be treated with the same degree of rigor with regard to symmetry.

A metal sandwich compound does not strictly fit our previous concept of an AB_n-type molecule, since the ligand atoms interact strongly with each other as well as with the central atom. It is desirable to extend the discussion to these molecules, however, since they provide clear and important examples of how to treat the situation in which the ligands in a complex are themselves polyatomic entities with an internal set of MO's perturbed by interaction with the AO's of the central atom.

Using ferrocene, $(C_5H_5)_2Fe$, as an example, we can demonstrate all of the basic ideas in the MO treatment for the whole class of molecules. Accordingly, we will first treat ferrocene in detail then briefly outline the application of the method to a few other selected cases.

Ferrocene

The basic strategy is to construct linear combinations of all of the $p\pi$ orbitals of the two C_5H_5 rings belonging to the irreducible representations of the molecular point group D_{5d},* to classify the orbitals in the valence shell of the metal atom according to their symmetry in the point group, and then to combine metal and ring orbitals into molecular orbitals of the entire molecule. The important difference from what has been done earlier in treating AB_n molecules arises from the strong interaction of the various pendent atoms among themselves. Because of this, it is best to first work out the "local" MO's and their energies and then formulate the A $\cdots$ B interactions in terms of the established local MO's for groups of ligand atoms—the C_5H_5 rings, in this case—rather than between A and each B as an individual atom.

In order to construct the proper linear combinations of $p\pi$ orbitals, we can make use of the results we have already obtained for a single C_5H_5 ring. For such a ring we have constructed LCAO-MO's transforming correctly under the rotations belonging to the group C_5 (see page 142). These are of A, E_1, and E_2 symmetry. The set of ten $p\pi$ orbitals provided by two such rings oriented as they are in the ferrocene molecule spans the following representations of D_{5d}:

D_{5d}	E	$2C_5$	$2C_5^2$	$5C_2$	i	$2S_{10}$	$2S_{10}^3$	$5\sigma_d$
Γ_π	10	0	0	0	0	0	0	2

$$\Gamma_\pi = A_{1g} + A_{2u} + E_{1g} + E_{1u} + E_{2g} + E_{2u}$$

Thus we see that for the system of two rings we require two A orbitals, one symmetric and the other antisymmetric to inversion in the center, two E_1 orbitals, one symmetric and the other antisymmetric, and finally, two E_2 orbitals, one symmetric and one antisymmetric to inversion in the center. It is rather easy to write down expressions for these by making appropriate combinations of the orbitals that we already have for the individual rings. In doing this, we will refer to the rings and orbitals as they are shown and labeled in Figure 8.16. It is very important here to note that we have chosen the directions of the $p\pi$ orbitals such that all of their positive lobes point in toward the metal atom. Thus the $+z$ axis for ring 1 is in the opposite direction to the $+z$ axis for ring 2.

Referring to the character table for group D_{5d}, we see that an A_{1g} orbital

---

* The energy difference between the staggered (D_{5d}) and eclipsed (D_{5h}) rotomers of ferrocene is apparently very small (≤ 1 kcal/mole), with the latter being perhaps the more stable [cf. R. K. Bohn and A. Haaland, *J. Organomet. Chem.*, **5**, 470 (1966)]. The two symmetries are equally suitable and convenient for a discussion of bonding. Since most of the research literature has used D_{5d}, we make the same choice here.

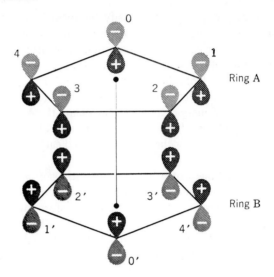

Figure 8.16   Sketch showing the $p\pi$ orbitals on the two rings used to construct MO's for
a bis(cyclopentadienyl)metal molecule.

must be symmetric to inversion in the center of symmetry. This requirement
will be satisfied by the following combination of the A orbitals of the two
rings:

$$\psi(A_{1g}) = \frac{1}{\sqrt{2}} [\psi_1(A) + \psi_2(A)]$$

where we use subscripts 1 and 2 to refer to the rings. It can also easily be seen
that the $\psi(A_{1g})$ so obtained satisfies all other symmetry requirements. To
obtain an orbital of the system of two rings which is antisymmetric to inversion we take the linear combination

$$\psi(A_{2u}) = \frac{1}{\sqrt{2}} [\psi_1(A) - \psi_2(A)]$$

Again, a simple check will show that this orbital, $\psi(A_{2u})$, satisfies all the
symmetry requirements of the A_{2u} representation.

For the E_1 and E_2 orbitals we proceed in exactly the same way, choosing
normalized combinations of the E_1 and E_2 orbitals of the individual rings so
as to obtain functions which are symmetric and antisymmetric to inversion,
namely,

$$\left| \psi(E_{1g}a) = \frac{1}{\sqrt{2}} [\psi_1(E_1a) + \psi_2(E_1a)] \right.$$

$$\left| \psi(E_{1g}b) = \frac{1}{\sqrt{2}} [\psi_1(E_1b) + \psi_2(E_1b)] \right.$$

$$\left| \psi(E_{1u}a) = \frac{1}{\sqrt{2}} [\psi_1(E_1a) - \psi_2(E_1a)] \right.$$

$$\left| \psi(E_{1u}b) = \frac{1}{\sqrt{2}} [\psi_1(E_1b) - \psi_2(E_1b)] \right.$$

$$\left| \psi(E_{2g}a) = \frac{1}{\sqrt{2}} [\psi_1(E_2a) + \psi_2(E_2a)] \right.$$

$$\left| \psi(E_{2g}b) = \frac{1}{\sqrt{2}} [\psi_1(E_2b) + \psi_2(E_2b)] \right.$$

$$\left| \psi(E_{2u}a) = \frac{1}{\sqrt{2}} [\psi_1(E_2a) - \psi_2(E_2a)] \right.$$

$$\left| \psi(E_{2u}b) = \frac{1}{\sqrt{2}} [\psi_1(E_2b) - \psi_2(E_2b)] \right.$$

For the metal atom, iron, the valence shell orbitals are the five $3d$ orbitals, the $4s$ orbital, and the three $4p$ orbitals. The transformation properties of these orbitals may be ascertained immediately by inspection of the character table for D_{5d}, the results being

$$A_{1g}: \quad 4s, 3d_{z^2}$$
$$E_{1g}: \quad (3d_{xz}, 3d_{yz})$$
$$E_{2g}: \quad (3d_{xy}, 3d_{x^2-y^2})$$
$$A_{2u}: \quad 4p_z$$
$$E_{1u}: \quad (4p_x, 4p_y)$$

Thus we have a total, counting the degeneracies, of nineteen orbitals, but because of their symmetry properties, as shown in Section 7.1, we do not have to solve a 19 × 19 secular determinant. Instead we have only the following small determinants:

One 3 × 3 for the A_{1g} MO's
Two 2 × 2 for the E_{1g} MO's (same roots for both)
Two 2 × 2 for the E_{2g} MO's (same roots for both)
One 2 × 2 for the A_{2u} MO's
Two 2 × 2 for the E_{1u} MO's (same roots for both)

The E_{2u} MO's on the rings are in themselves E_{2u} MO's for the whole molecule, since there are no E_{2u} metal orbitals with which they might interact.

The problem is now reduced to one of evaluating the matrix elements. We shall not discuss the details of this process; the reading list, Appendix IX, cites several such calculations. Figure 8.17 shows the energy level diagram for ferrocene produced by one of them.

The energies of the ring orbitals relative to one another have already been estimated in the Hückel approximation in units of β (page 142); as explained in Appendix IV, we choose here the "spectroscopic" value of β, ~ 60 kcal/mole, for use in constructing the energy level diagram. Moreover, since the rings are about 4 Å apart in the molecule, the reasonable assumption is made that there is no significant direct interaction between them; the g and u orbitals of the same rotational symmetry are thus taken to have the same energy.

It will be seen that there are nine more or less bonding orbitals, which are just filled by the eighteen electrons originating in the π systems of the rings and in the valence shell orbitals of the metal atoms. The exact ordering of the levels varies in some respects from one calculation to another, depending on the approximations used in evaluating the matrix elements. For other $(C_5H_5)_2 M$ compounds, energy level diagrams having the same qualitative features would be anticipated, but the relative order of the least stable bonding MO's is subject to change because of variation in both the relative energies of metal and ligand orbitals and the relative magnitudes of the different interaction energies. Consequently, some caution must be exercised in attempting to predict from a diagram constructed specifically for one $(C_5H_5)_2M$ compound the electronic structure of another containing a different metal.

Dibenzenechromium

If this molecule is assumed to consist of two benzene rings placed on either side of the chromium atom so that their planes are parallel and their C_6 axes colinear, the molecular symmetry may be D_{6d} or D_{6h}, depending on whether the rings are staggered or eclipsed. X-ray study of the crystalline compound shows that the chromium atom is at a center of inversion so that in the crystalline state at least the molecular point group is D_{6h}. There is evidence from infrared and Raman spectra that the molecule has D_{6h} symmetry in solution.

Assuming D_{6h} symmetry we can easily develop an MO bonding scheme similar to that given above for ferrocene. Again we begin with the LCAO-MO π orbitals of a single ring and combine them to obtain symmetry orbitals appropriate to the entire molecule. Using the twelve carbon $p\pi$ orbitals as the

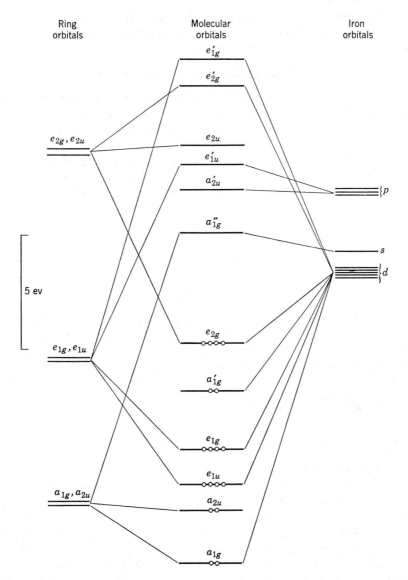

Figure 8.17 An energy level diagram for ferrocene. The MO energies are those calculated by Shustorovich and Dyatkina (*loc. cit*), using a self-consistent field procedure. The positions of the ring and iron orbitals on this diagram are only approximate.

basis for a representation of the group D_{6h}, and breaking this down into the irreducible representations, we find that the symmetry orbitals must belong to the following irreducible representations:

$$A_{1g}, A_{2u}, B_{2g}, B_{1u}, E_{1g}, E_{1u}, E_{2g}, E_{2u}$$

By combining the A, B, E_1, and E_2 orbitals for the individual rings into combinations which are symmetric and antisymmetric to inversion, we readily obtain expressions for symmetry orbitals of the required types. The relative energies of these symmetry orbitals in units of β (again using the spectroscopic value) are those already calculated (page 136) for benzene.

For the metal atom we find from the character table for the group D_{6h} that its valence shell orbitals belong to irreducible representations as follows:

$$A_{1g}: \quad s, d_{z^2}$$
$$E_{1g}: \quad (d_{xz}, d_{yz})$$
$$E_{2g}: \quad (d_{xy}, d_{x^2-y^2})$$
$$A_{2u}: \quad P_z$$
$$E_{1u}: \quad (p_x, p_y)$$

Figure 8.18 shows an energy level diagram for dibenzenechromium. It will be noted how similar the MO scheme is to that in ferrocene. This is not too surprising in view of the fact that the only qualitative difference between the two systems is in the presence of high energy orbitals of B symmetry in C_6H_6, which, however, do not participate in the ring-metal interactions since the metal has no valence shell orbitals of B symmetry.

Cyclopentadienylmanganese Tricarbonyl

This molecule provides a significant example of the type in which the true or overall molecular symmetry is very low but the bonding in parts of the molecule may be treated, at least qualitatively, in terms of relatively high *local* symmetries. In this case the $(C_5H_5)Mn$ part of the molecule may be considered to have C_{5v} symmetry and the $Mn(CO)_3$ part taken as having C_{3v} symmetry, although the molecule in its entirety can have no more than C_s symmetry, and that only for two particular orientations of the C_5H_5 ring relative to the $Mn(CO)_3$ grouping. In treating the $(C_5H_5)Mn$ and $Mn(CO)_3$ bonding separately, each in terms of its own ideal local symmetry, we make the assumption, *inter alia*, that degeneracies permitted in C_{5v} symmetry will not be greatly split by the presence of C_{3v} symmetry in the other part of the molecule and vice versa. Because of the particular shapes of the d orbitals such an assumption probably has some validity in this case, but it cannot always be taken as true.

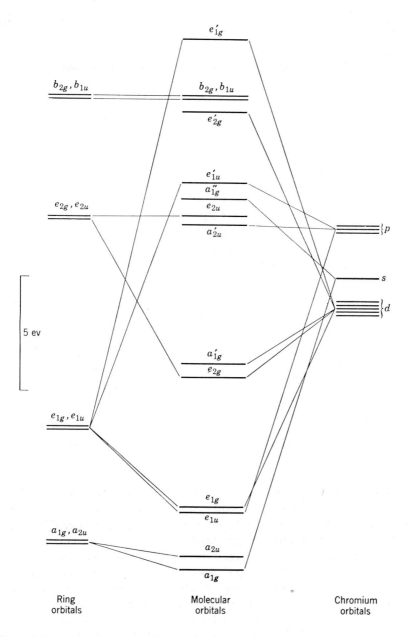

Figure 8.18   An energy level diagram for dibenzenechromium. The positions of the ring and chromium orbitals on this diagram are only approximate.

Figure 8.19 shows a schematic energy level diagram for $(C_5H_5)Mn(CO)_3$. In the center the valence shell orbitals of manganese are shown. They are labeled on the right with their symmetries in C_3 and on the left with their symmetries in C_5. At the extreme left are the π MO's of the C_5H_5 ring,

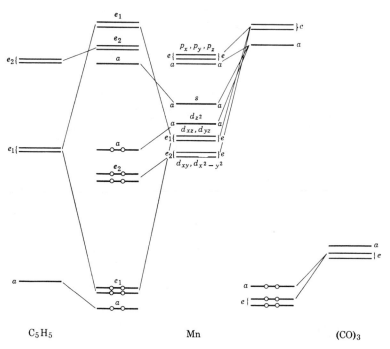

C_5H_5 Mn $(CO)_3$

Figure 8.19   A schematic MO energy level diagram for $(C_5H_5)Mn(CO)_3$.

labeled with their symmetries in C_5 and with the energy of the e_1 orbitals placed at about the same level as the d orbitals of manganese. On the extreme right are shown the energy levels of the σ orbitals for the three CO groups, with symmetry designations appropriate to C_3. These designations were obtained by taking the set of three σ orbitals on the carbon atoms as a basis for a representation of the group C_3 and decomposing the representation into its component irreducible representations. The energy of these symmetry orbitals constructed from the carbon σ orbitals has been assumed to be about the same as the energy of the a orbital of C_5H_5. No account has been taken of the interaction of the metal with the π orbitals of the CO groups. This could be done by finding the representations spanned by the six π orbitals and then permitting them to interact with metal orbitals of the same symmetry types. Although it is certain from the vibrational frequencies of the CO groups in this molecule that the M—C π interactions are substantial, we have chosen to omit them here, since they can be only crudely estimated and their inclusion in the energy level diagram would make it extremely unwieldy.

The only real justification for drawing an entirely schematic diagram such as this one is that it helps with the "bookkeeping." The diagram makes it somewhat easier to see how the orbitals of different components of the molecule may interact in order to produce a satisfactory set of bonding MO's than is possible by merely inspecting a list of these components. For example, the shapes of the metal e_1 orbitals, d_{xz} and d_{yz}, are such that their overlap with the ring e_1 orbitals should be fairly substantial and the diagram accordingly shows a sizable interaction of these orbitals. On the other hand, the metal e_2 orbitals are not strongly directed toward the ring e_2 orbitals and the energy difference is initially greater, so that a much weaker interaction would be expected, as indicated.

Cyclobutadiene Sandwich Compounds

As shown earlier (page 141), cyclobutadiene with a square ring would have two unpaired electrons, occupying E orbitals, and no net delocalization energy of stabilization. Therefore, C_4H_4 does not exist as a discrete molecule in this geometry. However, the possibility that overlap between the E orbitals of square C_4H_4 and the E orbitals of a metal atom could result in stabilization of the square configuration by formation of a strong delocalized M-C_4H_4 bond was recognized by theorists in 1956.* This idea was suggested, in part, by the pre-eminence of the E-E overlaps in stabilizing M-C_5H_5 bonds.

Since 1956 a number of presumed cyclobutadiene complexes have been prepared, and several, for example, $(C_6H_5C)_4Fe(CO)_3$ and $[(CH_3C)_4NiCl_2]_2$, shown in Figure 8.20, have been shown by X-ray study to be genuine, with local C_{4v} symmetry in the $M(RC)_4$ portions of the molecules.

In the case of the nickel compound each nickel atom is bound to two chlorine atoms which bridge the two nickel atoms and also to a third, non-bridging atom. These three chlorine atoms form a roughly triangular array, and the structure can be idealized by thinking of the nickel atom as sandwiched between the parallel planes of the C_4 ring and the Cl_3 triangle, the ring and the triangle having colinear four-and threefold axes passing through the nickel atom. Similarly, in the iron compound, the $Fe(CO)_3$ moiety is at least approximately a triangular pyramid with its threefold axis colinear with the fourfold axis of the cyclobutadiene ring.

---

* H. C. Longuet-Higgins and L. E. Orgel, *J. Chem. Soc.*, 1969 (1956).

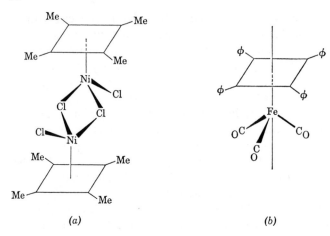

(a)   (b)

Figure 8.20   The structure of (a) tetramethylcyclobutadienenickel(II) chloride dimer and (b) tetraphenylcyclobutadieneiron tricarbonyl.

Thus, as in the $(C_5H_5)Mn(CO)_3$ molecule, there are no true symmetry axes in these molecules in their entirety but as an approximation we may think in terms of the local symmetry of the two halves. In Figure 8.21 we show in the center the orbitals of the metal atom, giving on the right their

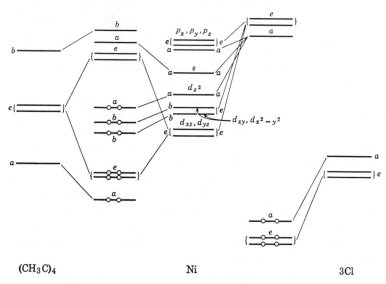

(CH₃C)₄   Ni   3Cl

Figure 8.21   Schematic MO energy level diagram for the bonding to one nickel atom in tetramethylcyclobutadienenickel dichloride dimer, $[(CH_3C)_4NiCl_2]_2$.

symmetries in C_3 and on the left their symmetries in C_4. It is seen that metal orbitals are available to form σ bonds to the three chlorine atoms or CO groups, leaving e orbitals which may interact with the e orbitals of the cyclobutadiene ring to create strong metal-ring bonding and leave no unshared electrons.

Exercises

8.1 Consider a trigonal prismatic AB$_6$ molecule. With s, p, and d orbitals available on atom A, what hydridization schemes are possible for σ bonds and π bonds?

8.2 Which s, p, and d orbitals of the central atom can be used to form σ hybrid orbitals in an AB$_8$ species having a square antiprism structure?

8.3 Give a qualitative analysis of the bonding to be expected in C$_4$H$_6$Fe (CO)$_3$, where C$_4$H$_6$ is trimethylenemethane, $(CH_2)_3C$, with intrinsic three-fold symmetry.

8.4 Carry out the symmetry analysis required to set up a factored secular equation for an MO calculation on the following molecules or complexes: PtCl$_4^{2-}$, PF$_5$, and Re$_2$Cl$_8^{2-}$.

Ligand Field Theory

9.1 Introductory Remarks

For conciseness, the title of this chapter is simply "Ligand Field Theory." However, many of the principles which will be developed are as much a part of crystal field theory and the molecular orbital theory of transition metal complexes as they are of ligand field theory. Indeed the three theories are very closely related, and hence it seems advisable to begin this chapter with a brief, historically oriented discussion of the nature of these theories.

The beginning of all three theories can be traced to the year 1929, when Hans Bethe published his classic paper entitled "Splitting of Terms in Crystals."[*] There are really two completely separate parts to Bethe's paper. The first is concerned purely with the *qualitative consequences of the symmetry* of the surroundings of a cation in a crystal lattice. In this part, Bethe showed that, in general, the states arising from a particular electronic configuration of an ion which are degenerate when the ion is free of perturbing influences must break up into two or more nonequivalent states when the ion is introduced into a lattice. He showed how it is possible, using the methods of group theory, to determine just what states will result when an ion of any given electronic configuration is introduced into a crystalline environment of definite symmetry.

The second part of Bethe's paper describes a method by which the magnitudes of the splittings of the free-ion states may be calculated, assuming that the surroundings effect these splittings by purely electrostatic forces. This assumption that all interactions between the ion and its surroundings may be treated as electrostatic interactions between point charges is the defining feature of the crystal field theory. It has the consequence that all electrons which are in metal-ion orbitals in the free ion are treated as though they remain in orbitals which are 100 per cent metal-ion orbitals.

It was first pointed out by Van Vleck that the symmetry part of Bethe's

---

[*] H. Bethe, *Ann. Physik*, **3**, 133–206 (1929). An English translation of this paper is available from Consultants Bureau Enterprises, Inc., 227 W. 17 Street, New York, N.Y. 10011.

approach will remain entirely valid if we change the computational part from a purely electrostatic approach to one which admits the existence of some chemical bonding between the metal ion and its neighbors. If we do so, the orbitals with which we must deal will be no longer pure metal orbitals, but only partly metal orbitals. This means that in principle we cannot write down the same fairly simple expressions for energies, because the orbitals involved are no longer simple. In practice, however, if the covalence of the metal ligand bonds is relatively small, energies will be given by equations identical in form to the equations derived in the crystal field theory. This modified crystal field theory, which admits that there is some covalent as well as electrostatic interaction between the ion and its neighbors, is called ligand field theory.

Van Vleck also pointed out that even in the case of very highly covalent bonding [as in, e.g., $Ni(CO)_4$ or $Fe(CN)_6{}^{4-}$], which is best treated by using molecular orbital theory, the symmetry properties and requirements remain exactly the same as for the crystal field model and the ligand field model.

Thus, in order to gain an understanding of any one of these theories, the same symmetry considerations are required at the outset.

9.2 Electronic Structures of Free Atoms and Ions

We intend in this chapter to consider the manner in which the symmetry of the chemical surroundings of an ion determines the effect of this environment on the energy levels of the ion. In the crystal field and ligand field theories we often wish to regard the effect of the environment as a small perturbation on the states of the free ion. For the benefit of readers not acquainted with certain general features of the electronic structures of free atoms and ions, a brief résumé of the subject is given in this section.

Wave Functions and Quantum Numbers for a Single Electron

The wave function Ψ for a single electron, in a hydrogen atom for example, may be written as a product of four factors. These are the radial function $R(r)$, which is dependent only on the radial distance r from the nucleus; two angular functions $\Theta(\theta)$ and $\Phi(\phi)$, which depend only on the angles θ and ϕ (cf. Figure 8.1); and a spin function ψ_s, which is independent of the spatial coordinates r, θ, and ϕ. Thus we write

$$\Psi = R(r) \cdot \Theta(\theta) \cdot \Phi(\phi) \cdot \psi_s \qquad (9.2\text{-}1)$$

This overall wave function and each of its factors separately have a parametric dependence on certain quantities called quantum numbers, of which there are four: n, l, m, s.

The principal quantum number, n, takes all integral values from 1 to infinity. It determines the nature of the radial part, $R(r)$, of the wave function only.

The quantum number l occurs in the $\Theta(\theta)$ factor of the wave function. It may be thought of as representing the angular momentum of the electron, in units of $h/2\pi$, due to its orbital motion, and we shall call it the orbital momentum quantum number.* It may take all values 0, 1, 2, ..., $n-1$, where n is the principal quantum number. Thus, in the first principal shell, there exist only wave functions with $l = 0$; in the second shell there are wave functions with $l = 0$ and 1; and so on. For historical reasons, letter symbols are given to orbitals according to the value of l, as shown in the following scheme:

$$l = 0 \quad 1 \quad 2 \quad 3 \quad 4 \quad 5 \quad 6 \quad \cdots$$
$$\text{Letter symbol:} \; s \quad p \quad d \quad f \quad g \quad h \quad i \quad \cdots$$

If we continue to take the classical view of an electron as a discrete charged particle having angular momentum due to its orbital motion, we must also conclude that because of its charge its orbital motion will generate a magnetic dipole. The vector representing this magnetic dipole will be colinear with the vector representing its angular momentum (both are perpendicular to the plane of the orbit), and the value of this orbital magnetic dipole, μ_l, is directly proportional to the angular momentum.

The quantum number m occurs in both the $\Theta(\theta)$ and $\Phi(\phi)$ parts of the wave function. It indicates the tilt of the plane of orbital motion with respect to some reference direction. It can take all integral values from l to $-l$, or $2l + 1$ values in all. Its relation to the tilt of the orbital plane may be stated more explicitly as follows. If the plane is perpendicular to the reference direction, the length of the projection of the vector representing l on the reference line is numerically equal to l. The next largest angle of tilt permitted by quantum mechanics is such that the length of the projection equals $l-1$, the next such that the length of the projection is $l-2$, and so on, until the value of $-l$ is reached. The quantum number m is simply the length of the projection of l on the reference line. The situation is illustrated for the case of $l = 2$ in the sketch below.

* To be precise, the total orbital angular momentum is not $l(h/2\pi)$ but $\sqrt{l(l+1)}(h/2\pi)$, but this point need not concern us here.

$m = l = 2$ $m = 1$ $m = 0$

$m = -1$ $m = -l = -2$

For given n and $l = 0$ there is only one possible orbital, namely, one with $m = 0$. Thus there is only one s orbital of each principal quantum shell. For given n and $l = 1$ there are three possible m values. Hence each principal shell has three different p orbitals. Similarly, d orbitals come in sets of five, f orbitals in sets of seven, and so on. In the absence of any external forces, the energy of an orbital is independent of its m value. Hence all three np orbitals, all five nd orbitals, and so on, are of the same energy.

The electron spin quantum number s is the number on which ψ_s depends, and it may take only the values $+\frac{1}{2}$ and $-\frac{1}{2}$. It may be interpreted, classically, as a measure of the spin angular momentum of the electron, that is, as a measure of angular momentum due to the rotation of the electron about its own axis. We may think of the electron as intrinsically having spin angular momentum equal to $\frac{1}{2}(h/2\pi)$* and consider that this may be oriented with respect to a reference direction so as to produce components of $+\frac{1}{2}$ or $-\frac{1}{2}$ along the reference direction. Again, however, in the framework of this classical picture, we must also expect that, if a charged body is rotating, a magnetic dipole is generated. Thus every electron has associated with it a spin magnetic dipole, μ_s, the direction of which depends on s.

Although these four quantum numbers are always sufficient to specify completely the state of an electron, there is another quantum number, j, which is useful in accounting for the energy of the state. In units of $h/2\pi$, j gives the total angular momentum of the electron, which is a vector sum of the orbital angular momentum and the spin angular momentum. Quantum mechanics requires, however, that the vector sum can be made only in certain

* Again, strictly $\sqrt{\frac{1}{2}(\frac{1}{2} + 1)}(h/2\pi)$.

ways. The value of j may be either $l + \frac{1}{2}$ or $l - \frac{1}{2}$. So long as we regard the spin wave function, ψ_s, as entirely independent of the orbital wave function, $R(r)\Theta(\theta)\Phi(\phi)$, these two j states have the same energy.

Actually, spin and orbital magnetic moments do interact, so that the state with $j = l - \frac{1}{2}$ is of lower energy than the state with $j = l + \frac{1}{2}$. This would be expected classically, since in the $j = l - \frac{1}{2}$ state the orbital and spin moments are opposed. In the hydrogen atom, in all hydrogen-like ions, and in all atoms and ions having one electron outside of a closed core, the splittings due to this phenomenon of spin-orbit coupling are very small compared to the energy differences between orbitals differing in their l values. Thus the effect of spin-orbit coupling in such cases is justifiably regarded as a small perturbation on an energy level pattern which is basically determined only by the values of n and l.

Quantum Numbers for Many-Electron Atoms

Although various general cases come under this heading, we need consider only one: that in which most of the electrons in the atom or ion are in closed shells and the others are in the *same* partly filled shell. The closed shells are spherically symmetric, and the only effect they have on the other electrons is to diminish the strength of the nuclear attraction for these electrons. This means that the wave function for the one electron in a partly filled shell containing only a single electron will have the same angular functions, $\Theta(\theta)$ and $\Phi(\phi)$, as it would if this electron were the only one in the atom, but its radial function will be different according to the "effective" nuclear charge which it feels.

To a first approximation each of several electrons in such a partly filled shell may be assigned its own private set of one-electron quantum numbers, n, l, m, and s. However, there are always fairly strong interactions among these electrons which make this approximation unrealistic. In general the nature of these interactions is not easy to describe, but the behavior of real atoms often approximates closely to a limiting situation called the L-S or Russell-Saunders coupling scheme.

In L-S or Russell-Saunders coupling, a quantum number L, which gives the total orbital angular momentum of all the electrons, and a quantum number S, which gives the total spin angular momentum of all the electrons, are used. Note the use of capital letters for quantum numbers characteristic of the entire configuration, in contrast to the use of lower-case letters for quantum numbers of individual electrons. This practice is carried further in assigning letter symbols to states of different L. We have the following scheme, which is completely analogous to the scheme for single electrons:

$$L = 0 \quad 1 \quad 2 \quad 3 \quad 4 \quad 5 \quad 6 \quad \cdots$$

Letter symbol: $S \quad P \quad D \quad F \quad G \quad H \quad I \quad \cdots$

In the Russell-Saunders coupling scheme it is assumed that the angular momenta of the individual electrons are coupled to one another and the spins of the individual electrons are also coupled to one another to give, respectively, the L and S values for the configuration. Consider, for example, an atom which has, outside of closed shells, a $3d$ electron and a $4d$ electron. Each of these has an l value of 2 and an s value of $\frac{1}{2}$. Now the two l vectors may combine to give integral vector sums as shown in Figure 9.1, where l_1

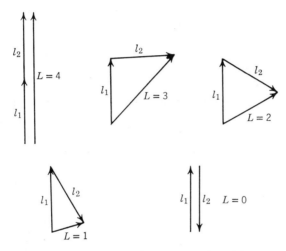

Figure 9.1　Integral vector sums, L, of two vectors, l_1 and l_2, each of length 2.

refers to the $3d$ electron and l_2 to the $4d$ electron, or vice versa. Thus L values of 0, 1, 2, 3, 4 are possible and hence states symbolized S, P, D, F, and G may arise from two nonequivalent d electrons. By "nonequivalent" we mean differing in the principal shell to which they belong. Similarly, the s vectors may combine to give S values of 1 or 0. Since the electrons differ in their principal quantum number, any L value may be combined with any S value without violating the exclusion principle. Thus there are altogether ten states or, as they are called, *terms*, which may arise when two d electrons are present but are in different principal shells.

With a single electron the total angular momentum, j, is the vector sum of l and s. Here there are only two possibilities, namely, $j = l + s = l + \frac{1}{2}$ and $j = l - s = l - \frac{1}{2}$. More generally, the total angular momentum, denoted J for a multielectron configuration, may take all of the values $L + S, L + S - 1,$

$L + S - 2, \ldots, |L - S|$, or altogether $2S + 1$ different values. The number $2S + 1$ is called the multiplicity of a term, and it is placed as a left superscript to the term symbol. Just as with one electron, states with the same value of L and S but different values of J will differ somewhat in energy. If these differences in J value are important and must be specified, this is done by putting the J value as a right subscript to the term symbol. Thus all of the different terms, including those differing in their J values, for an $ndmd$ configuration are:

$$
\begin{array}{ccccc}
{}^1S_0 & {}^1P_1 & {}^1D_2 & {}^1F_3 & {}^1G_4 \\
{}^3S_1 & {}^3P_0 & {}^3D_1 & {}^3F_2 & {}^3G_3 \\
 & {}^3P_1 & {}^3D_2 & {}^3F_3 & {}^3G_4 \\
 & {}^3P_2 & {}^3D_3 & {}^3F_4 & {}^3G_5
\end{array}
$$

For the ${}^3S$ state the triplet character is not actually realized because, when $L = 0$, J can only equal S. There is no finite value of L with which S may combine vectorially.

Now that we have seen how two d electrons in different principal shells couple to give terms in the Russell-Saunders or L-S coupling scheme, let us turn to the more directly interesting problem of what terms may arise when the two d electrons belong to the *same* principal shell. Straightforward but lengthy procedures for making certain that we do not violate the exclusion principle are necessary, since now we cannot count on a difference in n values to prevent this.* It is found that for a d^2 configuration only the following states may arise:

$$
{}^1S_0 \quad {}^3P_{0,1,2} \quad {}^1D_2 \quad {}^3F_{2,3,4} \quad {}^1G_4
$$

This L-S coupling scheme may be considered a useful approximation when the components of a given multiplet term, that is, states with the same S and L values but different J values, differ in energy by amounts which are small compared to the differences between one multiplet term as a whole and another. Among the transition elements, to which we wish to apply ligand field theory, the L-S coupling scheme works well enough for many purposes in the ions of elements in the first and second transition series. In the third transition series, however, it is not a very good approximation, although it can serve as a starting point for more elaborate treatments. In general, the adequacy of the L-S coupling approximation diminishes steadily as the atomic number increases. For the actinide elements it is of no use at all.

For convenience, the states which may arise by Russell-Saunders coupling

---

* These considerations are treated in many books on atomic structure. See, for example, H. E. White's *Introduction to Atomic Spectra*, McGraw-Hill, New York, 1934, Sections 12.1 and 13.11.

from all d^n configurations are listed in Table 9.1. For quantitative applications of ligand field theory we must know not only the nature of the states but also their relative energies in a particular ion. For a great number of the ions of practical interest these energies are known from experimental measurements. The standard tabulation of such data is C. E. Moore's "Atomic Energy Levels."*

Table 9.1   States for d^n Systems in Russell-Saunders Coupling

d^1		$^2(D)$					
d^2	$^1(S, D, G)$		$^3(P, F)$				
d^3		$^2(D)$		$^2(P, D, F, G, H)$	$^4(P, F)$		
d^4	$^1(S, D, G)$		$^3(P, F)$	$^1(S, D, F, G, I)$	$^3(P, D, F, G, H)$	$^5(D)$	
d^5		$^2(D)$		$^2(P, D, F, G, H)$	$^4(P, F)$ $^2(S, D, F, G, I)$ $^4(D, G)$ $^6(S)$		
d^6	Same as d^4						
d^7	Same as d^3						
d^8	Same as d^2						
d^9	Same as d^1						
d^{10}	$^1(S)$						

9.3   Splitting of Levels and Terms in a Chemical Environment

We may use the set of five d wave functions as a basis for a representation of the point group of a particular environment and thus determine the manner in which the set of d orbitals is split by this environment. Let us choose an octahedral environment for our first illustration. In order to determine the representation for which the set of d wave functions forms a basis, we must first find the elements of the matrices which express the effect upon the set of wave functions of each of the symmetry operations in the group; the characters of these matrices will then be the characters of the representation we are seeking.

Although the full symmetry of the octahedron is O_h, we can gain all required information about the d orbitals by using only the pure rotational

* *Circular* 467, National Bureau of Standards, for sale by the Superintendent of Documents, U.S. Government Printing Office, Washington 25, D. C., Volume I, Hydrogen-Vanadium; Volume II, Chromium-Niobium; Volume III, Molybdenum-Lanthanum and Hafnium-Actinium.

subgroup O because O_h may be obtained from O by adding the inversion, i. However, we already know that d orbitals are even to inversion, so that it is only the pure rotational operations of O which bring us new information.

We assume that the wave functions of a set of d orbitals are each of the general form specified by Equation 9.2-1. We shall further assume that the spin function, ψ_s, is entirely independent of the orbital functions and shall pay no further attention to it for the present. Since the radial function, $R(r)$, involves no directional variables, it is invariant to all operations in a point group and need concern us no further. The function $\Theta(\theta)$ depends only upon the angle θ. Therefore, if all rotations are carried out about the axis from which θ is measured (the z axis in Figure 8.1), $\Theta(\theta)$ will also be invariant. Thus, by always choosing the axes of rotation in this way (or, in other words, always quantizing the orbitals about the axis of rotation), only the function $\Phi(\phi)$ will be altered by rotations. The explicit form of the $\Phi(\phi)$ function, aside from a normalizing constant, is

$$\Phi(\phi) = e^{im\phi} \tag{9.3-1}$$

and the five d orbitals are those in which m takes the values $l, l-1, \ldots, 0, \ldots, 1-l, -l$, namely, 2, 1, 0, -1, -2.

If we take the function $e^{im\phi}$ and rotate by an angle α, the function becomes $e^{im(\phi+\alpha)}$. Thus we can easily see that the set of $\Phi(\phi)$ wave functions, I, becomes II on rotation by α:

$$
\begin{bmatrix}
e^{2i\phi} \\
e^{i\phi} \\
e^{0} \\
e^{-i\phi} \\
e^{-2i\phi}
\end{bmatrix}
\xrightarrow[\text{by }\alpha]{\text{rotation}}
\begin{bmatrix}
e^{2i(\phi+\alpha)} \\
e^{i(\phi+\alpha)} \\
e^{0} \\
e^{-i(\phi+\alpha)} \\
e^{-2i(\phi+\alpha)}
\end{bmatrix}
$$

$$\qquad \text{I} \qquad\qquad\qquad\qquad \text{II}$$

The matrix necessary to produce this transformation is

$$
\begin{bmatrix}
e^{2i\alpha} & 0 & 0 & 0 & 0 \\
0 & e^{i\alpha} & 0 & 0 & 0 \\
0 & 0 & e^{0} & 0 & 0 \\
0 & 0 & 0 & e^{-i\alpha} & 0 \\
0 & 0 & 0 & 0 & e^{-2i\alpha}
\end{bmatrix}
$$

This five-dimensional matrix is only a special case for a set of d functions, and clearly in the $(2l+1)$-fold set of functions ($l=0$ for an s level, 1 for a p level, 3 for an f level, and so on) we shall have

$$\begin{bmatrix} e^{li\alpha} & 0 & \cdots & \cdot & 0 \\ \cdot & e^{(l-1)i\alpha} & \cdots & \cdot & \cdot \\ \cdot & \cdot & \cdots & \cdot & \cdot \\ \cdot & \cdot & \cdots & \cdot & \cdot \\ \cdot & \cdot & \cdots & e^{(1-l)i\alpha} & 0 \\ 0 & 0 & \cdots & \cdot & e^{-li\alpha} \end{bmatrix}$$

The sum of the diagonal elements, $\chi(\alpha)$, can be shown to be*

$$\chi(\alpha) = \frac{\sin(l + \tfrac{1}{2})\alpha}{\sin \alpha/2} \quad (\alpha \neq 0) \tag{9.3-2}$$

We now have the necessary formula to determine the characters of the representation we seek. Let us proceed to work them out.

For a twofold rotation, $\alpha = \pi$ and hence

$$\chi(C_2) = \frac{\sin 5\pi/2}{\sin \pi/2} = \frac{1}{1} = 1$$

Similarly for the threefold and fourfold rotations:

$$\chi(C_3) = \frac{\sin 5\pi/3}{\sin \pi/3} = \frac{-\sin \pi/3}{\sin \pi/3} = -1$$

$$\chi(C_4) = \frac{\sin 5\pi/4}{\sin \pi/4} = -1$$

The general formula above is inapplicable if $\alpha = 0$; however, it is obvious that in this case each diagonal element is equal to 1 and the character is equal, in the general case, to $2l + 1$; in the present instance, $\chi(E) = 5$. Referring to the character table for the group O and using the methods developed in Chapter 4, we easily see that the representation we have derived is reducible to $E + T_2$. In the group O_h we will have, since the d wave functions are inherently g in their inversion property:

$$\Gamma_d = E_g + T_{2g}$$

Thus we have shown that the set of five d wave functions, degenerate in the free atom or ion (or, more precisely, under conditions of spherical symmetry) does not remain degenerate when the atom or ion is placed in an environment with O_h symmetry. The wave functions are split into a triply degenerate set, T_{2g}, and a doubly degenerate set, E_g.

---

* The proof of this will not be given here, but is suggested as a useful exercise. The quantities being summed form a geometric progression.

Table 9.2   Splitting of One-Electron Levels in an Octahedral Environment

TYPE OF LEVEL	l	$\chi(E)$	$\chi(C_2)$	$\chi(C_3)$	$\chi(C_4)$	IRREDUCIBLE REPRESENTATIONS SPANNED
s	0	1	1	1	1	A_{1g}
p	1	3	-1	0	1	T_{1u}
d	2	5	1	-1	-1	$E_g + T_{2g}$
f	3	7	-1	1	-1	$A_{2u} + T_{1u} + T_{2u}$
g	4	9	1	0	1	$A_{1g} + E_g + T_{1g} + T_{2g}$
h	5	11	-1	-1	1	$E_u + 2T_{1u} + T_{2u}$
i	6	13	1	1	-1	$A_{1g} + A_{2g} + E_g + T_{1g} + 2T_{2g}$

It is easy to apply the same treatment to electrons in other types of orbitals than d orbitals. The results obtained are collected in Table 9.2. It will be seen that an s orbital is totally symmetric in the O_h environment. The set of p orbitals remains unsplit, transforming as t_{1u}; this same conclusion could have been obtained directly from the O_h character table, where it is seen that (x, y, z) form a basis for the t_{1u} representation of O_h. All orbitals with higher values of the quantum number l, however, are split into two or more sets; this must be so, since the group O_h cannot allow any state to be more than threefold degenerate.

In a similar manner, we could determine the splitting of various sets of orbitals in environments of other symmetries which we may encounter in complexes, such as T_d, D_{4h}, D_{2d}, and C_{2v}, and indeed for any sort of symmetry we may encounter. An alternative and simpler way of obtaining this

Table 9.3   Splitting of One-Electron

TYPE OF LEVEL	O_h	T_d
SYMMETRY OF		
s	a_{1g}	a_1
p	t_{1u}	t_2
d	$e_g + t_{2g}$	$e + t_2$
f	$a_{2u} + t_{1u} + t_{2u}$	$a_2 + t_1 + t_2$
g	$a_{1g} + e_g + t_{1g} + t_{2g}$	$a_1 + e + t_1 + t_2$
h	$e_u + 2t_{1u} + t_{2u}$	$e + t_1 + 2t_2$
i	$a_{1g} + a_{2g} + e_g + t_{1g} + 2t_{2g}$	$a_1 + a_2 + e + t_1 + 2t_2$

information is to use the results we have obtained for the octahedral case in conjunction with the correlation table given in Appendix IIIB. In Table 9.3 are the results for a few point groups of particular interest.

The results we have obtained so far for single electrons in various types of orbitals apply also to the behavior of terms arising from groups of electrons. For example, just as a single d electron in a free atom has a wave function which belongs to a fivefold degenerate set corresponding to the five values which m may take in the $\Phi(\phi)$ factor of the wave function, so a D state arising from any group of electrons has a completely analogous fivefold degeneracy because of the five values which the quantum number M may take. Moreover, the splitting of a D term will be just the same as the splitting of the set of one-electron d orbitals. This is so because the $\Phi(\phi)$ factor of the wave function for a D term is $e^{iM\phi}$ in exact analogy to the $\Phi(\phi)$ factor, $e^{im\phi}$, in the wave function for a single d electron. Exactly the same relationship exists between f orbitals and F states, p orbitals and P states, and so on. Thus all of the results given in Table 9.2 for the splitting of various sets of one-electron orbitals apply to the splitting of analogous Russell-Saunders terms.

In Table 9.3 we have used small letters to represent the states for a single electron in the environments of various symmetries, corresponding with the use of the small letters, $s, p, d, f, \ldots$ to represent their states in the free atom. Similarly, we shall use capital letters to represent the states into which the environment splits terms of the free ion. Thus, for example, an F state of a free ion will be split into the states A_2, T_1, and T_2 when the ion is placed in the center of a tetrahedral environment.

In Table 9.3 the use of subscripts g and u is governed by the following rules. If the point group of the environment has no center of symmetry, then no subscripts are used, since they cannot have any meaning. When the

Levels in Various Symmetries

ENVIRONMENT

D_{4h}	D_3	D_{2d}
a_{1g}	a_1	a_1
$a_{2u} + e_u$	$a_2 + e$	$b_2 + e$
$a_{1g} + b_{1g} + b_{2g} + e_g$	$a_1 + 2e$	$a_1 + b_1 + b_2 + e$
$a_{2u} + b_{1u} + b_{2u} + 2e_u$	$a_1 + 2a_2 + 2e$	$a_1 + a_2 + b_2 + 2e$
$2a_{1g} + a_{2g} + b_{1g} + b_{2g} + 2e_g$	$2a_1 + a_2 + 3e$	$2a_1 + a_2 + b_1 + b_2 + 2e$
$a_{1u} + 2a_{2u} + b_{1u} + b_{2u} + 3e_u$	$a_1 + 2a_2 + 4e$	$a_1 + a_2 + b_1 + 2b_2 + 3e$
$2a_{1g} + a_{2g} + 2b_{1g} + 2b_{2g} + 3e_g$	$3a_1 + 2a_2 + 4e$	$2a_1 + a_2 + 2b_1 + 2b_2 + 3e$

environment does have a center of symmetry, the subscripts are determined by the type of orbital, all AO's for which the quantum number l is even $(s, d, g, \ldots)$ being centrosymmetric and hence of g character, and all AO's for which l is odd $(p, f, h, \ldots)$ being antisymmetric to inversion and thus of u character. In using Table 9.3 for term splittings the following rules apply. Again, if the environment does not have a center of symmetry, the g and u subscripts are inapplicable. For point groups having a center to which the inversion operation may be referred, the g or u character will be determined by the nature of the one-electron wave functions of the individual electrons making up the configuration from which the term is derived. We shall be interested only in terms derived from d^n configurations, and all of these will give g states in point groups possessing a center of symmetry.

One other point needs to be mentioned regarding the splitting of terms of the free ion in chemical environments, and this concerns the spin multiplicity. The chemical environment does not interact directly with the electron spins, and thus all of the states into which a particular term is split have the same spin multiplicity as the parent term.

In order to illustrate the splitting of terms of a d^n configuration, the states for a d^2 ion in several point groups are shown below. The free-ion terms have been given on page 248.

FREE-ION TERMS	STATES IN POINT GROUPS				
	O_h		T_d		D_{4h}
^1S	$^1A_{1g}$		$^1A_1$		$^1A_{1g}$
^1G	$^1A_{1g}$ $\quad ^1T_{2g}$		$^1A_1$ $\quad ^1T_2$		$2\,^1A_{1g}$ $\quad ^1B_{2g}$
	^1E_g		^1E		$^1A_{2g}$ $\quad 2\,^1E_g$
	$^1T_{1g}$		$^1T_1$		$^1B_{1g}$
^3P	$^3T_{1g}$		$^3T_1$		$^3A_{2g}$
					^3E_g
^1D	^1E_g		^1E		$^1A_{1g}$ $\quad ^1E_g$
	$^1T_{2g}$		$^1T_2$		$^1B_{1g}$
					$^1B_{2g}$
^3F	$^3A_{2g}$		$^3A_2$		$^3A_{2g}$ $\quad 2\,^3E_g$
	$^3T_{1g}$		$^3T_1$		$^3B_{1g}$
	$^3T_{2g}$		$^3T_2$		$^3B_{2g}$

9.4   Construction of Energy Level Diagrams

We have seen in Section 9.3 that all free-ion terms having $L > 1$ are split by chemical environments of symmetry O_h, T_d, or lower symmetries into two or more states which we label according to the representation of the point group describing their transformation properties. We now turn to the question of what the relative energies of these states are and how these energies depend on the strength of the chemical interaction of the ion with its surroundings. Obviously, these energies can be straightforwardly calculated by setting up and solving the requisite secular equations. It is also possible, however, to obtain a great deal of information about the energies, especially the relative energies, *almost* entirely by use of arguments based on the symmetry properties of the states, and this is the subject of this section.

Of course, from symmetry arguments *alone*, quantitative information on energies cannot be obtained. The procedures we are about to describe require one piece of quantitative information obtainable only by a calculation. In Section 9.5 we shall show how this item of information is obtained, but for the present we will accept it without proof and proceed to the construction of energy level diagrams.

It will be demonstrated in Section 9.5 that the relative energies of the doubly and triply degenerate sets of d orbitals into which the set of five d orbitals is split in a tetrahedral or octahedral environment are as shown in Figure 9.2. Thus, when there is a single d electron in an ion in an octahedral environment, it will occupy one of the t_{2g} orbitals and the energy required to promote it to an e_g orbital is Δ_0; for the same ion in a tetrahedral environment, the electron will occupy an e orbital and the energy required to promote it to a t_2 orbital will be Δ_t.

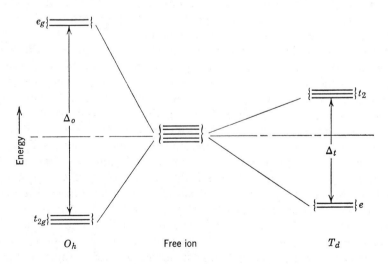

Figure 9.2   Diagram showing the relative energies of e and t_2 orbitals resulting from the splitting of the set of d orbitals by octahedral and tetrahedral environments.

The energy level diagram we wish to construct will show how the energies of the various states into which the free-ion terms are split depend on the strength of the interaction of the ion with its environment. The separation of the two sets of orbitals into which the group of five d orbitals is split can be taken as our measure of this interaction. Thus our diagram will have the magnitude of Δ_0 or Δ_t as abscissa and energy as ordinate. At the extreme left, where Δ_0 or Δ_t is zero, we shall have the free-ion term energies. At the right side of the diagram we shall have the energies of states which will exist when the interaction produces such a great separation of the e and t_2 orbitals that the energies due to interelectronic interactions become negligible by comparison.

We will now explain the method of constructing an energy level diagram by treating the particular case of a d^2 ion in an octahedral environment. For this system the free-ion terms, in order of increasing energy, are

$$^3F \quad ^1D \quad ^3P \quad ^1G \quad ^1S$$

In the limit of an extremely large splitting of the d orbitals, the following three configurations, in order of increasing energy, will be possible:

$$t_{2g}{}^2 \quad t_{2g}e_g \quad e_g{}^2$$

The usage of the symbols here is the same as that for showing free-atom configurations. Thus these are (1) the configuration in which both electrons are in t_{2g} orbitals, $t_{2g}{}^2$; (2) the configuration in which one electron is in a t_{2g} orbital while the other is in an e_g orbital, $t_2 e_g$; and (3) the highest energy configuration, $e_g{}^2$, in which both electrons are in e_g orbitals. The energy increase from one of these to the next highest is Δ_0.

Now let us consider what will happen as we begin to relax the strong interaction of the environment with the ion, so that the electrons start to feel one another's presence. They will begin to couple in certain ways, giving rise to a set of states of the entire configuration. The symmetry properties of these states can be determined by taking the direct products of the representations of the single electrons. Thus for the configuration $t_{2g}{}^2$ we take the direct product $t_{2g} \times t_{2g}$, and then decompose it into $A_{1g} + E_g + T_{1g} + T_{2g}$. Similarly the direct product $t_{2g} \times e_g$ gives $T_{1g} + T_{2g}$, and the direct product $e_g{}^2$ gives $A_{1g} + A_{2g} + E_g$. These are the symmetries of the orbital states produced by the interaction of the electrons. However, we have yet to determine the spin multiplicities of these states. Clearly, with two electrons involved, they must be either singlets or triplets, and we must be careful to observe any restrictions placed on the multiplicities by the exclusion principle.

Later in this section we shall describe a rigorous and completely general method of determining the multiplicities of the strong-field states, but for the present we shall work out the d^2 case by means of a less elegant but instructive method. Consider first the configuration $t_{2g}{}^2$. We may regard the t_{2g} levels as a set of six boxes as shown below:

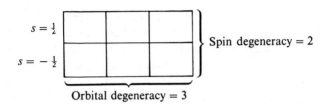

The number of ways in which two electrons may occupy these six boxes is given by $(6 \cdot 5)/2$, where the 2 in the denominator takes account of the indistinguishability of the electrons. Thus the total degeneracy of the $t_{2g}{}^2$ configuration is 15. Now as the field is decreased, giving rise to the separate orbital states, A_{1g}, E_g, T_{1g}, and T_{2g}, the total degeneracy must remain 15. Thus, if we write

$$t_{2g} \times t_{2g} \rightarrow {}^a A_{1g} + {}^b E_g + {}^c T_{1g} + {}^d T_{2g}$$

we may say that the total degeneracy equals 15 by writing

$$1 \cdot a + 2 \cdot b + 3 \cdot c + 3 \cdot d = 15$$

where a, b, c, and d must each be either 1 or 3. It is not difficult to see that with this restriction the equation has only three solutions:

	a	b	c	d
I	1	1	1	3
II	1	1	3	1
III	3	3	1	1

Similarly for the $e_g{}^2$ configuration we have two electrons to place in four equivalent boxes, and this may be done in $(4 \cdot 3)/2 = 6$ distinguishable ways. Thus we may write

$$e_g \times e_g \rightarrow {}^a A_{1g} + {}^b A_{2g} + {}^c E_g$$

and the equation

$$1 \cdot a + 1 \cdot b + 2 \cdot c = 6$$

which admits of only two solutions:

	a	b	c
I	1	3	1
II	3	1	1

Now for the $t_{2g}e_g$ configuration we may place one electron in any of six boxes while we *independently* place the second electron in any of four boxes, giving a total of 24 possible arrangements. We also note that there is no possibility of two electrons being in the same box, so that for all arrangements their spins may be either paired or unpaired. Thus both the T_{1g} and T_{2g} states derived from the $t_{2g}e_g$ configuration may be both triplet and singlet. We thus get the unique answer that the configuration $t_{2g}e_g$ gives $^1T_{1g}$, $^3T_{1g}$, $^1T_{2g}$, and $^3T_{2g}$. The total degeneracy of these four states is 24, in agreement with our count of the number of arrangements in the boxes.

Now we can determine which of the possible assignments of the multi-plicities in the $t_{2g}{}^2$ and $e_g{}^2$ configurations are correct by proceeding to correlate the states on the two sides of the diagram. To do this we shall use two prin-ciples, neither of which will be proved, but both of which are important and rather easily remembered. As we go from the weak to the strong interaction with the environment, we do not in any way change the symmetry properties of the system. Thus there must be the same number of each kind of state throughout, and we may accept, almost as an axiom, this principle:

There must exist a one-to-one correspondence between the states at the two extremes of the abscissa.

The second principle, which has been used earlier (page 184) in construct-ing the correlation diagrams for the Woodward-Hoffman rules, and which has its ultimate origin in the phenomenon of configuration interaction (page 165) is called the non-crossing rule:

As the strength of the interaction changes, states of the same spin degeneracy and symmetry cannot cross.

In Figure 9.3 we have shown on the extreme left the states of the free ion. Immediately to the right we have shown the states into which these free-ion states split under the influence of the octahedral environment. Here we know the spin multiplicities of all states. Now at the extreme right are the states in the (hypothetical) case of an infinitely strong interaction with the environ-ment, and immediately to the left of them are the distinct states which we have just shown to exist in the case of a very strong, but not infinitely strong, interaction. In order that each state on the left go over into a state of the

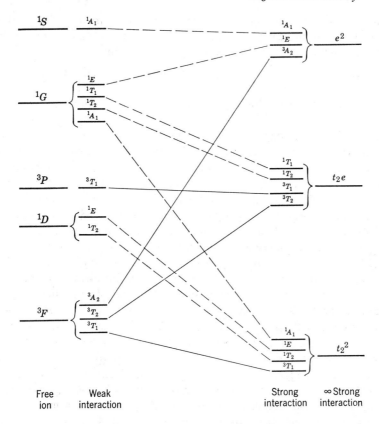

Figure 9.3   A correlation diagram for a d^2 ion in an octahedral environment. All states and orbitals are of g type, and this subscript has therefore been omitted.

same kind on the right without violation of the noncrossing rule, the connecting lines can be drawn only in the manner shown. The way in which this was done may be briefly recapitulated.

We note that there are two $^1A_{1g}$ states on the left and no $^3A_{1g}$ states. Thus both A_{1g} states on the right must be singlets. This immediately settles the multiplicities of the states coming from the $e_g{}^2$ configuration and rules out possibility III for the $t_{2g}{}^2$ configuration. Now we note that there are two $^3T_{1g}$ states at the left. The higher one must connect to the $^3T_{1g}$ state coming from the $t_{2g}e_g$ configuration. There is only one T_{1g} state below this, namely, that from $t_{2g}{}^2$, so this state must be a triplet, and this settles the assignment of spin multiplicities of the states coming from the $t_{2g}{}^2$ configuration. The remaining connections are now drawn in accord with the noncrossing rule.

The complete diagram is often called a *correlation diagram*. It shows how the energy levels of the ion behave as a function of the strength of the chemical interaction with an octahedrally symmetric environment.

The Method of Descending Symmetry

In the foregoing pages we have constructed the complete correlation diagram for a d^2 ion in O_h symmetry. In the course of doing so we ran into the problem of determining the spin multiplicities of the orbital states as they arise from interelectronic interactions in the configurations e_g^2 and t_{2g}^2, but we were able to obtain a solution by the somewhat oblique procedure of requiring the states to be just those necessary to correlate with the set of weak-field states. Such a procedure would be rather impractical in more complicated cases, but there is a straightforward and general method, due to Bethe and known as the method of descending symmetry, which can always give us the required information. We shall explain this method by showing its application in the d^2 case, and it should then be obvious how the same procedure can be used in any case.

Let us begin with the e_g^2 configuration. We have shown that this must go over into the states A_{1g}, A_{2g}, and E_g as the electron interactions take effect. Suppose now that we take the environment of the ion, which has O_h symmetry, and lower this symmetry to D_{4h}, say by taking a *trans* pair of ligands in an octahedral MX_6 complex and moving them out to a greater distance than the other four. The degeneracy of the one-electron e_g orbitals is now lifted, and, as may be seen from the correlation table in Appendix IIIB, we get two nondegenerate levels of symmetries a_{1g} and b_{1g}:

$$e_g \{ \overline{} \rightarrow \quad \begin{array}{l} \underline{}\ a_{1g} \\[2ex] \underline{}\ b_{1g} \end{array}$$

$$O_h \qquad D_{4h}$$

Now the number of ways in which we can place the two electrons in the two levels, a_{1g} and b_{1g}, are the following:

	DIRECT PRODUCT	POSSIBLE SPIN MULTIPLICITIES
a_{1g}^2	A_{1g}	$^1A_{1g}$
$a_{1g}b_{1g}$	B_{1g}	$^1B_{1g}, {}^3B_{1g}$
b_{1g}^2	A_{1g}	$^1A_{1g}$

Clearly the exclusion principle requires that the A_{1g} states resulting from the configurations a_{1g}^2 and b_{1g}^2 must be singlets, that is, the spins of the electrons must be different; for the configuration $a_{1g}b_{1g}$, however, since the two electrons have different orbital states, there is no restriction on their spins, and the resulting states $^1B_{1g}$ and $^3B_{1g}$ are both permitted. It should be noted that the total number of arrangements of electrons in the four e_g-type boxes was six and the total is still six in D_{4h} symmetry, as it must be.

Now just as the one-electron level, e_g, in O_h symmetry goes over into the levels a_{1g} and b_{1g} when the symmetry is lowered to D_{4h}, so the states deriving from the e_g^2 configuration in O_h symmetry, namely, A_{1g}, A_{2g}, and E_g, must go over into states appropriate to D_{4h} symmetry. Inspection of the correlation table shows that the relationship is

$$
\begin{array}{cc}
O_h & D_{4h} \\
A_{1g} \rightarrow & A_{1g} \\
A_{2g} \rightarrow & B_{1g} \\
E_g \rightarrow & \begin{cases} A_{1g} \\ B_{1g} \end{cases}
\end{array}
$$

Lowering the symmetry cannot change the spin degeneracies; hence, if the A_{1g} state in O_h is a singlet, then the corresponding A_{1g} state in D_{4h} must also be a singlet, and so on. Moreover, whatever is the multiplicity of the E_g state in O_h, both the A_{1g} and B_{1g} states which arise from it on lowering symmetry to D_{4h} must have that same spin multiplicity. Since the only A_{1g} states available in D_{4h} are $^1A_{1g}$, it immediately follows that the correlation between the O_h states and the D_{4h} states must be with the spin multiplicities, as shown below:

$$
\begin{array}{cc}
O_h & D_{4h} \\
^1A_{1g} \rightarrow & ^1A_{1g} \\
^3A_{2g} \rightarrow & ^3B_{1g} \\
^1E_g \rightarrow & \begin{cases} ^1A_{1g} \\ ^1B_{1g} \end{cases}
\end{array}
$$

This fixes the multiplicities of the states in O_h and, of course, gives the same assignments of spin multiplicities that we previously deduced.

Let us now proceed to the states arising from the t_{2g}^2 configuration, namely, $A_{1g} + E_g + T_{1g} + T_{2g}$. We found previously, it will be recalled, that only three assignments of multiplicities were possible, consistent with the total degeneracy (15) of the t_{2g}^2, which must be conserved. These are enumerated again in the chart below for convenient reference.

$t_{2g} \times t_{2g} =$	A_{1g}	E_g	T_{1g}	T_{2g}
Possible spin multiplicity assignments	$\left\{\begin{matrix} 1 \\ 1 \\ 3 \end{matrix}\right.$	1 1 3	1 3 1	3 1 1
Corresponding representations in C_{2h}	A_g	A_g B_g	A_g B_g B_g	A_g A_g B_g

It is now necessary to look for a subgroup of O_h such that each of the representations A_{1g}, E_g, T_{1g}, and T_{2g} of O_h goes over into a *different* one-dimensional representation or sum of one-dimensional representations of the subgroup. Unless these are all different, it will not be possible to obtain a complete and unambiguous result. Inspection of the correlation table for O_h in Appendix IIIB shows that the subgroups C_{2h} and C_{2v} will be satisfactory. We shall use C_{2h} here; the reader may obtain practice in applying the method by using C_{2v} to verify the results. In the chart above we have listed under each of the O_h representations the C_{2h} representations which correspond to it as obtained from the correlation table.

Since t_{2g} in O_h goes over to $a_g + a_g + b_g$ in C_{2h}, the direct product $t_{2g} \times t_{2g}$ goes over into the sum of the six direct products of $a_g + a_g + b_g$, namely,

$$a_g \times a_g = A_g$$
$$a_g \times a_g = A_g$$
$$a_g \times b_g = B_g$$
$$a_g \times a_g = A_g$$
$$a_g \times b_g = B_g$$
$$b_g \times b_g = A_g$$

The first of these represents the occupation of one a_g orbital by both electrons and so must be a singlet, ^1A_g. The second corresponds to the placement of each electron in a different a_g orbital and can therefore give rise to both singlet and triplet states, $^1A_g + {}^3A_g$. The third and the fifth also correspond to placing the electrons in different orbitals, and these too give rise to both triplet and singlet states, viz., $2^1B_g + 2^3B_g$. The fourth and the sixth correspond to placing the two electrons in the same orbital and thus give rise only to singlet states, 2^1A_g. In summary, then, the direct product $t_{2g} \times t_{2g}$, which goes over into $(a_g + a_g + b_g) \times (a_g + a_g + b_g)$ in C_{2h}, gives rise to the following states in C_{2h}:

$$4^1A_g + {}^3A_g + 2^1B_g + 2^3B_g$$

It may be noted that the sum of all the degeneracies, $4(1 \times 1) + (3 \times 1) + 2(1 \times 1) + 2(3 \times 1)$, equals 15, as it must if no errors have been made.

We can now obtain the result we desire. We can immediately make a unique assignment of multiplicities to the states listed in the lower part of our chart by noting that there is only one ^3A_g state and two ^3B_g states. These must therefore be assigned to the A_g and two B_g states arising from T_{1g}, thus establishing that the T_{1g} state is a spin triplet and, hence, that the A_{1g}, E_g, and T_{2g} states are all spin singlets.

Energy Level Diagrams in Tetrahedral Environments

Energy level diagrams for ions in tetrahedral environments can be constructed by the same procedures as those described in the preceding pages for the octahedral case. We shall briefly outline here the procedure for d^2.

To obtain the left (weak interaction) side we look up each of the free-ion terms in Table 9.3 and find that these terms split as follows:

$$^3F \rightarrow {}^3A_2 + {}^3T_1 + {}^3T_2$$
$$^1D \rightarrow {}^1E + {}^1T_2$$
$$^3P \rightarrow {}^3T_1$$
$$^1G \rightarrow {}^1A_1 + {}^1E + {}^1T_1 + {}^1T_2$$
$$^1S \rightarrow {}^1A_1$$

Turning our attention now to the right (strong interaction) side of the diagram, we observe in Figure 9.2 that the one-electron e orbitals are more stable (by Δ_t) than the one-electron t_2 orbitals. Hence the three configurations under the influence of a strong interaction with the tetrahedral environment will be, in increasing order of energy; e^2, et_2, $t_2{}^2$. Taking the direct product representations of these and obtaining their constituent irreducible representations, we see that interelectronic coupling will cause the following states to arise:

$$e^2 \rightarrow A_1 + A_2 + E$$
$$et_2 \rightarrow T_1 + T_2$$
$$t_2{}^2 \rightarrow A_1 + E + T_1 + T_2$$

We can now assign the correct spin multiplicity, 1 or 3, to each of these states by the same methods described in detail for the octahedral case and then correlate the states on the two sides of the diagram to obtain the complete correlation diagram shown in Figure 9.4.

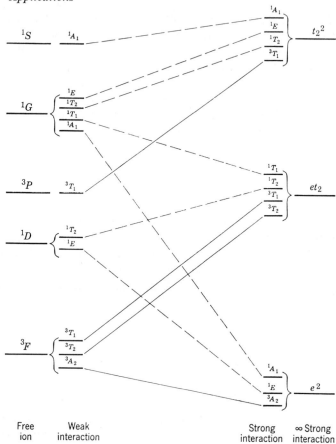

Figure 9.4   A correlation diagram for a d^2 ion in a tetrahedral environment.

The Hole Formalism

The methods illustrated above for working out the correlation diagrams of d^2 ions in octahedral and tetrahedral environments can be applied to all d^n configurations for $2 \leq n \leq 9$. However, the labor involved increases extraordinarily fast as the number of electrons increases. Fortunately several kinds of relationships make it possible to obtain certain diagrams from others which are more readily constructed.

One of these kinds of relationship is the hole formalism. According to this principle, which is thoroughly rigorous, a d^{10-n} configuration will behave, at all points along the abscissa of the energy level diagram, in the same way as the corresponding d^n configuration, *except* that all energies of interaction with the environment will have the opposite sign. A physical way of looking at the problem is to say that n holes in the d shell may be treated as n positrons.

In their interactions with one another, n positrons will behave the same as n electrons. However, where the environment tends to repel an electron, it will with the same force attract a positron and vice versa.

The consequences of this relationship are most easily seen by considering the right side of the correlation diagrams. For n electrons in an octahedral environment the configuration containing as many electrons as possible in the t_2 orbitals is the most stable, whereas for n positrons such a configuration is the least stable one. It should be easy indeed to see that quite generally the states at the extreme right of the diagram for a d^n configuration (in either octahedral or tetrahedral environments) will be inverted for the corresponding d^{10-n} configuration. Since the energy order of the states which arise from the free-ion terms of a d^n configuration is dictated by the energy order in which they go into the various $e^p t_2^q$ configurations on the right, it follows that the splitting patterns of the free-ion terms of a d^n system will be just inverted for the corresponding d^{10-n} system.

Thus, for example, the correlation diagram for a d^8 system in an octahedral field is obtainable from Figure 9.3 simply by inverting the vertical arrangement of the configurations t_2^2, et_2, e^2 and redrawing the connecting lines.

More General Relations

The prescription given in the preceding paragraph for obtaining the correlation diagram of a d^8 configuration in an octahedral environment from that for a d^2 configuration in an octahedral environment is exactly the same as one for obtaining the diagram of the d^2 ion in a tetrahedral environment from that for the d^2 ion in an octahedral environment. Changing the environment from octahedral to tetrahedral inverts the energies of e and t_2 orbitals, and so also does the change of n electrons to n positrons while keeping the symmetry of the environment the same.

We may thus state the following very general rule, in which we use the symbol d^n(oct) to mean the energy level order for a d^n system in an octahedral field, with the other symbols having analogous meanings:

$$d^n(\text{oct}) \equiv d^{10-n}(\text{tetr}) \text{ are inverse to } d^n(\text{tetr}) \equiv d^{10-n}(\text{oct})$$

Thus for the eighteen possible cases, that is, $d^1 - d^9$ each in tetrahedral and octahedral environments, correlation diagrams can be obtained by explicitly working out only those for the following cases, which are the simplest:

$$d^1(\text{oct}) \quad d^2(\text{oct}) \quad d^3(\text{oct}) \quad d^4(\text{oct}) \quad d^5(\text{oct})$$

The d^5 case is special in that all of the four related diagrams here are identical. The d^n and d^{10-n} are of course the same configurations when $n = 5$ and $d^5(\text{oct}) \equiv d^5(\text{tetr})$.

In Figure 9.5 are shown energy level diagrams for all of the d^n configurations

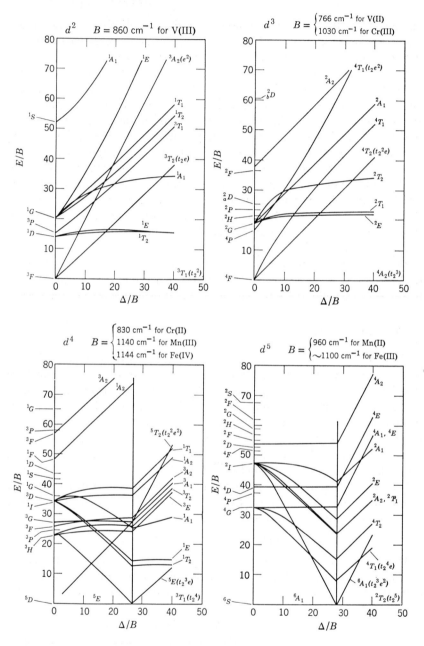

Figure 9.5   Energy level diagrams [after Tanabe and Sugano, *J. Phys. Soc. Japan*, **9**, 753 (1954)] for the d^2-d^8 configurations, in octahedral symmetry.

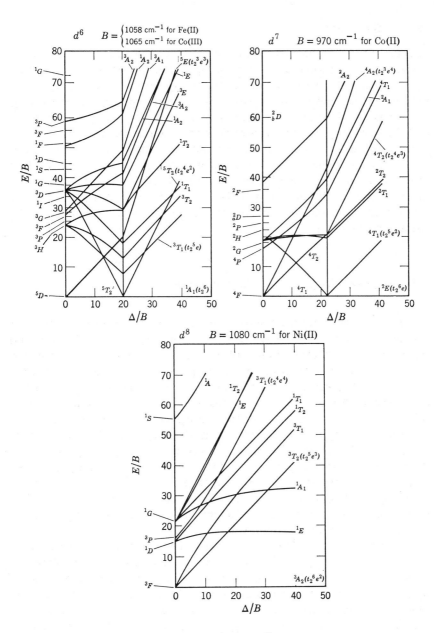

Figure 9.5 (*Continued*)

in octahedral environments. These diagrams are plotted in a manner requiring some comment. Instead of using absolute units for the ordinate and abscissa scales, which would restrict the application of each diagram to just the one case in which the separations of the free-ion terms matched those in the diagram, the energy unit is the interelectronic repulsion parameter B. On each diagram are given the values of B for the common free ions of the corresponding d^n and d^{10-n} configurations. In addition, the diagrams are so drawn that the energy of the ground state is taken as the zero of energy for all values of Δ_0. Thus, in cases where the ground state changes there are sharp changes in the slopes of all lines. It is to be emphasized that these "kinks" are artifacts of the diagrams and do not represent real discontinuities in the energies of the states.

Energy Level Diagrams for Lower Symmetries

We have so far considered only the most highly symmetrical situations of common occurrence. We have seen that for octahedral and tetrahedral symmetry the d orbitals split into only two sets, and thus only one parameter, Δ_0 or Δ_t, is required to describe the energy pattern. In cases in which the symmetry is high and the number of free parameters is small, symmetry considerations are, as we have seen, highly informative. When the symmetry is lower, there are more splittings and more parameters, and hence less may be learned from symmetry considerations alone regarding the relative order of the levels. Symmetry considerations then tend to become less an end in themselves and more a necessary preliminary to setting up the equations for a calculation in the simplest and most expedient form. With regard to the interpretation of spectral intensity measurements and especially the polarization of absorption bands, pure symmetry considerations remain immensely useful, as will be seen in Section 9.6.

One of the most commonly occurring of the lower symmetries in coordination compounds is D_{4h}. This is the point group of atoms surrounding and directly interacting with the metal atom in square planar complexes, of octahedral complexes which are distorted by elongation or compression along one of the fourfold axes, and of *trans*-disubstituted octahedral complexes such as *trans*-[Co(NH$_3$)$_4$Cl$_2$]$^+$. We shall discuss this point group as one example to illustrate the general nature of the problems which arise when the symmetry is lower than cubic.

As shown in Table 9.3, the d orbitals form a basis for a representation of the group D_{4h} which contains the irreducible representations A_{1g}, B_{1g}, B_{2g}, and E_g. By referring to the character table for D_{4h}, we can obtain the more specific information that the d orbitals correspond with these representations in the following way:

$$A_{1g}: \quad d_{z^2}$$

$$B_{1g}: \quad d_{x^2-y^2}$$

$$B_{2g}: \quad d_{xy}$$

$$E_g: \quad d_{xz}, d_{yz}$$

Since there are four different symmetry species of orbitals, we now require three parameters to specify the energy differences between them. The relative values of these parameters must be known in order to find the relative energies of these four types of orbital, so that at least two *actual numbers*, ratios of two of the parameters to the third, perhaps, must be known before any sort of energy level diagram can be constructed. Where the deviation from perfect octahedral symmetry is small, it can be assumed that the a_{1g} and b_{1g} levels arising from the splitting of the e_g levels in O_h will be separated by an energy comparable in magnitude to the separation of the b_{2g} and e_g levels arising from the t_{2g} levels in O_h, and that both of these energies will be small relative to the energy difference, Δ_0, initially existing between the e_g and t_{2g} levels in the octahedral environment. Nevertheless, there remains the question of the relative energies of a_{1g} versus b_{1g} and of b_{2g} versus e_g.

Relation of Energy Level Diagrams to Spectral and Magnetic Properties of Complexes

Among the most important applications of energy level diagrams of the sort we have been discussing is their use in the interpretation of spectral and magnetic properties of complexes and other compounds of the transition elements. For detailed discussion of these applications the reader is referred to several books dealing more broadly with ligand field theory,* but it is appropriate to give here a summary account.

The visible and near ultraviolet spectra of transition metal ions in chemical environments are due to transitions from their ground states to the various excited states, as these are shown on the energy level diagrams in Figure 9.5. As will be discussed more fully in Section 9.6, these transitions are nominally forbidden by selection rules in first approximation but appear weakly because of breakdown of these selection rules in higher approximations. Transitions

---

* See reading list, Appendix IX.

to excited states with the same spin multiplicity as the ground state are some 10^2 times stronger than those to states differing in spin quantum number, as might be expected. Thus the spin-forbidden transitions cause absorption bands which nearly always are too weak to be observed in ordinary measurements.

By inspection of the energy level diagrams it is possible to see directly what sort of spectrum the ion should have in the given environment. For example, it can be seen from Figure 9.3 that a d^2 ion in an octahedral complex, say $[V(H_2O)_6]^{3+}$, should have three spin-allowed transitions, from the $^3T_1$ ground state to the upper states $^3T_2$, $^3T_1$, and $^3A_2$. Experimentally, two absorption bands have been found at $\sim 17,000$ cm^{-1} and $\sim 24,000$ cm^{-1}, and these may be assigned to the $^3T_1 \to {}^3T_2$ and $^3T_1 \to {}^3T_1$ transitions if Δ_0 is taken as $\sim 21,500$ cm^{-1}. The $^3T_1 \to {}^3A_2$ transition would be expected to be at still higher energy at this value of Δ_0 and has not been definitely observed.

As another example, let us consider the d^7 ion Co(II) in a tetrahedral environment, for example, in $[CoCl_4]^{2-}$. The energy level diagram is the same as that for a d^3 ion in an octahedral complex, and from Figure 9.5 we observe that three spin-allowed bands are to be expected. Again only two have been observed at ~ 5500 cm^{-1} and $\sim 14,700$ cm^{-1}. From the energy level diagram it follows that these must be assigned as the $^4A_2 \to {}^4T_1(F)$ and $^4A_2 \to {}^4T_1(P)$ transitions, and it would then be predicted that the $^4A_2 \to {}^4T_2$ transition should lie in the range 3000–3500 cm^{-1}. In a few other tetrahedral complexes such a transition has been observed but is extremely weak. The reason for the weakness is discussed in Section 9.6. It is to be noted that Δ_t here is only 3000–3500 cm^{-1}, as compared to the value of $\sim 21,500$ cm^{-1} for $[V(H_2O)_6]^{3+}$ found earlier. This great difference is due to the combination of two effects. One is that, for a particular metal ion and ligands, Δ_0 for the octahedral complex MX_6 is about twice as great as Δ_t for the tetrahedral complex MX_4. Second, for octahedral complexes, an increase of 1 unit in the oxidation number of the metal causes Δ_0 to rise by a factor of 1.5–2.0.

Insofar as magnetic properties are concerned, the energy level diagrams provide a ready explanation of the way in which the symmetry and strength of interaction of the environment determine the spin multiplicities of the metal ions in their compounds. Basically, two different situations arise. In one of these, exemplified by the octahedral d^1, d^2, d^3, d^8, and d^9 cases, the ground state is derived from the lowest term of the free ion for all values of the parameters Δ_0, however large. Hence the number of unpaired electrons must be the same as that in the free ion, however strongly the ion may interact with its environment. In the other cases, namely d^4, d^5, d^6, and d^7, the ground state is derived from the lowest free-ion term only out to a certain critical value of Δ_0, beyond which a state of lower spin multiplicity originating in a higher free-ion term drops below it and hence becomes the ground state. For these systems, therefore, we can predict that for complexes and other compounds in which the perturbing effect of the environment, as measured by Δ_0, is weak there will be the maximum number of unpaired electrons, whereas for compounds in which the perturbing effect of the environment is very strong, greater than the critical value of Δ_0, there will be fewer (two or four fewer) unpaired electrons. Predictions of this sort have been found to be in remarkable accord with experimental observations. Although similar predictions could be made for the various d^n systems in tetrahedral environments, they would be of little practical value, since in real tetrahedral systems the value of Δ_t never seems to exceed the critical value, and hence all tetrahedral complexes known have the highest possible spin multiplicity.

Another way of deducing whether a given d^n ion in an octahedral environment will have only one possible spin multiplicity or several, and what these multiplicities will be, is so simple that it does not require reference to any published energy level diagrams or indeed even the use of pencil and paper. We consider the set of one-electron d orbitals split as shown in Figure 9.2 into the lower-lying t_{2g} subset and the upper-lying e_g subset. When an electron is to be placed in this set of d orbitals, two energy terms must be considered. If the electron enters the t_{2g} subset, it will be more stable by an amount Δ_0 than if it enters the e_g subset. However, if in order to enter the t_{2g} subset it will have to enter an orbital which is already occupied by one electron, there will be a repulsive energy, usually called a pairing energy, P. If this pairing energy is greater than Δ_0, the electron will go into an e_g orbital despite the fact that this costs an energy Δ_0. Thus the critical value of Δ_0 can be equated (approximately) to the pairing energy P. The latter can be estimated from spectroscopic data on the free ion.

With these considerations in mind we can make the following statements about the d electron distributions for the various d^n configurations in octahedral environments. For the d^1, d^2, and d^3 cases the electrons can enter the t_{2g} orbitals without any need of double occupancy of any orbital. Hence these ions will have one, two, and three unpaired electrons, respectively, regardless of the magnitude of Δ_0. For the d^8 and d^9 ions all possible configurations require double occupancy of three and four orbitals, respectively, and the lowest energy configurations will always be those, $t_{2g}^6 e_g^2$ and $t_{2g}^6 e_g^3$, with the t_{2g} orbitals fully occupied regardless of the magnitude of Δ_0. Thus the ground states of d^1, d^2, d^3, d^8, and d^9 ions in octahedral environments must have the maximum number of unpaired electrons irrespective of the magnitude of Δ_0. For d^4, d^5, d^6, and d^7 ions, however, the electron distributions will depend on the magnitude of Δ_0 compared to P, as indicated in the following table. The numbers in parentheses indicate the numbers of unpaired electrons.

CONFIGURATION	$\Delta_0 < P$	$\Delta_0 > P$
d^4	$t_{2g}^3 e_g\ (4)$	$t_{2g}^4\ (2)$
d^5	$t_{2g}^3 e_g^2 (5)$	$t_{2g}^5\ (1)$
d^6	$t_{2g}^4 e_g^2 (4)$	$t_{2g}^6\ (0)$
d^7	$t_{2g}^5 e_g^2 (3)$	$t_{2g}^6 e_g\ (1)$

It will be seen that all these results are in accord with the conclusions which can be drawn from the energy level diagrams of Figure 9.5.

9.5   Estimation of Orbital Energies

The numerical evaluation of the energies of orbitals and states is fundamentally a matter of making quantum mechanical computations. As indicated in Chapter 1, quantum mechanics per se is not the subject of this book, and indeed we have tried in general to avoid any detailed treatment of methods for solving the wave equation, emphasis being placed on the properties which the wave functions must have purely for reasons of symmetry and irrespective of their explicit analytical form. However, this discussion of the symmetry aspects of ligand field theory would be artificial and unsatisfying without some brief outline of the various models which may be used to make computations and also to visualize the nature of the interaction between the metal ion and its chemical environment.

Our discussion here of computational procedures will be very superficial and aimed at bringing out the physical features of the models. For a full treatment of this subject with references to the original literature, and for further discussion of the interpretation of the chemical behavior of transition metal compounds in terms of ligand field theory, the reader is referred to the publications cited in Appendix IX.

As noted in Section 9.1, there are three closely related theories of the electronic structures of transition metal complexes, all making quite explicit use of the symmetry aspects of the problem but employing different physical models of the interaction of the ion with its surroundings as a basis for computations. These three theories, it will be recalled, are the crystal field, ligand field, and molecular orbital theories. There is also the valence bond theory, which makes less explicit use of symmetry but is nevertheless in accord with the essential symmetry requirements of the problem. We shall now briefly outline the crystal field and ligand field treatments and comment on their relationship to the MO theory.

The Crystal Field Theory

This model of a complex or of a crystalline salt of a metal ion in a compound such as a halide or oxide is of an electrostatic, point charge, or point dipole type. The ligands or neighbors of the metal ion are treated as structureless, orbital-less point charges which set up an electrostatic field. The effect of this field on electrons in the d orbitals of the metal ion is then investigated.

Group theory alone has shown us that a single d electron in an ion which is at the center of an octahedron may be in either of two states. In one of these it will have either of two wave functions or one which is a linear combination of both, which together provide a basis for the E_g representation of the group O_h. In the other state it may have one of three wave functions or some linear combination of these, this set of three being such as to provide a basis for the T_{2g} representation of the group O_h. If we refer to the character table for the group O_h, we note that the wave functions $d_{x^2-y^2}$ and d_{z^2} form a basis for the E_g representation. Hence we may consider these to be the orbitals which may be occupied by an e_g electron. Similarly a t_{2g} electron may be assumed to occupy one of the orbitals d_{xy}, d_{yz}, d_{xz}, since these form a basis for the T_{2g} representation of the group. We shall now show how the energies of electrons in these orbitals are estimated according to the electrostatic model of the crystal field theory.

We assume that each of the six ligands is either an anion such as O^{2-}, F^-, Cl^-, ..., or a dipolar molecule such as $N^\delta-H_3^{\delta+}$, $O^\delta-H_2^{\delta+}$, ..., having its negative end close to the cation. In either case we may then look upon the environment of the cation as being that shown in Figure 9.6 in relation to a specific set of Cartesian axes. Before considering how, on the basis of such an electrostatic model, the actual magnitude of the energy difference between the e_g and t_{2g} states of the d electron might be computed, let us first consider the simpler, qualitative question of which state, e_g or t_{2g}, is the more stable.

Group theory tells us that both e_g orbitals have the same energy and that all three t_{2g} orbitals have the same energy. Hence we need only compare either of the e_g orbitals with any one of the t_{2g} orbitals to obtain the answer. Let us choose the $d_{x^2-y^2}$ and d_{xy} orbitals for our comparison. The shapes of these orbitals are indicated in Figure 9.7 by drawings intended to show surfaces enclosing a major fraction, say 90 per cent, of the electron density. Since the electron has a negative charge and the ligands either are negative or appear so to the electron, it is apparent that because of electrostatic repulsive forces the electron is more stable in the d_{xy} than in the $d_{x^2-y^2}$ orbital. The interaction of the electronic charge with the charges of the ligands lying along the z axis will be the same in either case, but it is evident that the electronic charge is much more concentrated in the region of the other negative

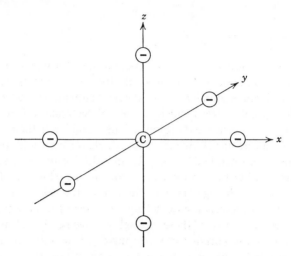

Figure 9.6   The electrostatic environment of a cation, C, surrounded by an octahedral array of anions or dipoles.

ligands in the $d_{x^2-y^2}$ orbital than it is in the d_{xy} orbital. Thus we know, qualitatively, that, insofar as the electrostatic model is a faithful representation of the true situation, the e_g orbitals are of higher energy than the t_{2g} orbitals.

It is easy to see the correctness of the group theoretical result that the d_{xy}, d_{xz}, and d_{yz} orbitals must all have the same energy. Each of these is identical in form to the other two, differing only in the plane in which its maxima lie. That the $d_{x^2-y^2}$ and d_{z^2} orbitals have the same energy is

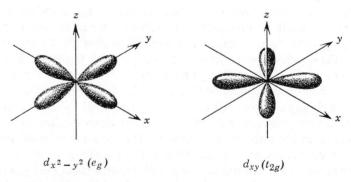

$$d_{x^2-y^2}\ (e_g)\qquad\qquad d_{xy}\ (t_{2g})$$

Figure 9.7   Sketches showing the $d_{x^2-y^2}$ and d_{xy} orbitals, representative of the e_g and t_{2g} orbitals respectively, in relation to the coordinate axes.

certainly not so obvious, merely on inspection. However, it is easy to grasp this equivalence by using the following line of reasoning. We know from wave mechanics that there can be only five linearly independent solutions to the wave equation with the same value of the quantum number n (≥ 3) and with $l = 2$. However, we may write an infinite number of solutions initially and select any five linearly independent combinations we desire (see page 94). Accordingly, let us consider the following *six* functions for the angular parts of nd orbitals:

$$\psi_1' \approx xy \quad \psi_4' \approx x^2 - y^2$$
$$\psi_2' \approx yz \quad \psi_5' \approx z^2 - y^2$$
$$\psi_3' \approx xz \quad \psi_6' \approx z^2 - x^2$$

Of these six the following set of five is usually selected:

$$\psi_1 = \psi_1' \approx xy$$
$$\psi_2 = \psi_2' \approx yz$$
$$\psi_3 = \psi_3' \approx xz$$
$$\psi_4 = \psi_4' \approx x^2 - y^2$$
$$\psi_5 = \frac{1}{\sqrt{2}}(\psi_5' + \psi_6') \approx 2z^2 - x^2 - y^2 \approx z^2$$

Thus we see that the d_{z^2} orbital can be regarded as a normalized linear combination of $d_{z^2-y^2}$ and $d_{z^2-x^2}$ orbitals. Now the $d_{x^2-y^2}$, $d_{z^2-y^2}$, and $d_{z^2-x^2}$ orbitals are obviously all of identical energy in an octahedral field, and, as shown on page 94, any linear combination of two degenerate wave functions has the same energy as each of its constituents. Thus we can see by geometrical reasoning that the d_{z^2} and $d_{x^2-y^2}$ orbitals are degenerate in an octahedral field, despite the fact that they may not clearly appear to be equivalent.

We may, on the basis of the foregoing argument, draw the simple energy level diagram, Figure 9.8, showing the relative energies of the e_g and t_{2g} orbitals. We know that the e_g levels are of higher energy than the t_{2g} levels, and we have denoted the magnitude of this difference by Δ_0 or $10Dq$, which are the symbols commonly used in the literature.

We may now consider whether it is possible to calculate the value of Δ_0 by using an electrostatic model. To do so we consider anion ligands as point charges and dipolar ligands as point dipoles. For the latter we should also

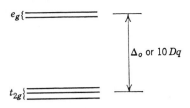

Figure 9.8 Splitting of d orbitals in an octahedral crystal field.

have to estimate the effective value of the dipole moment, which would be equal to the permanent moment plus that induced by the positive charge of the metal ion. Finally, we should require knowledge of the metal-ligand distance and a proper radial wave function for the d electron. It would then be possible to perform a calculation of the e_g-t_{2g} separation. Such calculations have been carried out in a number of cases. For dipolar ligands, it is necessary to assume very unrealistic values for the effective dipole moments in order to obtain correct values of Δ_0. For ionic ligands the model gives results which are of the right order of magnitude but not much better than that. However, it is now generally recognized that this purely electrostatic model is too simple to be taken literally, since in all cases the value of Δ_0 is determined by interactions other than purely electrostatic ones. Thus Δ_0 is best regarded as a phenomenological parameter to be determined from experiment rather than as one to be calculated from first principles using the crystal field model.

One further aspect of the splitting of the one-electron d orbitals in an octahedral field must be noted. Suppose that we consider the following *Gedanken* experiment. We surround an atom or ion by a concentric spherical shell of uniformly distributed negative charge, the total charge being $6q$ units. A set of ten d electrons in this ion will now have an energy, E_S, which is higher, because of repulsive forces between the electrons and the outer shell of negative charge, than its energy, E^0, in the free ion. However, since the charge distribution is spherical, the d electrons all have the same energy. Now suppose that the total charge on this spherical shell is redistributed, but moving only on the surface of the sphere, so as to place six point charges, each q units in magnitude, at the six apices of an octahedron. This redistribution cannot change the energy of the d^{10} configuration as a whole, and yet we know that now six electrons are in t_{2g} orbitals and four are in e_g orbitals and that these orbitals differ in energy by Δ_0. These relationships are depicted in Figure 9.9. In order for the total energy of the d^{10} configuration to be the same in the octahedral field as it is in the spherical field, the following equations must hold:

$$6(E_S - B) + 4(E_S + A) = 10E_S$$
$$A + B = \Delta_0$$

Solving; we obtain

$$3B = 2A$$

Hence

$$A = \tfrac{3}{5}\Delta_0$$
$$B = \tfrac{2}{5}\Delta_0$$

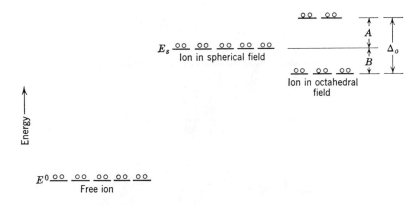

Figure 9.9   An energy diagram showing how the energies of the electrons in a d^{10} configuration are affected by spherical and octahedral electrostatic fields.

It should be noted that the splitting, Δ_0, is generally of the order of 1–3 electron volts, whereas the elevation of the set of d levels as a whole is of the order of 20–40 eV. Thus it should always be borne in mind that the crystal and ligand field theories focus attention on only one relatively small aspect of the overall energy of formation of a complex.

For a d electron in the electrostatic field of four anions or dipoles arranged tetrahedrally around it, the splitting pattern can be derived by an analogous line of reasoning. Group theory tells us that the fivefold degenerate state of the d electron in the free ion is split into two states, one twofold degenerate, E, and one threefold degenerate, T_2. Reference to the character table for the T_d group shows that the former state will be one of the two orbitals d_{z^2} and $d_{x^2-y^2}$ or a linear combination thereof, and that the T_2 state will be one of the orbitals d_{xy}, d_{xz}, and d_{yz} or some linear combination of these.

In order to find out the relative energies of the e and t_2 orbitals, let us place our tetrahedral complex in a set of coordinate axes as shown in Figure 9.10. Once more, we can make a comparison between either of the e orbitals and any one of the t_2 orbitals, and perhaps the best selection in order to visualize the relative electrostatic energies is again the pair d_{xy} and $d_{x^2-y^2}$. The difference here is much less striking than in the octahedral case, but it can be seen that qualitatively an electron in the d_{xy} orbital will have a higher potential energy than one in the $d_{x^2-y^2}$ orbital. If we call the energy separation between the e and t_2 orbitals Δ_t, then by an argument exactly analogous to the one used in the octahedral case we can show that the energies of the t_2 and e orbitals relative to their energies in a spherical shell of the same total charge will be as shown earlier in Figure 9.2.

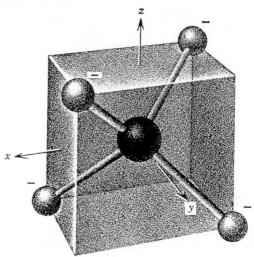

Figure 9.10   The electrostatic environment of a cation surrounded by a tetrahedral array of anions or dipoles.

It is worth noting that, when detailed expressions are written for Δ_0 and Δ_t in the purely electrostatic crystal field approximation, it turns out that for equal charges and metal-ligand distances these two quantities are in the ratio

$$\Delta_t/\Delta_0 = \tfrac{4}{9}$$

Even though the absolute values calculated by this model are, as noted before, grossly inaccurate, the ratio is in approximate agreement with experiment.

The Ligand Field Theory

This is a modification of the crystal field theory in which we drop the assumption that the partially filled electron shell is one consisting of pure d orbitals. Instead it is admitted that there is overlap between the d orbitals of the metal and the orbitals of the ligand atoms.

Various lines of evidence indicate that even in complexes in which the binding might be expected to be most ionic, for example, in hexafluoro complexes such as CoF_6^{3-} and in aquo ions such as $[Fe(H_2O)_6]^{2+}$, overlap of metal and ligand orbitals occurs to a small but significant extent. Thus there is direct evidence from ESR and NMR studies that spin density of the "d" electrons has a finite value at nuclei in the ligand molecules, and that in fact actual delocalization of the electron into ligand orbitals occurs.

Numerical estimates from these data of the time spent by a "d" electron in ligand atom orbitals in species such as $MnF_6{}^{4-}$ (in MnF_2) and $IrCl_6{}^{2-}$ are about 2–5 per cent per ligand atom. From this it can be concluded that the "d" orbitals have only about 80 ± 10 per cent d character and 10–30 per cent ligand orbital character. In reasonable accord with these observations is the fact that in complexes the interelectronic repulsion energies, which are responsible for the energy differences between the different terms of a d^n configuration, are only about 70 percent of their magnitudes in the free ions.

There are two practical consequences of the recognition of overlap between metal and ligand orbitals. First, we abandon all hope of making a priori calculations of orbital splittings by a pure point-charge electrostatic model using pure d wave functions. The electrostatic expressions for the "d" orbital energies remain valid in their general form, but the crystal field parameters, the effective charge or dipole of the ligands, the metal-ligand distance, and the radial part of the d orbital wave functions now lose their literal physical significance and must be considered as fictitious adjustable parameters.

Second, in constructing the energy level diagrams for d^n configurations, we must leave the separations between the various free-ion terms as functions of *adjustable* interelectronic repulsion parameters, rather than simply setting them down at the free-ion values. Practical energy level diagrams such as those in Figure 9.5 are interpreted using free-ion term separations equal to about 75 per cent of the separations spectroscopically observed for the free gaseous ions.

Comparison with Molecular Orbital Theory

Inspection of Figures 8.13 and 8.15, the MO energy level diagrams for octahedral and tetrahedral complexes, respectively, shows how the crystal field or ligand field treatment of the electronic structure of a complex fits into the complete picture which an MO treatment provides. The d orbitals of the metal ion provide the major contribution to the e_g and t_{2g} (octahedral) or e and t_2 (tetrahedral) orbitals lying in the center of the diagrams. Moreover, all MO's lower in energy than these are filled with electrons which originate in ligand orbitals, and indeed the MO's in which these "ligand" electrons now reside are made up mainly of the original ligand orbitals. The e and t_2 orbitals contain the electrons which were originally in the pure d orbitals of the uncomplexed metal ion.

The crystal and ligand field theories were developed to deal with only those properties of the complexes which are derived directly from the set of electrons originally occupying the d orbitals of the metal ion. Since these orbitals are the principal parents of the e and t_2 MO's of the complex, it is not unreasonable to treat the latter as though they were nothing more than split (crystal field theory) or split and somewhat diluted (ligand field theory) metal d orbitals. It is clear, however, that such a view can be only an approximation—indeed, a fairly ruthless one. Yet, with judicious empirical choice of one or more disposable parameters (the splitting energy, at least, and perhaps also the interelectronic repulsion energy and the spin-orbit coupling constant) the approximation has great practical utility for certain purposes. Specifically, it is rather good for fitting electronic spectra when only "*d-d*" transitions are involved and for interpreting magnetic behavior when the unpaired electrons occupy only "*d*" orbitals. The crystal and ligand field theories are, however, formalisms and cannot ever give complete and literal descriptions of the entire electronic structures of complexes.

9.6   Selection Rules and Polarizations

In Section 5.3 the general symmetry restrictions on transitions occurring by dipolar interaction with electromagnetic radiation were discussed. Here they will be invoked with particular reference to electronic transitions on metal ions in ligand fields.

Centrosymmetric Complexes, Vibronic Coupling

In a complex which possesses a center of symmetry, all states arising from a d^n configuration have the g character inherent in the d orbitals. Since the dipole moment vectors belong to odd representations, all of the integrals such as $\int \psi_g' x \psi_g \, d\tau$ are identically zero because the direct product of two g functions can never span any u representations. On this basis alone, we would predict that transitions between the various states arising from d^n configurations in octahedral environments would have zero absorption intensity. In fact, these transitions do take place but the absorption bands are only $\sim 10^{-3}$ times the intensity expected for symmetry allowed electronic transitions. Thus the prediction we have made is substantially correct, but at the same time there is obviously some intensity giving mechanism which has been overlooked.

It is generally accepted, following Van Vleck, that this mechanism is one called vibronic coupling—that is, a coupling of *vib*rational and elect*ronic* wave functions. In a qualitative sense we may say that some of the vibrations of the complex distort the octahedron in such a way that the center of symmetry is destroyed as the vibration takes place. The states of the d^n configuration then no longer retain rigorously their g character, and the transitions become "slightly allowed." Figure 9.11 shows the approximate nature of several of the modes of vibration of an octahedron which do destroy the symmetry center.

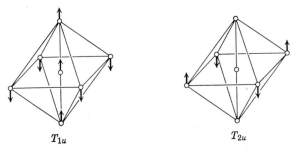

T_{1u} T_{2u}

Figure 9.11   Two of the normal vibrations of an octahedral AB_6 molecule in which the displacements of the atoms destroy the center of symmetry. Another type of T_{1u} vibration, not shown here, has the same property.

This phenomenon of vibronic coupling can be treated very effectively by using group theoretical methods. As will be shown in Chapter 10, the vibrational wave function of a molecule can be written as the product of wave functions for individual modes of vibration called *normal modes*, of which there will be $3n - 6$ for a nonlinear, n-atomic molecule. That is, we can express the complete vibrational wave function, ψ_v, as a product of $3n - 6$ functions each pertaining to one of the normal modes, viz.,

$$\psi_v = \prod_{i=1}^{3n-6} \psi_i$$

It is further shown in Chapter 10 that, when each of the normal modes is in its ground state, each of the ψ_i is totally symmetric and hence ψ_v is totally symmetric. If one of the normal modes is excited by one quantum number, the corresponding ψ_i may then belong to one of the irreducible representations other than the totally symmetric one, say Γ_i, and thus the entire vibrational wave function ψ_v will belong to the representation Γ_i. Simple methods for finding the representations to which the first excited states of the normal modes belong are explained in Chapter 10. In this section we will quote without proof results obtained by these methods.

To a first approximation, and usually a rather good one, the complete wave function, Ψ, for a molecule can be written as a product of an electronic wave function, ψ_e, a vibrational wave function, ψ_v, and a rotational wave function, ψ_r:

$$\Psi = \psi_e \psi_v \psi_r$$

It is then assumed that none of these factors of the complete wave function are interdependent, so that instead of having to solve one large wave equation

$$\mathscr{H}\Psi = E\Psi$$

it is possible to solve three simpler ones:

$$\mathscr{H}\psi_e = E_e\psi_e$$
$$\mathscr{H}\psi_v = E_v\psi_v$$
$$\mathscr{H}\psi_r = E_r\psi_r$$

and write the total energy as a simple sum of the electronic, vibrational, and rotational energies, namely,

$$E = E_e + E_v + E_r$$

This is, of course, only an approximation. Although it works well for many purposes, one of its limitations is that it cannot explain the low but finite intensity of the transitions between states of d^n configurations in centrosymmetric environments as we have shown above.

The way out of the difficulty is to drop the assumption that ψ_e and ψ_v are entirely independent, though retaining the approximation that ψ_r can be treated as independent of these other two. Thus it is not the values of integrals like

$$\int \psi_e' x \psi_e \, d\tau$$

which we must consider, but rather the values of the integrals

$$\int (\psi_e' \psi_v') x (\psi_e \psi_v) \, d\tau$$

It is easy to show by symmetry arguments that the latter do not in general vanish. First, we note that, if we assume that in the lower state, $\psi_e\psi_v$, the molecule is in its vibrational ground state, ψ_v is totally symmetric and we can ignore it. Our problem then is to decide whether there are any vibrational wave functions belonging to representations such that, although the direct product representation of $\psi_e' x \psi_e$ does not contain the totally symmetric representation, the direct product representation of $\psi_e' \psi_v x \psi_e$ does. Whenever this is so the transition will be vibronically allowed. According to the results of Section 5.2, the integral will be nonzero if there is any normal mode of vibration whose first excited state, ψ_v', belongs to one of the representations spanned by $\psi_e' x \psi_e$.

In order to show how this is done let us take a simple example. For the ion $[Co(NH_3)_6]^{3+}$, the ground state, ψ_e, transforms as $^1A_{1g}$. There are two excited states with the same spin $(S = 0)$ which belong to the representations T_{1g} and T_{2g}. In the group O_h the coordinates x, y, z jointly form a basis for the T_{1u} representation. Thus, for the $^1A_{1g} \rightarrow {}^1T_{1g}$ transition the direct product representation of $\psi_e'(x, y, z)\psi_e$ is given by

$$\Gamma[\psi_e'(x, y, z)\psi_e] = T_{1g} \times T_{1u} \times A_{1g}$$
$$= T_{1g} \times T_{1u}$$

This can be reduced to

$$A_{1u} + E_u + T_{1u} + T_{2u}$$

Thus, if there are any normal vibrations whose first excited states belong to any of these representations, there will be nonvanishing intensity integrals. By the methods of Chapter 10 it is easily found that the symmetries of the normal modes of an octahedral AB_6 molecule are

$$A_{1g}, E_g, 2T_{1u}, T_{2g}, T_{2u}$$

Thus, while the pure electronic transition $^1A_{1g} \rightarrow {}^1T_{1g}$ is not allowed, all the transitions in which there is simultaneous excitation of a vibration of T_{1u} or T_{2u} symmetry are allowed.

Similarly, for an $^1A_{1g} \rightarrow {}^1T_{2g}$ transition, we find

$$\Gamma[\psi_e'(x, y, z)\psi_e] = T_{2g} \times T_{1u} \times A_{1g}$$
$$= T_{2g} \times T_{1u}$$
$$= A_{2u} + E_u + T_{1u} + T_{2u}$$

Thus the $^1A_{1g} \rightarrow {}^1T_{2g}$ transition can also occur so long as there is simultaneous excitation of a T_{1u} or T_{2u} vibration.

Vibronic Polarization

For an octahedral complex we see that the direction of vibration of the electric vector of the light makes no difference, for the directions $x, y,$ and z are equivalent in the sense that they are interchangeable by the symmetry operations of the molecule. However, in less symmetrical complexes in which $x, y,$ and z do not all belong to the same representation, we encounter the phenomenon of *polarization*.

Let us suppose that we place a polarizing prism between the light source and the sample. If the sample is a single crystal in which all of the molecules have the same orientation relative to the crystallographic axes, we can so orient the crystal that the direction of the electric vector of the light will correspond to the *x*, *y*, or *z* direction in a coordinate system for the molecule. It is then possible that some transition may occur for only one or two of these orientations but not for all three.

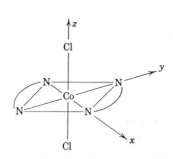

Figure 9.12   A sketch of the *trans*-dichlorobisethylenediaminecobalt-(III) ion showing a set of reference axes.

As an example, let us consider a *trans*-substituted octahedral complex such as *trans*-dichlorobisethylenediaminecobalt-(III), which is pictured in Figure 9.12 with a set of coordinate axes. Although the symmetry is no longer cubic, there is still a center of inversion so that *d-d* transitions can have finite intensity only if there is vibronic coupling. Since it is only vibrations in the part of the molecule including and immediately surrounding the cobalt ion which would be expected to have any appreciable interaction with the electronic wave functions of the metal ion, we shall restrict attention to the normal vibrations of a *trans*-$[CoCl_2N_4]$ group, which has local symmetry D_{4h}. The methods of Chapter 10 will tell us that for such a group the first excited states of the normal modes have the following symmetries:

$$2A_{1g}, B_{1g}, B_{2g}, E_g, 2A_{2u}, B_{1u}, 3E_u$$

For the *trans*-dichloro complex the ground state will be a $^1A_{1g}$ state, as in a strictly octahedral complex, but the excited singlet states $^1T_{1g}$ and $^1T_{2g}$ will be split as follows (cf. the correlation table, Appendix IIIB):

$$T_{1g} : A_{2g} + E_g$$
$$T_{2g} : B_{2g} + E_g$$

Thus the possible transitions from the ground state to excited states will be of the following types so far as the symmetry of the electronic states is concerned:

$$(1) \ A_{1g} \rightarrow A_{2g}$$
$$(2) \ A_{1g} \rightarrow B_{2g}$$
$$(3) \ A_{1g} \rightarrow E_g$$

For these transitions we obtain the following results for the representations of the purely electronic dipole integrals:

	$A_{1g} \rightarrow A_{2g}$	$A_{1g} \rightarrow B_{2g}$	$A_{1g} \rightarrow E_g$
$\int \psi_e' z \psi_e \, d\tau$	A_{1u}	B_{1u}	E_u
$\int \psi_e'(x, y) \psi_e \, d\tau$	E_u	E_u	$A_{1u} + A_{2u} + B_{1u} + B_{2u}$

Comparing these results with the list of the symmetries of the first excited states of the normal vibrations, we can immediately write the following predictions for the polarizations of the transitions:

	POLARIZATION WITH VIBRONIC COUPLING	
TRANSITION	z	(x, y)
$A_{1g} \rightarrow A_{2g}$	Forbidden	Allowed
$A_{1g} \rightarrow B_{2g}$	Allowed	Allowed
$A_{1g} \rightarrow E_g$	Allowed	Allowed

These results have actually been used to analyze experimental data. Figure 9.13 shows the experimental observations of Yamada et al.* on *trans*-[Co(en)$_2$Cl$_2$]Cl $\cdot$ HCl $\cdot$ 2H$_2$O. It can be seen that there are three regions of

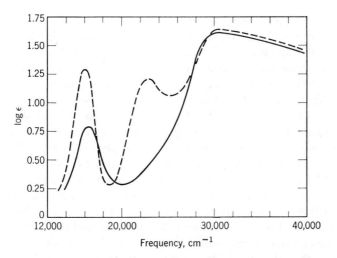

Figure 9.13   Dichroism of *trans*-[Co(en)$_2$Cl$_2$]$^+$ after the results of Yamada et al. (*loc. cit*) on *trans*-[Co(en)$_2$Cl$_2$]Cl$\cdot$2HCl$\cdot$2H$_2$O. The full line shows the spectrum with light polarized parallel (or nearly parallel) to the Cl—Co—Cl axis, and the dashed line shows the spectrum with light polarized perpendicular to the Cl—Co—Cl axis.

* S. Yamada et al., *Bull. Chem. Soc. Japan*, **25**, 127 (1952); **28**, 222 (1955).

absorption. From about 27,000 to about 40,000 cm^{-1} the absorption exhibits no significant polarization. At $\sim 22,000$ cm^{-1} there is a strongly polarized band which is absent for light parallel to z but present for light perpendicular to z, while at $\sim 16,000$ cm^{-1} there is a band which shows some difference in intensity for the two directions of polarization but is clearly present in both. The results of the above analysis permit a simple interpretation of these observations. The bands at $\sim 16,000$ and $\sim 22,000$ cm^{-1} are assigned, respectively, to transitions to the E_g and A_{2g} components of the T_{1g} state existing in O_h symmetry, since an $A_{1g} \to E_g$ transition is vibronically allowed for both directions, but an $A_{1g} \to A_{2g}$ transition is vibronically forbidden for z polarization. The broad absorption above $\sim 27,000$ cm^{-1} may be assigned to the unresolved transitions from the A_{1g} ground state to the E_g and B_{2g} states coming from the T_{2g} state for O_h symmetry. An interpretation along these lines was first given by Ballhausen and Moffitt,* who also showed that independent calculations and experimental evidence would lead to the expectation that the order of the excited levels should be as postulated in order to explain the polarization data.

Noncentrosymmetric Complexes

When the complex lacks a center of symmetry even in its equilibrium configuration, the "d-d" transitions become allowed as simple changes in the electronic wave functions. The breadth of the absorption bands shows that the electronic transitions are still accompanied by vibrational changes, but these vibrational changes are not in themselves essential to the occurrence of the transition. It is believed that the way in which the noncentrosymmetric ligand field alters the d wave function so that the states of the "d^n" configuration no longer have rigorous g character is by making possible the mixing of d and p orbitals. We have already shown on page 201 that in a tetrahedral field the p orbitals and the d_{xy}, d_{xz}, and d_{yz} orbitals belong to the same (T_2) representation, and that to some finite degree each of the two sets of T_2 orbitals which will be found among the orbitals arising from the valence shell atomic orbitals of a transition metal atom must have both p and d character. Then, if two different electronic states of the "d^n" configuration of a metal ion contain different amounts of p character, a transition from one to the other will be to a certain extent a $d \to p$ or $p \to d$ transition, which is highly allowed even in the free atom since the d orbitals are even to inversion and the p orbitals odd. The exact extent to which this mixing occurs and the resulting intensity of the transition must, of course, be computed

* C. J. Ballhausen and W. Moffit, *J. Inorg. Nucl. Chem.*, 3, 178 (1956).

using explicit wave functions, but symmetry considerations alone can tell us whether it is possible for a particular transition to acquire any intensity at all in this way.

As a first illustration let us consider the optical transitions in a tetrahedral complex of Co(II). The ground state belongs to the A_2 representation of the tetrahedral point group T_d, and there are two excited states of T_1 symmetry and one of T_2 symmetry. The character table for T_d tells us that the coordinates x, y, and z form a basis for the T_2 representation. For the $A_2 \rightarrow T_1$ transitions we then see that the intensity integral will span the representations in the direct product of $A_2 \times T_1 \times T_2$, and this reduces as follows:

$$A_2 \times T_1 \times T_2 = A_1 + E + T_1 + T_2$$

Since the A_1 representation is present, it follows that these transitions are allowed by the symmetry of the purely electronic wave functions. For the $A_2 \rightarrow T_2$ transition we must consider the direct product $A_2 \times T_2 \times T_2$, which reduces as follows:

$$A_2 \times T_2 \times T_2 = A_2 + E + T_1 + T_2$$

We see that the $A_2 \rightarrow T_2$ transition is not allowed by the symmetry of the pure electronic wave functions and that whatever intensity it may have must be attributed to vibronic interaction. In agreement with this prediction, it has been found that the $A_2 \rightarrow T_2$ transition is observed but with an intensity 10–100 times smaller than the intensities of the $A_2 \rightarrow T_1$ transitions in the systems [for example, Co(II) in ZnO] which have been studied.

Polarization of Electronically Allowed Transitions

Just as with vibronically allowed transitions, in symmetry groups in which all Cartesian axes are not equivalent (noncubic groups), it is found that, in general, transitions will be allowed only for certain orientations of the electric vector of the incident light. One class of compounds in which this phenomenon has been studied both theoretically and experimentally consists of trischelate compounds such as tris(acetylacetonato)M(III) and tris(oxalato)M(III) complexes. In these complexes the six ligand atoms form an approximately octahedral array but the true molecular symmetry is only D_3. Since there is no center of symmetry in these molecules, the pure electronic selection rules might be expected to be dominant.

For the tris(oxalato)Cr(III) ion, $[Cr(C_2O_4)_3]^{3-}$, the electronic states are those into which the states of a complex of O_h symmetry are reduced when the symmetry is reduced to D_3. From the correlation table (Appendix IIIB) we see that the correlation of the O_h and D_3 states is as follows:

O_h	D_3
A_{2g}	A_2 (ground state)
T_{1g}	$A_2 + E$
T_{2g}	$A_1 + E$

Thus in D_3 symmetry we want to know the polarizations of the following types of transitions: $A_2 \rightarrow A_1$, $A_2 \rightarrow A_2$, and $A_2 \rightarrow E$. Noting in the character table for D_3 that z belongs to the A_2 representation and (x, y) to the E representation, we obtain the following results for the irreducible representations spanned by the dipole integrals for each of these transitions:

	$A_2 \rightarrow A_1$	$A_2 \rightarrow A_2$	$A_2 \rightarrow E$
$\int \psi'_e z \psi_e \, d\tau$	A_1	A_2	E
$\int \psi'_e (x, y) \psi_e \, d\tau$	E	E	$A_1 + A_2 + E$

Thus the selection rules are:

TRANSITION	POLARIZATION OF INCIDENT RADIATION	
	z	(x, y)
$A_2 \rightarrow A_1$	Allowed	Forbidden
$A_2 \rightarrow A_2$	Forbidden	Forbidden
$A_2 \rightarrow E$	Forbidden	Allowed

It can be seen that these are very powerful selection rules indeed. On the other hand, we might have assumed that the symmetry of the environment of the metal ion could have been adequately approximated by considering only the six coordinated oxygen atoms. In this case, the symmetry would be D_{3d}, in which there is a center of inversion and the transitions would be governed by vibronic selection rules. When these are worked out, it is found that all of the transitions are vibronically permitted. Thus, experimental study of the polarizations should provide clear-cut evidence as to the correct effective symmetry and selection rules. Such a study has been reported* and shows conclusively that the selection rules followed are those given above for pure electronic transitions in D_3 symmetry.

* T. S. Piper and R. L. Carlin, *J. Chem. Phys.*, **35**, 1809 (1961). These authors also give selection rules and experimental data for the oxalato complexes of the trivalent ions of Ti, V, Mn, Fe, and Co.

9.7 Double Groups

In Section 9.3 we showed that for an orbital or state wave function having angular momentum quantum number l (or L) the character of the representation for which this forms a basis, under a symmetry operation which consists of rotation by an angle α, is given by

$$\chi(\alpha) = \frac{\sin (l + \frac{1}{2})\alpha}{\sin \alpha/2} \qquad (9.3\text{-}2)$$

We then applied this formula to various types of single-electron wave functions for example, s, p, d, f, g, and to wave functions for various Russell-Saunders terms characterized by integral values of the quantum number L.

There are, however, many cases of interest in which we may want to determine the splitting of a state that is well characterized by its total angular momentum, J. This will in fact be the only thing of importance in the very heavy elements, for example, the rare-earth ions, where states of particular L cannot be used since the various free-ion states of different J are already separated by much greater energies than the crystal field splitting energies.

Now $J = L + S$, and for ions with an odd number of electrons S and hence J must be half integral numbers. For states in which J is an integer, the characters can be obtained by using the above formula with l replaced by J. However, when J is half integral a difficulty arises. We know that a rotation by 2π is an identity operation and therefore it should be true that

$$\chi(\alpha) = \chi(\alpha + 2\pi)$$

It can easily be seen that this is true when L or J is an integer. However, when J is half integral we have

$$\chi(\alpha + 2\pi) = \frac{\sin (J + \frac{1}{2})(\alpha + 2\pi)}{\sin (\alpha + 2\pi)/2} = \frac{\sin [(J + \frac{1}{2})\alpha + 2\pi]}{\sin [\alpha/2 + \pi]}$$

$$= \frac{\sin (J + \frac{1}{2})\alpha}{-\sin \alpha/2}$$

$$= -\chi(\alpha)$$

Since the characters of a representation must be uniquely defined, we see that those we would obtain by the above procedure when J is half integral cannot belong to true representations.

A simple device for avoiding this difficulty was proposed by Bethe. We introduce the fiction (mathematically possible but not physically significant) that rotation by 2π be treated as a symmetry operation but not as an identity operation. We must then expand any ordinary rotation group by taking the product of this new operation, which we shall call R, with all of the existing rotations. The new group will therefore contain twice as many operations and more classes and representations (though not twice as many) than the simple rotation group with which we start. This new group is called a *double group*.

In working out the various products $C_n{}^m R$ and $RC_n{}^m$, we first note that two rotations about the same axis commute so that $C_n{}^m R = RC_n{}^m$. If $n = 2$ we have RC_2, which is a special case since

$$\chi(\pi) = \chi(3\pi) = 0$$

For rotation by any other angle, viz., $m2\pi/n$, it is not difficult to show that the following equality holds generally:

$$\chi(m2\pi/n + 2\pi) = \chi[(n - m)2\pi/n]$$

In order to evaluate the characters of E and R, that is $\chi(0)$ and $\chi(2\pi)$, we must evaluate the limit of an indeterminate form, for as $\alpha \to 0$ or $\alpha \to 2\pi$,

$$\frac{\sin (J + \frac{1}{2})\alpha}{\sin \alpha/2} \to \frac{0}{0}$$

This may easily be done by using l'Hospital's rule, and the results are

$$\chi(0) = 2J + 1$$

$$\chi(2\pi) = \begin{cases} 2J + 1 & \text{when } J \text{ is an integer} \\ -(2J + 1) & \text{when } J \text{ is a half integer} \end{cases}$$

After we have worked out the characters for all of the new operations, $C_n{}^m R$, of the double group, we will then collect them into classes, using the same rule as for simple groups, namely, that all operations having the same characters are in the same class. Thus in general we shall find the following classes in double rotation groups:

(1) E
(2) R
(3) C_2 and $C_2 R$
(4) C_n and $C_n{}^{n-1} R$
(5) $C_n{}^m$ and $C_n{}^{n-m} R$

We can then determine the number and dimensions of the irreducible representations by using the familiar rules that there are as many irreducible representations as there are classes and that the sum of the squares of the dimensions of the irreducible representations must equal the group order.

In order to illustrate this procedure let us consider the group D_4 and the corresponding double group D'_4. The eight operations of D_4 are E, C_4, $C_4{}^3$, C_2, $2C'_2$, $2C''_2$. According to the general results given above, the sixteen operations of D'_4 may be arranged into the following classes:

$$E \quad R \quad C_4 \quad C_4{}^3 \quad C_2 \quad 2C'_2 \quad 2C''_2$$
$$C_4{}^3R \quad C_4R \quad C_2R \quad 2C'_2R \quad 2C''_2R$$

Since there are seven classes, there must be seven irreducible representations, and their orders, l_i, must satisfy the equation

$$l_1{}^2 + l_2{}^2 + l_3{}^2 + l_4{}^2 + l_5{}^2 + l_6{}^2 + l_7{}^2 = 16$$

It is easy to convince oneself that the only combination of positive integers satisfying this equation is 1, 1, 1, 1, 2, 2, 2, Thus there are four one-dimensional and three two-dimensional irreducible representations of the double group D'_4.

Double groups are of greatest importance for transition metal complexes. In Appendix VII we give the character tables for the double groups D'_4 and O' corresponding to the simple rotation groups D_4 and O. Several features of these tables should be noted. First, there are two systems for labeling the representations. One is an adaptation of the Mulliken system for simple groups, in which we use primed Mulliken symbols. The other is Bethe's original system, in which we use a serially indexed set of Γ_i's. Second, it will be noted that among the representations of the double groups are all the representations of the simple group. Whenever we form a representation using a wave function having an integral value of angular momentum, l, L, S, or J, it will either be one of these representations or it will be reducible to a sum of only these representations. In other words, when the angular momentum quantum number used is integral, we have no need of the double group. However, when the angular momentum quantum number, s, S, or J, is half integral, we will obtain one of the new representations not occurring in the simple group, or a representation which can be reduced to a sum containing only these new irreducible representations. It will be noted that all of these new representations have an even order, 2, 4, and so on. Thus all wave functions of a system must be at least twofold degenerate. This is a manifestation of Kramer's theorem that in the absence of an external magnetic field the spin degeneracy of a system having an odd number of electrons must always persist even when the low symmetry of the environment lifts all other degeneracies.

The direct products of representations of double groups can be taken in the usual way and reduced to sums of irreducible representations.

In order to illustrate the utility of double groups let us consider several examples. Suppose that we have an ion with one d electron in a planar complex. Real examples of this case are represented by complexes of Cu(II) and Ag(II) (where we virtually have one positron, but this behaves as one electron except in the signs of the energies). In each case there will be two states with J values of $l \pm \frac{1}{2} = 2 \pm \frac{1}{2}$, namely, $J = \frac{3}{2}$ and $J = \frac{5}{2}$. Using Equation 9.3-2, we find that these form bases for the following representations:

D_4'	E	R	$2C_4$	$2C_4 R$	$2C_2$	$4C_2'$	$4C_2''$
$\Gamma_{3/2}$	4	-4	0	0	0	0	0
$\Gamma_{5/2}$	6	-6	$-\sqrt{2}$	$\sqrt{2}$	0	0	0

These can be reduced in the standard way, giving

$$\Gamma_{3/2} = \Gamma_6 + \Gamma_7$$
$$\Gamma_{3/2} = \Gamma_6 + 2\Gamma_7$$

The procedure we have used would be particularly appropriate in the case of Ag(II), where the two J states are already well separated in the free ion because of a very large spin-orbit coupling. An energy level diagram of the following sort could then be drawn:

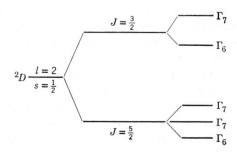

This illustrates how the two J states would be further split up by an environment of D_{4h} symmetry.

If the spin-orbit coupling is relatively small, we might wish to consider first the splitting of the ^2D state by the environment and then the further splitting of the resulting states by spin-orbit coupling. To do this we first use the D_{4h} character table to find that the ^2D state splits into A_{1g}, B_{1g}, B_{2g}, and E_g. These representations in D_{4h} can easily be seen to correspond, respectively, to A_1, B_1, B_2, and E in D_4, and to Γ_1, Γ_3, Γ_4, and Γ_5 in D_4'. This analysis has dealt with the orbital part of the wave function (see Section 9.2, especially Equation 9.2-1). For the spin part, ψ_s in Equation 9.2-1, we find the representation for which the spin angular momentum, $s = \frac{1}{2}$, forms a basis in the double group D'_4. Using the formulae given above, we easily obtain

D'_4	E	R	$2C_4$	$2C_4 R$	$2C_2$	$4C'_2$	$4C''_2$
$\Gamma_{1/2}$	2	-2	$\sqrt{2}$	$-\sqrt{2}$	0	0	0

Thus

$$\Gamma_{1/2} = \Gamma_6$$

Now, to find the representation of a wave function which is a product of two other functions we must obtain the representation of the direct product of the two functions. The characters of this representation are the products of the characters of the representations of the two functions. Thus, using the character table for D'_4, we obtain the results

$$\Gamma_1 \times \Gamma_6 = \Gamma_6$$
$$\Gamma_3 \times \Gamma_6 = \Gamma_7$$
$$\Gamma_4 \times \Gamma_6 = \Gamma_7$$
$$\Gamma_5 \times \Gamma_6 = \Gamma_6 + \Gamma_7$$

Of course the final results are the same as those previously obtained. The manner in which we obtain them is unimportant so far as pure symmetry considerations are concerned. Normally, however, we would choose the first method in a case where we expected the splitting between the free-ion states with $J = \frac{3}{2}$ and $J = \frac{5}{2}$ to be greater than the further splittings caused by the environments, and the second method when we expected the splitting of the Russell-Saunders term, ^2D, by the environment to be much larger than the splittings due to spin-orbit coupling. In the latter case our energy level diagram for one electron might look somewhat as follows:

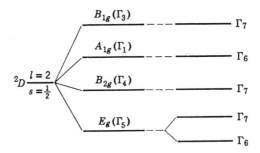

In this example the relative order of the orbitals is somewhat arbitrary. Only by solving an appropriate wave equation could the actual order be determined in a particular case, but the result would have to correspond with this diagram in regard to the types and the number of each type of wave functions obtained.

Exercises

9.1 Use the subgroup C_{2v} in the method of descending symmetry to verify the correlation of Figure 9.3.

9.2 Select an appropriate subgroup of T_d and verify Figure 9.4 by the method of descending symmetry.

9.3 The selection and polarization rules for magnetic dipole-allowed transitions are given by a set of integrals of the type 5.3-3 in which the transition moment operators have the same symmetry properties as the rotations, R_x, R_y, R_z. The magnetic dipole mechanism is weaker by several orders of magnitude than the electronic dipole one, but in centrosymmetric situations, where *d-d* transitions are not electric dipole allowed, magnetic dipole and vibronic mechanisms both need to be considered to explain the weak bands. Work out the magnetic dipole polarizations for *trans*-$[Coen_2Cl_2]^+$. Are they consistent with the observations or not?

9.4 How will the *d* orbitals split in a trigonal bipyramidal environment? Using a crystal field (pure point charge) model, determine what the relative energies of the orbitals will be.

9.5 Assume that in the orbital energy diagram obtained in Exercise 9.4 the *a'* and *e'* orbitals are separated by three times as much distance as are the *e'* and *e"* orbitals. Work out a weak-field to strong-field correlation diagram for the case of a d^2 ion in a trigonal bipyramidal environment.

Molecular Vibrations

10.1 Introductory Remarks

A molecule possesses three types of internal energy. These are, in the usual decreasing order of their magnitudes, electronic, vibrational, and rotational energies. In preceding chapters we have dealt with the use of symmetry properties for understanding the electronic states of various kinds of molecules. The rotational energy states have no symmetry properties of importance in ordinary chemical processes and will not concern us directly in this book. We are left, then, with the subject of molecular vibrations, to which symmetry arguments may be very fruitfully applied.

Every molecule, at all temperatures, including even the absolute zero, is continually executing vibrational motions, that is, motions in which its distances and internal angles change periodically without producing any net translation of the center of mass of the molecule or imparting any net angular momentum (rotatory motion) to the molecule. Of course the molecule may, and if free certainly will, be traveling through space and rotating, but we may divorce our attention from these motions by supposing that we are seated at the center of gravity of the molecule and that we travel and rotate with it. Then it will not appear to us to be undergoing translation or rotation, and our full attention may be focused on its internal or *genuine* vibrations.

Although a cursory glance at a vibrating molecule might suggest that its vibratory motion is random, close inspection and proper analysis reveal a basic regularity and simplicity. It is the underlying basis for this simplicity which we shall formulate in this chapter. We shall also develop working methods by which all of the analysis of molecular motions which symmetry alone allows may be rapidly and reliably performed.

10.2 The Symmetry of Normal Vibrations

The complex, random, and seemingly aperiodic internal motions of a vibrating molecule are the result of the superposition of a number of relatively simple vibratory motions known as the *normal vibrations* or *normal modes of vibration* of the molecule. Each of these has its own fixed frequency. Naturally, then, when many of them are superposed, the resulting motion must also be periodic, but it may have a period so long as to be difficult to discern.

The first question to be considered regarding the normal modes is that of their number in any given molecule. This, fortunately, is a very easy question, and doubtless many readers know the answer already. An atom has three degrees of motional freedom. It may move from an initial position in the x direction independent of any displacement that it may or may not undergo in the y or z direction, in the y direction independent of whether or not it moves in the x or z direction, and so on. In the molecule consisting of n atoms there will thus be $3n$ degrees of freedom.

Let us now suppose that all n atoms move simultaneously by the same amount in the x direction. This will displace the center of mass of the entire molecule in the x direction without causing any alteration of the internal dimensions of the molecule. The same may of course be said of similar motions in the y and z directions. Thus, of the $3n$ degrees of freedom of the molecule, three are not genuine vibrations but only translations. Similarly, concerted motions of all atoms in circular paths about the x, y, and z axes do not constitute vibrations either but are instead, molecular rotations. Thus, of the $3n$ degrees of motional freedom, only $3n - 6$ remain to be combined into genuine vibratory motions.

We make note here of the special case of a linear molecule. In that instance rotation of the molecule may occur about each of two axes perpendicular to the molecular axis, but "rotation" of nuclei about the molecular axis itself cannot occur since all nuclei lie on the axis. Thus an n-atomic linear molecule has $3n - 5$ normal modes.

Let us now look at the normal modes of vibration of a molecule which is as simple as possible and yet exemplifies all general features ordinarily encountered. The planar ion CO_3^{2-} will serve for this purpose. As a nonlinear four-atomic species, it must have $3(4) - 6 = 6$ normal modes. In Figure 10.1 we have depicted these vibrations. In each drawing the length of an arrow relative to the length of another arrow in the same drawing shows how much the atom to which it is attached is displaced at any instant relative to the simultaneous displacement of the atom to which the other arrow is attached. The lengths of the arrows relative to interatomic distances in the drawings, however, are exaggerated.

As may be seen in the particular case of CO_3^{2-}, which we are using as an illustration, normal modes have two important properties:

1. Each of the vectors representing an instantaneous atomic displacement may be regarded as the resultant of a set of three basis vectors.

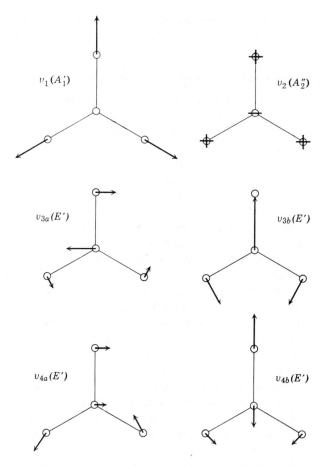

Figure 10.1   The six normal modes of vibration of the carbonate ion.

2. Each of the normal modes forms a basis for or "belongs to" an irreducible representation of the molecule.

Let us first consider the ways in which we might regard the displacement vectors in the normal modes as resultants of some set of basis vectors. There are many ways in which the set of basis vectors might be chosen, but only two are of interest. In the first, we attach a separate Cartesian coordinate system to each atom, with the atom at the origin and all x axes, all y axes, and all z axes parallel and pointing in the same direction (cf. Figure 10.3). In each small coordinate system we place unit vectors along the x, y, and z axes. Now, the vector representing the displacement of a given atom, the ith atom, may be expressed as the vector sum of Cartesian displacement vectors of this atom, x_i, y_i, and z_i. We may call this process resolution of a general displacement into Cartesian displacements. It will be noted that the three translational motions and the three (or two) rotational motions may also be resolved into vector sums of Cartesian displacements. Thus all $3n$ degrees of motional freedom of the molecule may be represented by suitable combinations of the $3n$ Cartesian displacements.

The second important way of resolving the displacement vectors of the normal modes is to use basis vectors related to the internal coordinates of the molecule, that is, the interatomic distances and bond angles. There is no unique way of doing this, Normally, however, we choose first the changes in interatomic distances between bonded atoms and then as many changes in bond angle (taking care that those chosen are all independent) as are necessary to provide a set of $3n - 6$ internal displacement vectors. For example, in the carbonate ion we require six internal displacement vectors to represent the six normal modes. We choose first changes in the three C—O distances. Next we may choose changes in two of the three OCO angles. Our sixth choice might be a change in the remaining OCO angle or a change in the angle between a C-O bond axis and the molecular plane.

Let us consider now the second important property of the normal modes, namely, their symmetry. It is easy to see by comparison of the diagrams in Figure 10.1 and the character table for the group D_{3h}, to which the carbonate ion belongs, that each normal mode (or pair of normal modes) transforms exactly as required by the characters of the representation to which it belongs. These representations are noted in parenthesis on Figure 10.1. Clearly the set of vectors representing v_1 is carried into itself by all operations; hence it belongs to the A_1' representation. It is equally obvious that the set of vectors representing v_2 is carried into itself by the operations E, C_3, and σ_v but into the negative of itself by C_2, S_3, and σ_h. Thus this mode belongs to the A_2'' representation, as stated in Figure 10.1.

The v_{3a} and v_{3b} modes together form the basis for the E' representation of the group D_{3h}. Clearly the identity operation takes each component into itself, as required by the character of 2. We can express this symbolically as follows:

$$E(v_{3a}) = v_{3a} + 0v_{3b}$$
$$E(v_{3b}) = 0v_{3a} + v_{3b}$$

The matrix of the coefficients on the right side of this set of equations is

$$\begin{bmatrix} 1 & 0 \\ 0 & 1 \end{bmatrix}$$

It is, of course, a two-dimensional unit matrix with the character 2. The effect of a threefold rotation on either v_{3a} or v_{3b} is to take the mode into a linear combination of both v_{3a} and v_{3b}. Figure 10.2 illustrates this for the case of a

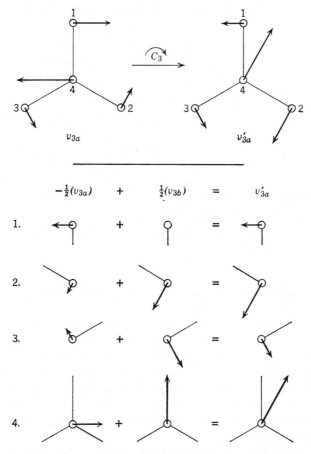

Figure 10.2   Vector diagrams showing how a threefold rotation transforms ν_{3a} into ν'_{3a} and how the latter is a linear combination of ν_{3a} and ν_{3b}; specifically, $\nu'_{3a} = -\frac{1}{2}\nu_{3a} + \frac{1}{2}\nu_{3b}$.

clockwise rotation of ν_{3a} by $2\pi/3$, to give a mode which we have labeled ν'_{3a}. The lower part of Figure 10.2 shows in detail how each of the displacement vectors in ν'_{3a} is the vector sum of $-\frac{1}{2}\nu_{3a}$ and $\frac{1}{2}\nu_{3b}$. Thus we can write

$$C_3(\nu_{3a}) = -\tfrac{1}{2}\nu_{3a} + \tfrac{1}{2}\nu_{3b}$$

It can similarly be shown that application of a clockwise rotation by $2\pi/3$ to ν_{3b} would produce a mode which could be expressed as the following linear combination of ν_{3a} and ν_{3b};

$$C_3(\nu_{3b}) = -\tfrac{3}{2}\nu_{3a} - \tfrac{1}{2}\nu_{3b}$$

Now the matrix of the coefficients of these two transformations is:

$$\begin{bmatrix} -\frac{1}{2} & \frac{1}{2} \\ -\frac{3}{2} & -\frac{1}{2} \end{bmatrix}$$

and its character is -1 as required by the character table.

It is easy to see that the operation C_2 transforms v_{3a} into the negative of itself and v_{3b} into itself. Thus a matrix is obtained which has only the diagonal elements -1 and 1 and the character 0 as required by the character table. It is equally easy to see that σ_h carries each component of v_3 into itself, so that the matrix of the transformation has only the diagonal elements 1 and 1 and hence a character of 2. We could carry out similar reasoning for the remaining operation applied to v_{3a} and v_{3b} and also with respect to the application of all of the operations in the group to v_{4a} and v_{4b}, and it would be found that they satisfy the requirements of the characters of the E' representation in every respect.

10.3 Determining the Symmetry Types of the Normal Modes

The two characteristic features of normal modes of vibration which have been stated and discussed above lead directly to a simple and straight-forward method of determining how many of the normal modes of vibration of any molecule will belong to each of the irreducible representations of the point group of the molecule. This information may be obtained entirely from knowledge of the molecular symmetry and does not require any knowledge, or by itself provide any information, concerning the frequencies or detailed forms of the normal modes.

We have illustrated in detail for the case of CO_3^{2-} how the normal modes of genuine vibration have symmetry corresponding to one or another of the irreducible representations of the molecule. This is true for every molecule, though we shall not offer proof here.* It is also true that the nongenuine vibrations, the translational and rotational motions, transform according to irreducible representations of the molecular point group. Moreover the entire set of $3n$ normal modes may be expressed as functions of a set of $3n$ Cartesian displacements, as described in the preceding section. It is also evident that we may use the $3n$ Cartesian displacement vectors as the basis for a reducible representation of the molecular symmetry group. This representation will contain (or, as is sometimes said, span) the set of irreducible representa-

---

* Rigorous proof, which involves more quantitative discussion of the mechanics of the normal vibrations and the use of explicit expressions for the kinetic and potential energies, may be found in more specialized texts, for example, *Molecular Vibrations* by E. B. Wilson, J. C. Decius, and P. C. Cross, McGraw-Hill, New York, 1955.

tions to which all of the normal modes, genuine and nongenuine, belong.

We shall now illustrate this, using CO_3^{2-} as an example. A number of further examples will be found in Section 10.7. The first step must be to determine the symmetry group to which the molecule belongs, as described in Chapter 3, especially Section 3.13. We find that CO_3^{2-} belongs to the D_{3h} group. Figure 10.3 shows the CO_3^{2-} ion with the sets of Cartesian displace-

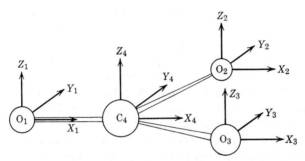

Figure 10.3 The set of $3n = 12$ Cartesian displacement vectors used in determining the reducible representation spanning the irreducible representation of the normal modes of CO_3^{-2}.

ment vectors attached to each atom. There are of course $3n = 12$ in all, and the representation will therefore be of dimension 12.

We turn now to the character table for the group D_{3h}. The first operation, naturally, is the identity operation. When this is applied to the set of vectors, each remains in place, that is, remains identical with itself. We may express this as shown in Figure 10.4. We assume that the symmetry operation is

	X_1	Y_1	Z_1	X_2	Y_2	Z_2	X_3	Y_3	Z_3	X_4	Y_4	Z_4
X_1'	1	0	0	0	0	0	0	0	0	0	0	0
Y_1'	0	1	0	0	0	0	0	0	0	0	0	0
Z_1'	0	0	1	0	0	0	0	0	0	0	0	0
X_2'	0	0	0	1	0	0	0	0	0	0	0	0
Y_2'	0	0	0	0	1	0	0	0	0	0	0	0
Z_2'	0	0	0	0	0	1	0	0	0	0	0	0
X_3'	0	0	0	0	0	0	1	0	0	0	0	0
Y_3'	0	0	0	0	0	0	0	1	0	0	0	0
Z_3'	0	0	0	0	0	0	0	0	1	0	0	0
X_4'	0	0	0	0	0	0	0	0	0	1	0	0
Y_4'	0	0	0	0	0	0	0	0	0	0	1	0
Z_4'	0	0	0	0	0	0	0	0	0	0	0	1

Figure 10.4 The matrix expressing the effect of the identity operation on the set of Cartesian displacement coordinates (Figure 10.3) for CO_3^{2-}.

applied only to the set of vectors, moving them but leaving the nuclei themselves fixed. Thus we may specify or label each vector before the operation by stating its direction and the number of the atom to which it is attached, viz., X_1 or Z_4. For the same vector after the symmetry operation we use the same symbol primed, whether the vector has moved in any way or not. The left vertical column in Figure 10.4 lists the vectors after application of the symmetry operation, and the top horizontal row lists the original set. The purpose of Figure 10.4 and a similar device for each symmetry operation is to express the composition of the primed vectors in terms of those in the original unprimed set. In this case the results are trivial: each primed vector is identical with its unprimed counterpart. The square array of numbers so obtained is a matrix describing the effect of the symmetry operation upon the set of vectors, and its character is the character corresponding to the particular operation in the reducible representation that we are seeking. Thus, for the identity operation, we have here a character of 12.

We now apply a threefold rotation to the set of Cartesian displacement vectors with the results pictured in Figure 10.5. Again we wish to construct

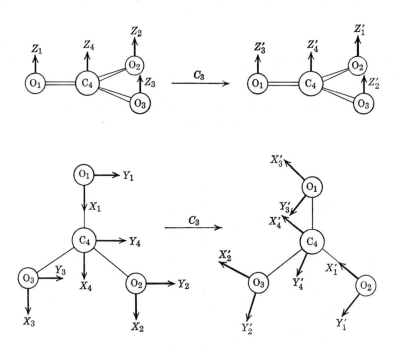

Figure 10.5   Diagrams showing the effect of a threefold rotation on the set of Cartesian displacement vectors.

the matrix expressing these results. This procedure is a trifle tedious but requires no more than the simplest trigonometry. For example, as Figure 10.6 shows, X'_1 can be expressed as $-\frac{1}{2}X_2 - (\sqrt{3}/2)\,Y_2$, and this result has been entered in the first row of the matrix, Figure 10.7. The reader should have no

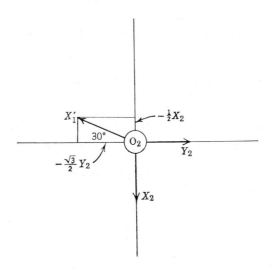

Figure 10.6   Diagram showing the resolution of the displacement vector X'_1 into its X_2 and Y_2 components.

	X_1	Y_1	Z_1	X_2	Y_2	Z_2	X_3	Y_3	Z_3	X_4	Y_4	Z_4
X'_1	0	0	0	$-\frac{1}{2}$	$-\sqrt{\frac{3}{2}}$	0	0	0	0	0	0	0
Y'_1	0	0	0	$\sqrt{\frac{3}{2}}$	$-\frac{1}{2}$	0	0	0	0	0	0	0
Z'_1	0	0	0	0	0	1	0	0	0	0	0	0
X'_2	0	0	0	0	0	0	$-\frac{1}{2}$	$-\sqrt{\frac{3}{2}}$	0	0	0	0
Y'_2	0	0	0	0	0	0	$\sqrt{\frac{3}{2}}$	$-\frac{1}{2}$	0	0	0	0
Z'_2	0	0	0	0	0	0	0	0	1	0	0	0
X'_3	$-\frac{1}{2}$	$-\sqrt{\frac{3}{2}}$	0	0	0	0	0	0	0	0	0	0
Y'_3	$\sqrt{\frac{3}{2}}$	$-\frac{1}{2}$	0	0	0	0	0	0	0	0	0	0
Z'_3	0	0	1	0	0	0	0	0	0	0	0	0
X'_4	0	0	0	0	0	0	0	0	0	$-\frac{1}{2}$	$-\sqrt{\frac{3}{2}}$	0
Y'_4	0	0	0	0	0	0	0	0	0	$\sqrt{\frac{3}{2}}$	$-\frac{1}{2}$	0
Z'_4	0	0	0	0	0	0	0	0	0	0	0	1

Figure 10.7   The matrix expressing the effect of the operation C_3 on the set of Cartesian displacement coordinates for $CO_3{}^{2-}$.

difficulty in verifying the other entries in Figure 10.7. The character of this matrix has the value zero.

Before proceeding further, we take note of a labor-saving procedure. The operation C_3 shifts all of the vectors originally associated with atom 1 to atom 2. Thus, when we write down the components of X_1', Y_1', and Z_1', we find that they come entirely from the set X_2, Y_2, Z_2. This being so, all of the diagonal elements in the first three rows are zero. Since all of the vectors initially on atom 2 are rotated to atom 3, there are no nonzero diagonal elements in the next three rows. For the same reason, there are none in the next three rows either. Only the vectors X_4, Y_4, and Z_4 which the operation C_3 merely shuffles among themselves but does not shift to a different atom, contribute nonzero diagonal elements. Thus we could have determined the character of the C_3 matrix by ignoring all the vectors which are shifted from one atom to another when the molecule is rotated and taking account of only those which remain associated with the same atom. Henceforth we shall approach the task in this way.

When the molecule is subjected to a twofold rotation, we see that vectors on two of the oxygen atoms, let us say numbers 2 and 3, are shifted. Thus we know that these vectors will contribute nothing to the character of the matrix. For each of the other two atoms, O_1 and C_4, however, the Z vectors are transformed into the negatives of themselves, two Y vectors go over into their own negatives, while the X vectors remain unchanged. These results are expressed in the abbreviated matrix of Figure 10.8, in which only the elements

	X_1	Y_1	Z_1	X_4	Y_4	Z_4
X_1'	1	0	0			
Y_1'	0	-1	0			
Z_1'	0	0	-1			
X_4'				1	0	0
Y_4'				0	-1	0
Z_4'				0	0	-1

Figure 10.8 Abbreviated matrix for the operation C_2 on the Cartesian displacement vectors of CO_3^{2-}.

relating to the vectors on atoms 1 and 4 are given. The value of the character is seen to be -2.

The character of the matrix corresponding to the operation σ_h may be found without writing out any part of the complete matrix itself. We note that the operation σ_h does not shift any vectors from one atom to another. Hence no set of vectors may be summarily ignored. We note further that each set of vectors will be affected by σ_h in exactly the same way. Thus whatever contribution to the character is made by one of the four sets may simply be multiplied by 4 in order to get the total value of the character. In any one set, σ_h transforms the X and Y vectors into themselves and the Z vector into its own negative. Thus the submatrix for this set of vectors will be diagonal with the elements 1, 1, and -1 and hence a character of 1. The character of the entire matrix corresponding to the operation σ_h is thus 4.

The operation S_3 shifts all the vectors on oxygen atoms, so we know that these nine vectors can be ignored. The effect of S_3 on the vectors of the carbon atom can be found quickly by recalling the effect of C_3 on the same set, as shown in the lower right portion of Figure 10.7. Since S_3 is simply C_3 followed by σ_h, its effect on X_4 and Y_4 is the same as that of C_3. However, whereas C_3 left Z_4 unaffected, S_3 transforms it into its negative. Thus S_3 acting on the displacement vectors of the carbon atom causes the following elements to appear along the diagonal of the matrix: $-\frac{1}{2}$, $-\frac{1}{2}$, -1. The character of the entire matrix is therefore -2.

Finally, we have to consider the operation σ_v. Let us choose the plane passing through O_1 and C_4. The vectors on atoms 2 and 3, which are shuffled by reflection through this plane, may be dismissed from consideration. It is obvious by inspection that X_1, Z_1, X_4, and Z_4, go into themselves, while Y_1 and Y_4 go into the negative of themselves by reflection of these sets of vectors through the σ_v considered. Thus the diagonal elements generated are 1, 1, 1, 1, -1, and -1, and the character of the entire matrix corresponding to σ_v has the value 2.

In Figure 10.9 we reproduce the character table of the group D_{3h} and append to it the results just obtained for the characters of the operations in the reducible representation for which the twelve Cartesian displacement coordinates form a basis. This may be reduced by the methods of Section 4.3 with the following result:

$$\Gamma_t = A_1' + A_2' + 3E' + 2A_2'' + E''$$

D_{3h}	E	$2C_3$	$3C_2$	σ_h	$2S_3$	$3\sigma_v$		
A_1'	1	1	1	1	1	1		x^2+y^2, z^2
A_2'	1	1	-1	1	1	-1	R_z	
E'	2	-1	0	2	-1	0	(x, y)	(x^2-y^2, xy)
A_1''	1	1	1	-1	-1	-1		
A_2''	1	1	-1	-1	-1	1	z	
E''	2	-1	0	-2	1	0	(R_x, R_y)	(xz, yz)
Γ_t	12	0	-2	4	-2	2		

Figure 10.9 The character table for the group D_{3h} with the characters for the representation generated from the twelve Cartesian displacement coordinates appended.

We know that of the twelve normal modes of the molecule only six are genuine vibrations while three are translations and three are rotations. We can easily strike the nongenuine ones from the above list by reference to the information on the right side of the character table. The translatory motions must belong to the same representations as do the coordinates x, y, and z. Thus we delete from our list one of the E' and one of the A_2'' species. We see that rotation about the z axis is a motion having A_2' symmetry and that rotations about the x and y axes are a degenerate pair having E'' symmetry. We therefore strike the A_2' and the E'' from our list. This then leaves the following list of the representations to which the six genuine normal modes of the molecule belong:

$$\Gamma_g = A_1' + 2E' + A_2''$$

These results are, of course, in agreement with information given in Figure 10.1.

We conclude this section by summarizing the procedure developed and the meaning of the results that it provides. In order to find out how many genuine normal modes of vibration of a molecule there are belonging to each irreducible representation of the molecular symmetry group, we merely have to form the reducible representation, which is generated on applying the operations of the group (actually just one operation from each class) to the set of Cartesian displacement vectors. This gives directly the symmetry types of both the genuine and the nongenuine modes, and from this list those of the translations and rotations may be removed immediately by use of information explicit in the character table. The process of finding the characters of the reducible representation is greatly simplified by recognizing that none of the vectors which are shifted to different atoms by a symmetry operation makes any contribution to the value of the character of the matrix corresponding to this operation. The final result is a list of irreducible representations of the molecular symmetry group such that the sum of their dimensions equals the number of internal or genuine modes of vibration of the molecule. There is a genuine normal mode having symmetry corresponding to each of the irreducible representations in the list.

10.4   Contributions of Particular Internal Coordinates to Normal Modes

We noted in Section 10.3 that the normal modes could be expressed not only as functions, that is, vector sums, of a set of Cartesian displacement vectors but also as functions of a set of internal displacement vectors. We have seen how the former relationship is utilized in determining the total number of normal modes of each symmetry type. We use the second relationship mainly to gain information on how the stretching of bonds and the bending of bond angles contribute to the normal modes according to their symmetry types. For instance, inspection of Figure 10.1 reveals that the A'_1 vibration involves purely the stretching of C-O bonds while the A''_2 vibration involves only bending or deformation of the molecule out of the equilibrium plane. The E' vibrations, however, are clearly not of this "purity"; both involve a mixture of C—O stretching and in-plane deformation of the OCO angles. We may ask: Could these facts have been deduced from symmetry considerations alone without knowledge of the actual forms of the normal modes? The answer to this question is in the affirmative, and it is our purpose in this section to explain how such information is obtained.

Suppose that we choose as a set of three internal displacement coordinates stretchings of the three C-O bonds, and use these as the basis for a three-dimensional representation of the symmetry group. The irreducible representations spanned by this representation will include only those to which belong normal modes involving C—O stretching. From Figure 10.1, we anticipate that the answer will be $A'_1 + E'$. Let us carry through the procedure outlined and see whether we do obtain this representation. In so doing we will make use of the short cut developed above in handling Cartesian displacement vectors. Any vector which is shifted to a different position by a given symmetry operation will contribute nothing to the value of the character of the corresponding matrix.

The various symmetry operations will affect our set of C—O stretchings in the same way as they will affect the set of C-O bonds themselves. With this in mind we can determine the desired characters very quickly as follows. For the operation E the character equals 3, since each C-O bond is carried into itself. The same is true for the operation σ_h. For the operations C_3 and S_3 the characters are zero because all C-O bonds are shifted by these operations. The operations C_2 and σ_v have characters of 1, since each carries one C-O bond into itself but interchanges the other two. The set of characters, listed in the same order as are the symmetry operations at the top of the D_{3h} character table, is thus as follows: 3 0 1 3 0 1. The representation reduces to $A'_1 + E'$, as it should according to our previous discussion. Thus we have shown that normal modes of symmetry types A'_1 and E' must involve some degree of C—O stretching. Since there is only one A'_1 mode, we can state further that this mode must involve entirely C—O stretching.

As a second set of internal displacement coordinates we may choose the increments or decrements to the three OCO angles. Before using this set to form a representation which will tell us the normal coordinates involving in-plane OCO bends, we must be careful to note that all of the coordinates in the set are *not* independent. If all three of the angles were to increase by the same amount at the same time, the motion would have A_1' symmetry. It is obviously impossible, however, for all three angles simultaneously to expand within the plane. Thus the A_1' representation which we shall find when we have reduced the representation is to be discarded as spurious.

This problem of spurious or, as they are conventionally called, *redundant* coordinates always arises when there are sets of angles which form a closed group, as in the case just considered, in planar cyclic molecules and also in three dimensions (e.g., tetrahedral and octahedral molecules). Several of the examples discussed in Section 10.7 will illustrate the point further. Redundant coordinates can usually be recognized without much difficulty, though troublesome cases sometimes arise.

A set of three increases in the three OCO angles will be affected by the symmetry operations in the same way as the three angles themselves, therefore we may make use of the angles in determining the characters. The operation E transforms each angle into itself, giving a character of 3 for this operation. So also does the operation σ_h. Since the operations C_3 and S_3 shift all of the angles, these operations have characters of zero. The operations C_2 and σ_v each leave one angle unshifted but interchange the other two, thus having characters of 1. The complete set of characters, again in the order in which the symmetry operations are listed, is as follows: 3 0 1 3 0 1. This representation reduces to $A_1' + E'$. Disregarding the A_1' for the reason explained above, we have the result that in-plane bending of the OCO angles contributes to vibrations of E' symmetry. This result is seen to be in agreement with the actual nature of the normal modes as shown in Figure 10.1.

Since there remains only one other kind of internal coordinate, namely, changes in the angles between the C-O bonds and the plane of the ion, and only one normal mode is unaccounted for, we may naturally conclude that this remaining mode, of A_2'' symmetry, must consist entirely of out-of-plane bending of the ion, which is, of course, correct. However, we may bring more positive reasoning to bear on the problem. It is evident that normal modes involving C—O stretching, OCO in-plane angle bending, or both must by nature be symmetric with respect to reflection through the plane of the molecule. All vectors used to indicate the displacements in such vibrations will lie in the plane and cannot be affected by reflection through the plane in which they lie. Since a mode of A_2'' symmetry must be antisymmetric to σ_h, according to the character table, we might have seen that the A_2'' mode would not involve C—O stretching or in-plane OCO bending without having any knowledge as to which vibrations did involve such internal displacements. Moreover, it is obvious that only a vibration in which all displacements are perpendicular to the molecular plane can belong to the A_2'' representation.

10.5 How to Calculate Force Constants: the F and G Matrix Method

Qualitative ways of analyzing a problem in molecular vibrations, that is, methods for determining the number of normal modes of each symmetry type which will arise in the molecule as a whole and in each set of equivalent internal coordinates, have been developed. There is also the quantitative problem of how the frequencies of these vibrations, which can be obtained by experiment, are related to the masses of the atoms, the bond angles and bond lengths, and, most particularly the force constants of the individual bonds and interbond angles. In this section we shall show how to set up the equations which express these relationships, making maximum use of symmetry to simplify the task at every stage.

For this purpose we shall adopt, without proof of its validity, Wilson's method of F and G matrices*. All of the required relations are combined in the master equation

$$|FG - E\lambda| = 0 \qquad (10.5\text{-}1)$$

in which F, G, and E are matrices and the entire left-hand side of the equation is a determinant. F is a matrix of force constants and thus brings the potential energies of the vibrations into the equation. G is a matrix which involves the masses and certain spatial relationships of the atoms and thus brings the kinetic energies into the equation. E is a unit matrix, and λ, which brings the frequency, v, into the equation, is defined by

$$\lambda = 4\pi^2 c^2 v^2 \qquad (10.5\text{-}2)$$

If the atomic masses in the G matrix elements are expressed in atomic mass units rather than in grams and the frequencies are expressed in Cm^{-1}, Equation 10.5-2 may be written

$$\lambda = 5.8890 \times 10^{-2} v^2$$

The plausibility of 10.5-1 may be appreciated by comparing it with the equation obtained by treating a diatomic molecule, AB, as a harmonic oscillator, viz.,

$$f \cdot \mu^{-1} - \lambda = 0 \qquad (10.5\text{-}3)$$

in which λ is defined as above, f is the force constant, and μ is the reduced mass [i.e., $M_A M_B / (M_A + M_B)$]. Equation 10.5-3 is evidently the limiting case

* For a full exposition of this method see *Molecular Vibrations*, by Wilson, Decius, and Cross.

of the *F-G* matrix equation when the F, G, and E matrices are one-dimensional.

Once the elements of the F and G matrices are known for the molecule in question, the determinantal equation (10.5-1) may be written out explicitly. This is the secular equation for the vibrational problem, analogous to the secular equation for energy, which has already been examined in connection with LCAO-MO theory in Chapter 7. Using known frequencies, we may apply the secular equation to calculate force constants, or, utilizing force constants, to calculate the frequencies of the normal modes. However, for a nonlinear molecule of N atoms this equation will involve a $3N \times 3N$ determinant, equivalent to a polynomial of order $3N$ of which only six roots will equal zero. It is then obviously advantageous to be able to factor this equation so that several smaller determinantal equations, rather than one large one, are to be solved. The great virtue of the F and G matrix method is that it affords a convenient and systematic means of employing the symmetry properties of the molecules to achieve maximum factorization of the secular equation.

Symmetry Coordinates

In order to use the F-G matrix method to maximum advantage, it is first necessary to set up linear combinations of internal coordinates which provide the proper number of functions, transforming according to each of the irreducible representations spanned by the $3N - 6$ normal modes of genuine vibration. To do this, the methods of Sections 10.3 and 10.4 are used to obtain the full list of irreducible representations involved and the internal coordinates which contribute to each. Then, using the sets of internal coordinates as basis functions, we set up the SALC's employing the projection operator technique developed in Chapter 6. The SALC's should be normalized.

The details of the *FG* matrix procedure are best explained by working through a simple example, such as the water molecule. This belongs to the point group C_{2v}. The nine Cartesian displacement vectors, three on each atom, gives rise to the representation.

	E	C_2	$\sigma_v(xz)$	$\sigma_v(yz)$
Γ	9	-1	3	1

where we have placed the molecule in the xz plane. This reduces to

$$\Gamma = 3A_1 + A_2 + 3B_1 + 2B_2$$

From the character table we can tell at once that translations and rotations account for $A_1 + A_2 + 2B_1 + 2B_2$. Thus, the three genuine internal vibrations span the irreducible representations $2A_1 + B_1$.

Displacements of the internal coordinates of the water molecule, the two O—H distances (d_1 and d_2), and the HOH angle ($\Delta\theta$) are bases for the following irreducible representations:

$$\Delta d_1, \Delta d_2: \quad A_1 + B_1$$
$$\Delta\theta: \quad A_1$$

Writing symmetry coordinates as SALC's of the internal coordinates is very easy in this simple case, where $\Delta\theta$ itself is an A_1 symmetry coordinate. Applying projection operators to a member of the set $\Delta d_1, \Delta d_2$, we obtain

$$\hat{P}^{A_1} \Delta d_1 \approx (1)\hat{E} \Delta d_1 + (1)\hat{C}_2 \Delta d_1 + (1)\hat{\sigma}_v \Delta d_1 + (1)\hat{\sigma}_v \Delta d_1$$
$$= \Delta d_1 + \Delta d_2 + \Delta d_1 + \Delta d_2$$
$$\approx \Delta d_1 + \Delta d_2$$

$$\hat{P}^{B_1} \Delta d_1 \approx (1)\hat{E} \Delta d_1 + (-1)\hat{C}_2 \Delta d_1 + (1)\hat{\sigma}_v \Delta d_1 + (-1)\hat{\sigma}_v \Delta d_1$$
$$= \Delta d_1 - \Delta d_2 + \Delta d_1 - \Delta d_2$$
$$\approx \Delta d_1 - \Delta d_2$$

The normalizing factor for each of these, assuming Δd_1 and Δd_2 to constitute an orthogonal set ($\Delta d_i \cdot \Delta d_j = \delta_{ij}$), is $1/\sqrt{2}$. Thus, the complete set of symmetry coordinates for vibrations is

$$A_1 \begin{cases} S_1 = \Delta\theta \\ S_2 = \dfrac{1}{\sqrt{2}}(\Delta d_1 + \Delta d_2) \end{cases}$$
$$B_1 \quad S_3 = \dfrac{1}{\sqrt{2}}(\Delta d_1 - \Delta d_2) \tag{10.5-4}$$

The F Matrix

We assume that the nuclei vibrate harmonically. Thus the potential energy, V, of the molecule may be written

$$2V = \sum_{i,k} f_{ik} s_i s_k \tag{10.5-5}$$

where s_i and s_k are changes in internal coordinates, $f_{ik} = f_{ki}$ and the sum extends over all values of i and k. A term such as $f_{ii} s_i^2$ represents the potential energy of stretching a given bond or bending a given angle, while the cross terms represent the energies of interaction between such motions. The f_{ik}'s are called force constants.

For H_2O there are three internal coordinates and hence nine force constants. A systematic way to list these is to make a square array, where the rows and columns are labeled by the internal displacements. Thus we have the f matrix, a matrix of the f_{ik}'s; note that this is symmetrical about its diagonal because $f_{ik} = f_{ki}$. For H_2O the f matrix is, explicitly,

	Δd_1	Δd_2	$\Delta\theta$
Δd_1	f_d	f_{dd}	$f_{d\theta}$
Δd_2	f_{dd}	f_d	$f_{d\theta}$
$\Delta\theta$	$f_{d\theta}$	$f_{d\theta}$	f_θ

where the force constant for stretching an O-H bond is f_d, the one for bending the HOH angle is f_θ, and the constants for interaction of one bond stretch with the other and with angle bending are f_{dd} and $f_{d\theta}$, respectively. This is equivalent to writing the appropriate form of Equation 10.5-5 as

$$2V = f_d(\Delta d_1)^2 + f_d(\Delta d_2)^2 + f_\theta(\Delta\theta)^2 + 2f_{dd}(\Delta d_1\,\Delta d_2) + 2f_{d\theta}(\Delta d_1\,\Delta\theta)$$
$$+ 2f_{d\theta}(\Delta d_2\,\Delta\theta) \tag{10.5-6}$$

It is also possible to express the potential energy in terms of symmetry coordinates. We write Equation 10.5-7 instead of 10.5-5, viz.,

$$2V = \sum_{j,l} F_{jl} S_j S_l \tag{10.5-7}$$

Here the F_{jl} are again force constants but pertain to vibrations described by the symmetry coordinates S_j, S_l, etc. From the standpoint of physical insight, it is the f_{ik} which have meaning for us, whereas mathematically the F_{jl} and the associated symmetry coordinates provide the easiest route to calculations because of symmetry factorization of the secular equation. Clearly, if we could express the F_{jl}'s in terms of the f_{ik}'s we would have an optimum situation. The following considerations will show how to do this.

Each of the two equations for the potential energy, 10.5-5 and 10.5-7, can be expressed in matrix notation:

$$2V = \mathbf{s'fs} \tag{10.5-6a}$$

$$2V = \mathbf{S'FS} \tag{10.5-7a}$$

by writing the s_i's as a column matrix (vector) $\mathbf{s}$, and the S_j's as a column matrix $\mathbf{S}$, and taking $\mathbf{s'}$ and $\mathbf{S'}$ as the corresponding row matrices. To illustrate, for H_2O we have

$$(\Delta d_1 \; \Delta d_2 \; \Delta \theta) \begin{bmatrix} f_d & f_{dd} & f_{d\theta} \\ f_{dd} & f_d & f_{d\theta} \\ f_{d\theta} & f_{d\theta} & f_{\theta\theta} \end{bmatrix} \begin{bmatrix} \Delta d_1 \\ \Delta d_2 \\ \Delta_\theta \end{bmatrix}$$

$$= (\Delta d_1 \; \Delta d_2 \; \Delta \theta) \begin{bmatrix} f_d \, \Delta d_1 + f_{dd} \, \Delta d_2 + f_{d\theta} \, \Delta \theta \\ f_{dd} \, \Delta d_1 + f_d \, \Delta d_2 + f_{d\theta} \, \Delta \theta \\ f_{d\theta} \, \Delta d_1 + f_{d\theta} \, \Delta d_2 + f_\theta \, \Delta \theta \end{bmatrix}$$

$$= f_d(\Delta d_1 \; \Delta d_1) + f_{dd}(\Delta d_2 \; \Delta d_1) + f_{d\theta}(\Delta \theta \Delta d_1)$$
$$+ f_{dd}(\Delta d_1 \; \Delta d_2) + \cdots + \cdots f_\theta(\Delta \theta \; \Delta \theta)$$

$$= f_d(\Delta d_1)^2 + f_d(\Delta d_2)^2 + f_\theta(\Delta \theta)^2 + 2 f_{dd}(\Delta d_1 \; \Delta d_2) + 2 f_{d\theta}(\Delta d_1 \; \Delta \theta)$$
$$+ 2 f_{d\theta}(\Delta d_2 \; \Delta \theta)$$

which is equivalent, as shown, to 10.5-6.

It is also possible to write the relationship between the internal coordinates and the symmetry coordinates, 10.5-4, in matrix form:

$$\mathbf{S} = \mathbf{Us} \qquad\qquad (10.5\text{-}8)$$

where the matrix **U** is

	Δd_1	Δd_2	$\Delta \theta$
S_1	0	0	1
S_2	$1/\sqrt{2}$	$1/\sqrt{2}$	0
S_3	$1/\sqrt{2}$	$-1/\sqrt{2}$	0

The inverse of the matrix **U** is simply its transpose, **U′**, because **U** describes a linear, orthogonal transformation. Thus, 10.5-8 may be rewritten as

$$\mathbf{s} = \mathbf{U}^{-1}\mathbf{S} = \mathbf{U}'\mathbf{S} \qquad\qquad (10.5\text{-}8a)$$

and we have

$$\mathbf{s}' = (\mathbf{U}'\mathbf{S})' = \mathbf{S}'\mathbf{U} \qquad\qquad (10.5\text{-}9)$$

since for matrices, as for group operations in general (page 5), the inverse (or transpose) of a product is the product of the inverses (or transposes) in reverse order.

We may now equate the right-hand side of Equations 10.5-6a and 10.5-7a and employ relations 10.5-8 and 10.5-9:

$$\mathbf{s}'\mathbf{fs} = \mathbf{S}'\mathbf{FS}$$

$$(\mathbf{S}'\mathbf{U})\,\mathbf{f}(\mathbf{U}'\mathbf{S}) = \mathbf{S}'\mathbf{FS}$$

$$\mathbf{S}'(\mathbf{U}\,\mathbf{f}\,\mathbf{U}')\mathbf{S} = \mathbf{S}'\mathbf{FS}$$

$$\mathbf{U}\,\mathbf{f}\,\mathbf{U}' = \mathbf{F}$$

We thus obtain a simple matrix equation for transforming the **f** matrix into the **F** matrix. Let us apply this to the water molecule.

$$\mathbf{F} = \mathbf{U f U'}$$

$$= \begin{bmatrix} 0 & 0 & 1 \\ 1/\sqrt{2} & 1/\sqrt{2} & 0 \\ 1/\sqrt{2} & -1/\sqrt{2} & 0 \end{bmatrix} \begin{bmatrix} f_d & f_{dd} & f_{d\theta} \\ f_{dd} & f_d & f_{d\theta} \\ f_{d\theta} & f_{d\theta} & f_\theta \end{bmatrix} \begin{bmatrix} 0 & 1/\sqrt{2} & 1/\sqrt{2} \\ 0 & 1/\sqrt{2} & -1/\sqrt{2} \\ 1 & 0 & 0 \end{bmatrix}$$

$$= \begin{bmatrix} 0 & 0 & 1 \\ 1/\sqrt{2} & 1/\sqrt{2} & 0 \\ 1/\sqrt{2} & -1/\sqrt{2} & 0 \end{bmatrix} \begin{bmatrix} f_{d\theta} & 1/\sqrt{2}(f_d + f_{dd}) & 1/\sqrt{2}(f_d - f_{dd}) \\ f_{d\theta} & 1/\sqrt{2}(f_{dd} + f_d) & 1/\sqrt{2}(f_{dd} - f_d) \\ f_\theta & 2/\sqrt{2}(f_{d\theta}) & 0 \end{bmatrix}$$

$$= \begin{bmatrix} f_\theta & \sqrt{2} f_{d\theta} & 0 \\ \sqrt{2} f_{d\theta} & f_d + f_{dd} & 0 \\ 0 & 0 & f_d - f_{dd} \end{bmatrix}$$

We see that the **F** matrix is symmetry-factored into a 2×2 block for the two A_1 vibrations and a one-dimensional block for the single B_1 vibration and that each element of it is expressed as a linear combination of the various internal force constants, $f_d, f_\theta, f_{dd}, f_{d\theta}$.

The G Matrix

The **G** matrix may be set up by a procedure formally analogous to that used for the **F** matrix. Thus, in general we have

$$\mathbf{G} = \mathbf{U g U'} \tag{10.5-11}$$

For the water molecule the explicit equation is:

$$\mathbf{G} = \begin{bmatrix} 0 & 0 & 1 \\ 1/\sqrt{2} & 1/\sqrt{2} & 0 \\ 1/\sqrt{2} & -1/\sqrt{2} & 0 \end{bmatrix} \begin{bmatrix} g_{11} & g_{12} & g_{13} \\ g_{12} & g_{22} & g_{23} \\ g_{13} & g_{23} & g_{33} \end{bmatrix} \begin{bmatrix} 0 & 1/\sqrt{2} & 1/\sqrt{2} \\ 0 & 1/\sqrt{2} & -1/\sqrt{2} \\ 1 & 0 & 0 \end{bmatrix}$$

where the **g** matrix elements are given general symbols except that g_{21} is replaced by g_{12}, g_{31} by g_{13}, and g_{32} by g_{23} because of the general requirement that the matrix be symmetrical about its diagonal. We may also make the substitutions

$$g_{11} = g_{22}, \quad g_{13} = g_{23}$$

because internal coordinates 1 and 2 are equivalent (Δd_1, Δd_2). With these further substitutions, the result of matrix multiplication is

$$\mathbf{G} = \begin{bmatrix} g_{33} & \sqrt{2}g_{13} & 0 \\ \sqrt{2}g_{13} & g_{11} + g_{12} & 0 \\ 0 & 0 & g_{11} - g_{12} \end{bmatrix}$$

It now remains to consider how the various g_{ij}'s are expressed in terms of the atomic masses and the dimensions of the molecule. The procedure, although not difficult or profound, is definitely tedious; it is explained in Wilson, Decius, and Cross and other specialized textbooks. Fortunately, it is possible to tabulate the most commonly used ones in the form of general expressions into which the specific parameters of any molecule may be inserted. Such a tabulation and directions for its use are given in Appendix VIII. It is found that

$$g_{11} = \mu_H + \mu_O$$

$$g_{12} = \mu_O \cos \theta$$

$$g_{13} = -(\mu_O/r) \sin \theta$$

$$g_{33} = 2(\mu_H + \mu_O - \mu_O \cos \theta)/r^2$$

where μ_H = the reciprocal of the mass of the H atom, etc.

The G matrix, which is symmetry-factored in conformity with the $\mathbf{F}$ matrix, then becomes

$$\begin{bmatrix} 2(\mu_H + \mu_O - \mu_O \cos \theta)/r^2 & -(\sqrt{2}\mu_O/r) \sin \theta & 0 \\ -(\sqrt{2}\mu_O/r) \sin \theta & \mu_H + \mu_O(1 + \cos \theta) & 0 \\ 0 & 0 & \mu_H + \mu_O(1 - \cos \theta) \end{bmatrix}$$

When the molecular structure parameters are known, or estimated, this matrix may be reduced to a set of numbers. Thus, for H_2O, using $\theta = 104°31'$ and $r = 0.9580$ Å, we obtain

$$G = \begin{bmatrix} 2.332 & -0.0893 & 0 \\ -0.0893 & 1.0390 & 0 \\ 0 & 0 & 1.0702 \end{bmatrix}$$

The problem of expressing, for the H_2O molecule, the relationship between the frequencies of the fundamental modes and a set of force constants has now been solved in such a way that the equations are as simple as symmetry will permit them to be. In this case, the secular equation is factored into one of second order for the two A_1 vibrations and one of first order for the sole B_1 vibration. The explicit forms of these separate equations are:

$$A_1: \begin{bmatrix} f_\theta & \sqrt{2}f_{d\theta} \\ \sqrt{2}f_{d\theta} & f_d + f_{dd} \end{bmatrix} \begin{bmatrix} 2.332 & -0.0893 \\ -0.0893 & 1.0390 \end{bmatrix} - \begin{bmatrix} \lambda & 0 \\ 0 & \lambda \end{bmatrix} = 0$$

$$B_1: \quad 1.0702(f_d - f_{dd}) = \lambda$$

This is the end of the symmetry analysis; from here on the problem is a purely computational one.

The computational problem is not a proper subject for discussion here, but a few remarks about it may be pertinent. In practice one is nearly always interested in computing force constants from observed frequencies. However, analytical expressions for force constants as explicit functions of the frequencies and geometric parameters cannot generally be obtained. The only efficient way to calculate force constants from frequencies is to use an iterative approach, implemented by a digital computer. To do this a starting set of assumed force constants is refined by successive approximations until the set which yields calculated frequencies in best agreement with the observed ones is obtained.

An additional problem in computing force constants is that usually—and H_2O as treated here is a case in point—a force field which includes the principal interaction constants, as well as the stretching and bending constants, will have more constants to evaluate than there are fundamental frequencies. The most generally applicable procedure for coping with this problem is to measure frequencies in isotopically substituted molecules, thereby providing new sets of equations involving different frequencies and some different G matrix elements, but the same force constants. For light molecules, which can be observed in the gas phase, centrifugal distortion effects in rotational fine structure afford another source of data on force constants.

10.6 Selection Rules for Fundamental Vibrational Transitions

The wave functions of normal modes can be written in the simplest possible way by using the normal coordinates as the variables. As is shown in many books on elementary quantum mechanics, these wave functions, assuming that the oscillations are harmonic (potential energy a function only of the square of the displacement coordinate), have the form

$$\psi_i(n) = N_i e^{-(a_i/2)\xi_i^2} H_n(\sqrt{\alpha_i}\,\xi_i) \tag{10.6-1}$$

where N_i is a normalizing constant, $\alpha_i = 2\pi v_i/h$ (in which v_i is the frequency of the ith normal mode), H_n is a Hermite polynomial of order n, and ξ_i is the ith normal coordinate expressed as a SALC of internal displacement coordinates. The number n is the vibrational quantum number; it is 0 in the ground state, 1 in the first excited state, and so forth.

The first few Hermite polynomials, $H_n(x)$, are

$$H_0(x) = 1 \qquad H_1(x) = 2x$$
$$H_2(x) = 4x^2 - 2 \quad H_3(x) = 8x^3 - 12x$$

(10.6-2)

As shown in Section 5.1, the wave functions must form bases for irreducible representations of the symmetry group of the molecule, and the same holds, of course, for all kinds of wave functions, vibrational, rotational, electronic, and so on. Let us now see what representations are generated by the vibrational wave functions of the normal modes. Inserting $H_0(\sqrt{\alpha_i}\,\xi_i)$ into 10.6-1, we obtain

$$\psi_i(0) = N_i\, e^{-(\alpha_i/2)\xi_i^2}$$

In the event that ξ_i represents a nondegenerate vibration, all symmetry operations change it into ± 1 times itself. Hence ξ_i^2 is unchanged by all symmetry operations. Thus $\psi_i(0)$ is invariant under all symmetry operations and forms a basis for the totally symmetric representation of the group. If ξ_i is a normal coordinate describing one of a set of degenerate vibrations, any symmetry operation will change it into ± 1 times itself or into a linear combination of all members of the set (cf. Section 10.2). Let us suppose that ξ_a and ξ_b are the normal coordinates of a pair of degenerate vibrations and that some symmetry operation, R, acting on ξ_a has the effect:

$$\hat{R}\xi_a = \xi_a' = r_a\xi_a + r_b\xi_b$$

If ξ_a, ξ_b are normalized, r_a and r_b will be such that ξ_a' will also be normalized. We then have, because normal coordinates are orthogonal and normalized,

$$\xi_a^2 = 1, \qquad \xi_a'^2 = 1.$$

Thus, in the degenerate case also, $\psi_i(0)$ is invariant to all symmetry operations.

We may therefore state the following important rule:

All wave functions for normal vibrations in their ground states, $\psi_i(0)$, are bases for the totally symmetric representation of the point group of the molecule.

For the excited vibrational states, $\psi_i(n)$, the wave functions are products of the same exponential as in the ground state, which we have seen is always totally symmetric, and the nth Hermite polynomial. Therefore $\psi_i(n)$ has the symmetry of the nth Hermite polynomial.

For the first excited state, we then find that the wave function, $\psi_i(1)$, has the same symmetry as ξ_i: for the second excited state, $\psi_i(2)$ has the symmetry of ξ_i^2, which means that it is totally symmetric, and so on.

Consider a molecule with k normal modes of vibration. At any time each of these modes will be in a certain quantum state; the wave function for the ith mode in the nth state is $\psi_i(n_i)$. For the total molecular vibrational wave function, ψ_v, we may use the product of the $\psi_i(n_i)$, since the variables in each, the ξ_i, are linearly independent. Thus we write

$$\psi_v = \psi_1(n_1) \cdot \psi_2(n_2) \cdot \psi_3(n_3) \ldots \ldots \psi_k(n_k) \qquad (10.6\text{-}3)$$

When each of the n_i is equal to 0, the molecule is in its vibrational ground state. If it absorbs radiation so that the ith normal mode is excited to the state with $n_i = 1$, while the remaining $k - 1$ normal modes remain in their lowest $(n = 0)$ states, the molecule is said to have undergone a *fundamental transition* in the ith normal vibration. The k different transitions of this kind are called the fundamentals of the molecule.

Since the fundamental transitions generally give rise to infrared absorption bands and Raman lines which are more intense by at least an order of magnitude than any other kinds* of transition, they are of the greatest interest and we shall deal only with the fundamentals here. Selection rules for other types of transition can also be obtained by arguments of the type we shall use, but the reader is cautioned that where degenerate modes are concerned subtle complications often arise.†

For a fundamental transition of the jth normal mode we may write

$$\prod_i \psi_i(0) \rightarrow \psi_j(1) \prod_{i \neq j} \psi_i(0)$$

Abbreviating the total molecular vibrational wave functions in the ground and excited states by ψ_v^0 and ψ_v^j, respectively, we can also write

$$\psi_v^0 \rightarrow \psi_v^j$$

For a fundamental transition to occur by absorption of infrared dipole radiation (see Section 5.3) it is necessary that one or more of the integrals

---

* If one mode alone is doubly excited, this transition is called the *first overtone* of the fundamental; if one mode alone is triply excited, this is called the *second overtone* and so on. Transitions in which two or more modes are simultaneously excited (by one or more units each) are called *combination tones*. If the molecule initially has one or more fundamentals in an excited $(n > 0)$ state, *any* transitions that it makes are called *hot transitions*.
† For details regarding the selection rules for other types of transitions, *Molecular Vibrations*, by Wilson, Decius, and Cross, should be consulted, especially pages 151–155, where complications arising because of degeneracy are treated.

$$\int \psi_v{}^0 x \psi_v{}^j \, d\tau$$

$$\int \psi_v{}^0 y \psi_v{}^j \, d\tau$$

$$\int \psi_v{}^0 z \psi_v{}^j \, d\tau$$

be nonzero. The x, y, and z in these integrals refer to the orientation of the oscillating electric vector of the radiation relative to a Cartesian coordinate system fixed in the molecule.

Now, by using the results of Section 5.3, it is extremely easy to determine whether or not such integrals vanish. Since $\psi_v{}^0$ belongs to the totally symmetric representation, the coordinate x, y, or z and $\psi_v{}^j$ must belong to the same representation in order that the representation of their direct product will contain the totally symmetric representation, whereby the representation given by the entire integral may contain the totally symmetric representation. Since $\psi_j(1)$ must have the same symmetry as the jth normal coordinate, ξ_j, and since all of the other wave functions, $\psi_i(0)$, are totally symmetric, $\psi_v{}^j$ belongs to the same representation as the normal mode which is undergoing its fundamental transition.

We therefore have the following very simple rule for the activity of fundamentals in infrared absorption:

A fundamental will be infrared active (i.e., will give rise to an absorption band) if the normal mode which is excited belongs to the same representation as any one or several of the Cartesian coordinates.

Of course the character tables show to which representations the Cartesian coordinates belong, so that this rule can be used with the utmost ease.

For Raman scattering, it is necessary that at least one integral of the type

$$\int \psi_v{}^0 P \psi_v{}^j \, d\tau$$

be nonzero. In these integrals P is one of the quadratic functions of the Cartesian coordinates, viz., x^2, y^2, z^2, xy, yz, zx, all of which (simply or in combinations such as $x^2 - y^2$) are listed opposite the representations that they generate in the character tables. These P's are components of the polarizability tensor, and the requirement that the above integrals be nonzero means physically that there must be a change in polarizability of the molecule when the transition occurs. By exactly the same reasoning as used above in regard to infrared absorption, we obtain the following simple rule for the Raman activity of fundamentals:

A fundamental transition will be Raman active (i.e., will give rise to a Raman shift) if the normal mode involved belongs to the same representation as one or more of the components of the polarizability tensor of the molecule.

We may illustrate these rules, using the carbonate ion. We see in the D_{3h} character table that (x, y) form a basis for the E' representation and z for the A_2'' representation. For the polarizability tensor components we see that one or more of these belong to the A_1', E', and E'' representations. Thus, for *any* molecule of D_{3h} symmetry, we have the following selection rules:

$$\begin{array}{rl} \text{Raman-active only:} & A_1', E'' \\ \text{Infrared-active only:} & A_2'' \\ \text{Both Raman- and infrared-active:} & E' \end{array}$$

In the particular case of the carbonate ion:

$$\begin{array}{rl} \nu_1(A_1'): & \text{Raman only} \\ \nu_2(A_2''): & \text{Infrared only} \\ \nu_3(E'), \nu_4(E'): & \text{Infrared and Raman} \end{array}$$

10.7   Illustrative Examples

The Pyramidal AB₃ Molecule

This type of molecule will be used to illustrate all of the procedures and techniques discussed earlier in this chapter, including setting up the vibrational secular equation by the *F-G* matrix method.

Symmetry Types of the Normal Modes. For this nonlinear four-atomic molecule there are $3(4) - 6 = 6$ genuine internal vibrations. Using a set of three Cartesian displacement coordinates on each atom, we obtain the following representation of the group C_{3v}:

C_{3v}	E	$2C_3$	$3\sigma_v$
Γ	12	0	2

which can be reduced, giving

$$\Gamma = 3A_1 + A_2 + 4E$$

The C_{3v} character table shows that the translations account for $A_1 + E$ and the rotations for $A_2 + E$, leaving, for the genuine, internal vibrations,

$$2A_1 + 2E$$

Selection Rules. The character table shows that A_1 and E vibrations are both infrared-active and Raman-active. Thus, all four fundamental modes for such a molecule should be observable in both the infrared and the Raman spectra.

Internal Coordinate Contributions. The nature of the vibrations in terms of changes in internal coordinates may be found by using the three A-B bond lengths and the three ABA angles as internal coordinates. In this case no spurious or redundant results will be obtained because all of these six internal coordinates may change independently. The representation Γ_{AB}, given by the bond lengths, and Γ_δ, given by the bond angles, are readily found to be the following:

C_{3v}	E	$2C_3$	$3\sigma_v$
Γ_{AB}	3	0	1
Γ_δ	3	0	1

The two representations are identical, each reduces to $A_1 + E$. Therefore it follows that each of the A_1 normal modes will involve both bond stretching and deformations of the bond angles, and so also will the E modes.

F and G Matrices. The internal coordinates are shown in the sketch below. The SALC's, which constitute symmetry coordinates, have already been derived for an AB_3 molecule in Section 6.3 and are reproduced here.

$$A_1 \begin{cases} S_1 = \dfrac{1}{\sqrt{3}}(\Delta r_1 + \Delta r_2 + \Delta r_3) \\[2mm] S_2 = \dfrac{1}{\sqrt{3}}(\Delta\theta_1 + \Delta\theta_2 + \Delta\theta_3) \end{cases}$$

$$E \begin{cases} S_{3a} = \dfrac{1}{\sqrt{6}}(2\Delta r_1 - \Delta r_2 - \Delta r_3) \\[2mm] S_{3b} = \dfrac{1}{\sqrt{2}}(\Delta r_2 - \Delta r_3) \\[2mm] S_{4a} = \dfrac{1}{\sqrt{6}}(2\Delta\theta_1 - \Delta\theta_2 - \Delta\theta_3) \\[2mm] S_{4b} = \dfrac{1}{\sqrt{2}}(\Delta\theta_2 - \Delta\theta_3) \end{cases}$$

It should be noted that the displacement coordinates have been labeled and then combined into the E-type symmetry coordinates in a particular way. The angles have been indexed so that θ_i is opposite to r_i. This assures that θ_i and the change therein, $\Delta\theta_i$, are related to the molecular symmetry elements in the same way as are r_i and Δr_i. Then, when the SALC's are written, $\Delta\theta_i$ and Δr_i occupy corresponding positions in the expressions. Unless this is done the symmetry factorization will not work out.

We can now write the **U** matrix. In doing so, the corresponding types of E symmetry coordinates, S_{3a}, S_{4a}, and S_{3b}, S_{4b}, are placed in successive rows. The **U** matrix is as follows:

	r_1	r_2	r_3	θ_1	θ_2	θ_3
S_1	$1/\sqrt{3}$	$1/\sqrt{3}$	$1/\sqrt{3}$	0	0	0
S_2	0	0	0	$1/\sqrt{3}$	$1/\sqrt{3}$	$1/\sqrt{3}$
S_{3a}	$2/\sqrt{6}$	$-1/\sqrt{6}$	$-1/\sqrt{6}$	0	0	0
S_{4a}	0	0	0	$2/\sqrt{6}$	$-1/\sqrt{6}$	$-1/\sqrt{6}$
S_{3b}	0	$1/\sqrt{2}$	$-1/\sqrt{2}$	0	0	0
S_{4b}	0	0	0	0	$1/\sqrt{2}$	$-1/\sqrt{2}$

The **f** matrix which will be adopted in this illustration is the following.

	Δr_1	Δr_2	Δr_3	$\Delta\theta_1$	$\Delta\theta_2$	$\Delta\theta_3$
Δr_1	f_r	$f_{rr'}$	$f_{rr'}$	0	$f_{r\theta}$	$f_{r\theta}$
Δr_2	$f_{rr'}$	f_r	$f_{rr'}$	$f_{r\theta}$	0	$f_{r\theta}$
Δr_3	$f_{rr'}$	$f_{rr'}$	f_r	$f_{r\theta}$	$f_{r\theta}$	0
$\Delta\theta_1$	0	$f_{r\theta}$	$f_{r\theta}$	f_θ	0	0
$\Delta\theta_2$	$f_{r\theta}$	0	$f_{r\theta}$	0	f_θ	0
$\Delta\theta_3$	$f_{r\theta}$	$f_{r\theta}$	0	0	0	f_θ

It may be noted that a complete **f** matrix (that is, one containing no zeros) would have six different force constants: f_r, $f_{rr'}$, f_θ, $f_{r\theta}$, $f_{r\theta'}$, and $f_{\theta\theta'}$. The last two, pertaining to the interaction between Δr_i and $\Delta\theta_i$ (interaction between stretching a bond and deforming an angle to which that bond does *not* belong) and the interaction between two angle deformations, respectively, have been omitted. Since there are only four fundamental frequencies, it would not be possible to solve for all six force constants by using the data for only a single isotopic species. In this case, there are two force constants which seem very likely to be smaller and less important than the others, and these have been omitted from the outset, thus bringing the number of force constants carried into line with the number of experimental data likely to be readily available.

Using the above U and f matrices, we find that

$$\mathbf{F} = \mathbf{U}\mathbf{f}\mathbf{U}' = \begin{bmatrix} f_r + 2f_{rr'} & 2f_{r\theta} & 0 & 0 & 0 & 0 \\ 2f_{r\theta} & f_\theta & 0 & 0 & 0 & 0 \\ 0 & 0 & f_r - f_{rr'} & -f_{r\theta} & 0 & 0 \\ 0 & 0 & -f_{r\theta} & f_\theta & 0 & 0 \\ 0 & 0 & 0 & 0 & f_r - f_{rr'} & -f_{r\theta} \\ 0 & 0 & 0 & 0 & -f_{r\theta} & f_\theta \end{bmatrix}$$

The first block, at the upper left, is for the two A_1 modes, and the two remaining blocks, which are identical, are for the a and b components of the E modes. Thus, the factorization is into two second-order equations, as expected. If the internal coordinates and the symmetry coordinates (SALC's) had not been selected and arranged so that the Δr_i and $\Delta\theta_i$ sets matched correctly, as specified earlier, the factorization of the 4×4 E block into two identical 2×2 blocks would not have been accomplished.

Finally, the G matrix must be set up. Using the notation defined in Figure AVIII.1 of the Appendix and omitting elements related by the symmetry about the diagonal to those given we write the g matrix as follows.

	r_1	r_2	r_3	θ_1	θ_2	θ_3
r_1	g_{rr}^2	$g_{rr'}^1$	$g_{rr'}^1$	$g_{r\theta}^1(2)$	$g_{r\theta}^2$	$g_{r\theta}^2$
r_2		g_{rr}^2	$g_{rr'}^1$	$g_{r\theta}^2$	$g_{r\theta}^1(2)$	$g_{r\theta}^2$
r_3			g_{rr}^2	$g_{r\theta}^2$	$g_{r\theta}^2$	$g_{r\theta}^1(2)$
θ_1				$g_{\theta\theta}^3$	$g_{\theta\theta'}^2(1)$	$g_{\theta\theta'}^2(1)$
θ_2					$g_{\theta\theta}^3$	$g_{\theta\theta'}^2(1)$
θ_3						$g_{\theta\theta}^3$

From this we obtain:

$$\mathbf{G} = \mathbf{U}\mathbf{g}\mathbf{U}' = \begin{bmatrix} g_{rr}^2 + 2g_{rr'}^1 & g_{r\theta}^1(2) + 2g_{r\theta}^2 & 0 & 0 & 0 & 0 \\ g_{r\theta}^1(2) + 2g_{r\theta}^2 & g_{\theta\theta}^3 + 2g_{\theta\theta'}^2(1) & 0 & 0 & 0 & 0 \\ 0 & 0 & g_{rr}^2 - g_{rr'}^1 & g_{r\theta}^1(2) - 2g_{r\theta}^2 & 0 & 0 \\ 0 & 0 & g_{r\theta}^1(2) - 2g_{r\theta}^2 & g_{\theta\theta}^3 - 2g_{\theta\theta'}^2(1) & 0 & 0 \\ 0 & 0 & 0 & 0 & g_{rr}^2 - g_{rr'}^1 & g_{r\theta}^1(2) - 2g_{r\theta}^2 \\ 0 & 0 & 0 & 0 & g_{r\theta}^1(2) - 2g_{r\theta}^2 & g_{\theta\theta}^3 - 2g_{\theta\theta'}^2(1) \end{bmatrix}$$

The F and G matrices may now be combined to give two two-dimensional equations, one for the A_1 modes and one for the E modes. To illustrate what these modes actually look like in a real case, they are depicted for ND_3 in Figure 10.10. These drawings are based on calculations (cf. Wilson, Decius, and Cross for details) from experimentally observed frequencies. The molecule ND_3 is used rather than NH_3 because the inverse mass dependence of the amplitudes would make the vectors on the nitrogen atom of NH_3 impractically small compared to those on the hydrogen atoms.

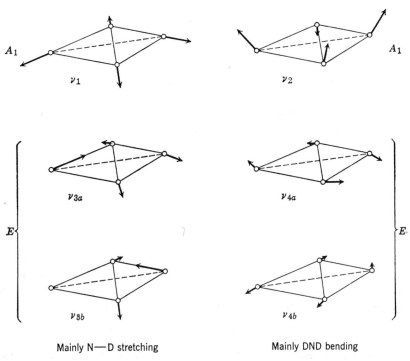

Mainly N—D stretching   Mainly DND bending

Figure 10.10   The normal vibrations of ND_3 (after Herzberg).

It will be seen that one each of the A_1 and E modes involves *almost* purely bond stretching, while each of the others involves almost purely angle bending. Such a situation is not uncommon and makes it easy to interpret qualitatively the shifts in observed frequencies when molecules or parts of molecules (e.g., CF_3 or CH_3 groups) change their chemical environment. It must be emphasized, however, that the occurrence of such convenient separations is a consequence of the particular values of the masses and force constants, and no such separation into "stretching modes" and "bending modes" is in general required. Quite often the normal modes involve complex mixtures of several internal coordinates; in that case, naive discussions in terms of pure "stretches" and "bends" can be quite misleading.

Trans-N_2F_2

This molecule has the following planar, nonlinear structure:

$$\overset{\displaystyle F}{\underset{\displaystyle F}{\ddot{N}=N}}$$

It belongs to the point group C_{2h}. Since it is a nonlinear four-atomic molecule it has $3(4) - 6 = 6$ degrees of internal freedom.

The set of twelve Cartesian displacement vectors for the entire molecule generates the following reducible representation:

C_{2h}	E	C_2	i	σ_h
Γ	12	0	0	4

This can be reduced in the following manner:

$$\Gamma = 4A_g + 2B_g + 2A_u + 4B_u$$

Inspection of the character table shows that the translation and rotations span the representations $A_g + 2B_g + A_u + 2B_u$. On deleting these from the total number constituting Γ, we are left with the following list of the representations spanned by the genuine normal vibrations:

$$3A_g + A_u + 2B_u$$

The selection rules for the fundamentals of these modes are obtained immediately from the right columns of the character table and are:

Infrared-active: A_u, B_u
Raman-active: A_g

The nature of these six vibrations may be further specified in terms of the contribution made to each of them by the various internal coordinates. We first note that A_g and B_u vibrations must involve only motions within the molecular plane, since the characters of the representations A_g and B_u with respect to σ_h are positive. The A_u vibration will, however, involve out-of-plane deformation, since the character of A_u with respect to σ_h is negative. Thus we may describe the normal mode of A_u symmetry as "the out-of-plane deformation." In order to treat the remaining five in-plane vibrations we need a set of five internal coordinates so chosen that changes in them may occur entirely within the molecular plane. A suitable set, related to the bonding in the molecule, consists of the two N—F distances, the two NNF angles, and the N=N distance.

It is found that the two N—F distances form the basis for the representation Γ_{NF}, the two NNF angles for Γ_{NNF}, and the N=N distance for Γ_{NN}, each of which is given below:

C_{2h}	E	C_2	i	σ_h
Γ_{NF}	2	0	0	2
Γ_{NNF}	2	0	0	2
Γ_{NN}	1	1	1	1

It is then easily shown that

$$\Gamma_{NF} = A_g + B_u$$

$$\Gamma_{NNF} = A_g + B_u$$

$$\Gamma_{NN} = A_g$$

It therefore follows that the three Raman-active vibrations (A_g) will be compounded of symmetric N—F stretching, symmetric NNF bending, and N=N stretching, the relative amounts of each involved in each normal mode depending, of course, on the actual values of the force constants and atomic masses. Similarly the two B_u vibrations in the infrared will involve asymmetric N—F stretching and NNF angle bending. Again the proportion of each of these in each of the true normal modes will depend on force constants and atomic masses.

Tetrahedral Molecules, Such as Methane

Such molecules have T_d symmetry. As nonlinear, five-atomic species they have $3(5) - 6 = 9$ degrees of internal freedom.

The set of fifteen Cartesian displacement vectors forms a basis for the following representation:

T_d	E	$8C_3$	$3C_2$	$6S_4$	$6\sigma_d$
Γ	15	0	-1	-1	3

This reduces as follows:

$$\Gamma = A_1 + E + T_1 + 3T_2$$

The character table shows that the rotations transform as T_1 and the translations as T_2; hence the symmetry types of the genuine vibrations are

$$A_1, E, 2T_2$$

The character table also shows that the activities of these fundamentals are as follows:

$$\text{Raman only:} \quad A_1 \text{ and } E$$
$$\text{Infrared and Raman:} \quad T_2$$

To find the contributions of the internal coordinates, C–H bond lengths and HCH angles, to these vibrational modes we first use the set of four C—H bond lengths as the basis for a representation, obtaining Γ_{CH} shown below.

T_d	E	$8C_3$	$3C_2$	$6S_4$	$6\sigma_d$
Γ_{CH}	4	1	0	0	2
Γ_{HCH}	6	0	2	0	2

This is easily found to reduce as follows:

$$\Gamma_{CH} = A_1 + T_2$$

Similarly, we can use the six interbond angles to generate a representation, obtaining Γ_{HCH}, which reduces thus:

$$\Gamma_{HCH} = A_1 + E + T_2$$

It will be seen that the total dimensionality of these two representations, ten, is one in excess of the correct number, and specifically, that there is an extra A_1 representation. It is easy to determine that the spurious or redundant representation is the one in Γ_{HCH}, for, although it is possible for all four of the C—H distances to change independently, it is not possible for all six angles to change independently. If any five are arbitrarily altered, the alteration of the sixth one is then automatically fixed. For an A_1 vibration all six angles would have to change in the same way at the same time (i.e., all increase or all decrease), and this is clearly impossible. Hence we obtain the results that the A_1 vibration of CH_4 consists purely of C—H stretching, and the E vibration purely of HCH angle deformations, while both bond stretching and angle bending contribute to each of the normal vibrations of T_2 symmetry.

Octahedral Molecules, Such as SF_6

Such molecules belong to the point group O_h. They have $3(7) - 6 = 15$ degrees of internal freedom. The twenty-one Cartesian displacement vectors generate the representation Γ given in the table below:

O_h	E	$8C_3$	$6C_2$	$6C_4$	$3C_2(= C_4{}^2)$	i	$6S_4$	$8S_6$	$3\sigma_h$	$6\sigma_d$
Γ	21	0	-1	3	-3	-3	-1	0	5	3
Γ_{SF}	6	0	0	2	2	0	0	0	4	2
Γ_{FSF}	12	0	2	0	0	0	0	0	4	2

The representation Γ reduces as follows:

$$\Gamma = A_{1g} + E_g + T_{1g} + 3T_{1u} + T_{2g} + T_{2u}$$

The character table shows that the rotations and translations belong, respectively, to the T_{1g} and T_{1u} representations. After deleting these, we obtain the following list of genuine normal modes, grouped according to the activities of their fundamentals:

$$\text{Infrared-active:} \quad 2T_{1u}$$
$$\text{Raman-active:} \quad A_{1g}, E_g, T_{2g}$$
$$\text{Inactive:} \quad T_{2u}$$

We encounter here for the first time the occurrence of a normal vibration which is completely inactive as a fundamental. This phenomenon is not commonplace but is encountered occasionally in relatively symmetrical molecules.

In the table above we also give the representations generated by the set of six S-F bonds, Γ_{SF}, and the set of twelve FSF angles, Γ_{FSF}. These representations can be reduced as follows:

$$\Gamma_{SF} = A_{1g} + E_g + T_{1u}$$

$$\Gamma_{FSF} = A_{1g} + E_g + T_{2g} + T_{1u} + T_{2u}$$

Obviously there is some redundancy here; since the S—F coordinates are wholly independent, the redundancy must be entirely in Γ_{FSF} (see the preceding discussion of CH_4). By comparing the total of Γ_{SF} and Γ_{FSF} with the correct list of genuine internal modes we see that the A_{1g} and E_g occurring in Γ_{FSF} are the spurious ones. Thus we conclude that each of the two T_{1u} modes will involve a combination of bond stretching and angle deformation, the A_{1g} and E_g modes will involve only bond stretching, and the T_{2g} and T_{2u} modes will involve only angle deformation.

10.8   Some Important Special Effects

The Exclusion Rule

Consider a molecule which has a center of symmetry. When one of the Cartesian coordinates, x, y, or z, is inverted through this center, it goes into the negative of itself. Hence all representations generated by x, y, or z or any set of these must belong to a u representation. On the other hand, a binary product of two Cartesian coordinates, say xy or z^2, does not change sign on inversion, since each coordinate separately does change sign and $-1 \times -1 = 1$. It therefore follows that all such binary products, which represent components of the polarizability tensor, belong to g representations.

From these rules, which can be verified by inspection of the character tables, we conclude that in centrosymmetric molecules only fundamentals of modes belonging to g representations can be Raman-active and only fundamentals of modes belonging to u representations can be infrared-active. It is also obvious that the same must be true for other transitions besides fundamentals, since the reasoning is completely general.

Another way of stating this result, the so-called *exclusion rule*, is as follows:

In a centrosymmetric molecule no Raman-active vibration is also infrared-active and no infrared-active vibration is also Raman active.

The reader may refer to the preceding section to see this rule exemplified for *trans*-N_2F_2 and SF_6.

Fermi Resonance

The vibrational secular equation as it is normally set up (e.g., by the *F–G* matrix method) deals only with the fundamental modes of vibration. The n fundamental modes with a given symmetry are governed by the appropriate $n \times n$ dimensional block factors of the F and G matrices. When $n \geq 2$ the frequencies of the various modes are mutually interdependent to a degree determined by the off-diagonal matrix elements. The usual secular equation does not take account of overtones or combinations and hence neglects any influence these may have on the fundamentals. In most cases this approximation is very satisfactory, but there are instances in which a combination or overtone interacts strongly with a fundamental. This may happen when the two excitations give states of the same symmetry. The phenomenon of interaction is called *Fermi resonance* because it was first recognized and explained by Enrico Fermi in the vibrational spectrum of carbon dioxide.

Fermi resonance is only a special example of the general phenomenon called configuration interaction, which has been discussed earlier in connection with electronic spectra (page 168). In the latter case each of two different excited electronic configurations gave rise to states which then mixed and interacted with each other to give the two actual states, each of which involved contributions from both electron configurations. Also, the energies of the two actual states were further separated than the energies calculated for each of the pure configurations. We will see now that exactly the same type of interaction is involved in Fermi resonance.

Let us suppose that the frequencies, v_i, v_j, and v_k, of three fundamental vibrational transitions are related as follows:

$$v_i + v_j \approx v_k$$

and the symmetry of the doubly excited state

$$\Psi_{ij} = \psi_i(1)\psi_j(1) \prod_{l \neq i, j} \psi_l(0)$$

is the same as that of the singly excited state

$$\Psi_k = \psi_k(1) \prod_{l \neq k} \psi_l(0)$$

This will be true when the direct product representation for the ith and jth normal modes is or contains the irreducible representation for which the kth normal mode forms a basis. These two excited states of the same symmetry will then interact in a way which can be represented by the usual sort of secular equation, viz.,

$$\begin{vmatrix} (v_i + v_j) - v & W_{ij,k} \\ W_{ij,k} & v_k - v \end{vmatrix} = 0$$

where the roots, v, will be the actual frequencies. The magnitude of the interactions is given by

$$W_{ij,k} = \int \Psi_{ij} W \Psi_k \, d\tau$$

The interaction operator, W, has a nonzero value because the vibrations are anharmonic. It is also totally symmetric, and for this reason the two states, ψ_{ij} and ψ_k, must belong to the same representation in order for the integral to have a finite value.

It is clear that one of the roots of the secular equation will be greater than either $(v_i + v_j)$ or v_k and that the other will be less than either $(v_i + v_j)$ or v_k. This is an obvious consequence of the nature of the quadratic equation represented in determinantal form. The closer $(v_i + v_j)$ and v_k are to begin with, the more the actual roots will diverge from these values. Thus, one effect of Fermi resonance is to displace the two expected frequencies away from each other.

Fermi resonance also has an effect on the intensities of the two transitions. Overtone or combination transitions generally have less than one tenth the intensity of fundamental transitions. However, the interaction between Ψ_{ij} and Ψ_k, which affects the actual energies, also leads to new wave functions which are linear combinations of these two, as explained for secular equations generally in Section 7.1. Thus, the actual excited state whose energy is closest to $(v_i + v_j)$ will be described by a wave function

$$\Psi'_{ij} = N(\Psi_{ij} + x\Psi_k)$$

where N is a normalizing factor and $x \leq 1$, while the actual excited state whose energy is closest to v_k will be described by a wave function

$$\Psi'_k = N(\Psi_k + x\Psi_{ij})$$

Therefore, even though a transition from the ground state, ψ_0 to ψ_{ij}, has an inherently low intensity, proportional to the small integral

$$\int \Psi_0(x, y, z)\Psi_{ij} \, d\tau$$

the transition to the actual excited state, Ψ'_{ij}, has an intensity proportional to

$$\int \Psi_0(x, y, z)\Psi_{ij} \, d\tau + x \int \Psi_0(x, y, z)\Psi_k \, d\tau$$

where the second integral must be large, because the $\Psi_0 \to \Psi_k$ transition is intense. The weak overtone transition "borrows" intensity from the strong fundamental transition because it is close to it and has an excited state with the same symmetry. The closer are the energies $(v_i + v_j)$ and v_k, the more the total intensity tends to be shared equally between them. The situation is illustrated schematically in Figure 10.11.

In the actual case of the CO_2 molecule, which provides an excellent example, the three fundamental transitions have frequencies of 667, 1300, and 2350 cm^{-1}. The first overtone of the 667 cm^{-1} vibration, which is doubly degenerate, has a frequency of 1334 cm^{-1}, which is quite close to that of the 1300 cm^{-1} fundamental. Now it can be shown that the excited state for the 1300 cm^{-1} fundamental and one component of the representation generated by the excited state corresponding to the first overtone of the 667 cm^{-1} vibration do belong to the same representation of the group $D_{\infty h}$, and hence a Fermi resonance occurs. Thus, in the Raman spectrum of CO_2 two strong bands at 1285 and 1388 cm^{-1} are observed, instead of just one at ~ 1300 cm^{-1}.

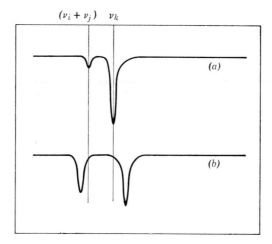

Figure 10.11 (*a*) The spectrum expected for the combination, $(\nu_i + \nu_j)$, and the fundamental ν_k, in the absence of Fermi resonance. (*b*) The actual spectrum where the two bands have diverged and shared intensity as a result of Fermi resonance.

This example also serves to emphasize that Fermi resonance occurs in Raman as well as infrared spectra, although, for the sake of definiteness, the immediately preceding intensity integrals were written as appropriate for electric dipolar activity in the infrared. Fermi resonance has been unambiguously recognized in various other thoroughly analyzed molecules, such as CCl_4, C_6H_6, and $CHCl_3$. Undoubtedly, it occurs frequently in very complex molecules, where there is considerable chance of a near coincidence of one of the many combinations of lower frequency modes with one of the higher frequency fundamentals of appropriate symmetry. Nevertheless, it should be noted that papers reporting hasty and superficial vibrational analyses of complex molecules display a deplorable tendency to employ "Fermi resonance" as an escape from difficulties or inconsistencies without adequate evidence.

Solid-State Effects

The vibrations of an individual molecule in the gas phase are subject only to the symmetry restrictions based on its own intrinsic point symmetry, and this chapter has so far been concerned exclusively with symmetry conditions of that kind. When a molecule resides in a crystal it is, in principal, subject only to the symmetry restrictions arising out of its crystalline environment.

To be *entirely* rigorous, the molecule cannot even be treated as a discrete entity: instead the entire array of molecules must be analyzed. However, such a completely rigorous approach is essentially impossible for practical reasons and unnecessary for most purposes, and therefore approximations are justifiably made. Two levels of approximation have frequently been used: (1) the site symmetry approximation, and (2) the correlation field (sometimes called "factor group") approximation. The first is conceptually very simple and very often is entirely adequate. Sometimes, however, it fails, and then the more abstruse correlation field treatment must be employed. Each of these approximations will now be explained and illustrated.

Site Symmetry Approximation. This approximation assumes that the detailed dynamical nature of the interactions between one molecule and the molecules surrounding it may be ignored. The surroundings are treated as a static environment whose only relevant property is its symmetry, as seen by the molecule occupying a site within it.

In general a crystal lattice has various elements of symmetry. Some of these are of the same kind as those occurring in molecules, that is, they generate symmetry operations which cause no translation of the center of mass of the object to which they are applied. Thus a given point in a crystal may be a center of inversion, it may lie on a proper or improper axis, or it may lie in a reflection plane. Of course several such symmetry elements may intersect at one point in the crystal lattice. The collection of all the operations made possible by the symmetry elements intersecting at such a point in the crystal constitute a point group which can be used to specify the *site symmetry* of a molecule whose center of mass is situated at this point. In a rigid, ordered crystal, the site symmetry can never be such as to contain symmetry elements not belonging to the free molecule,* and in general the site symmetry is lower than the molecular symmetry. In other words, of a number of symmetry elements present in the free molecule, only some will correspond with symmetry elements of the lattice. The symmetry elements of the free molecule which do not correspond with symmetry elements inherent in its crystal environment are not part of the site symmetry. In principal, the molecule loses these symmetry elements when it enters the crystal, and its vibrational

---

* There are some apparent exceptions. In the ammonium halides which have the rock salt lattice, the ammonium ions, which do not themselves have centers of symmetry, lie at lattice points which would, were the ammonium ions not present, be centers of inversion. Depending on the particular halide and the temperature, either the ammonium ions are freely rotating, so that the time-averaged symmetry is centric, or else ammonium ions at different sites have different orientations so that, although any particular site is not centro-symmetric, the crystal as a whole appears to be centrosymmetric. In this brief discussion we shall consider only cases in which the crystals are completely ordered and contain no rotating molecules.

behavior must be treated only in terms of the site symmetry. In general, then, the point group for the site symmetry will be a subgroup of the molecular point group, though occasionally the two groups are the same (e.g., for the $(SiF_6)^{2-}$ ion in the cubic form of K_2SiF_6).

In many cases molecules in a crystal occupy what is called a *general position* in the unit cell. This is a position through which *no* nontranslatory symmetry element passes. The site symmetry for a molecule on a general position is C_1 (i.e., nil), and in principle *all* selection rules and degeneracies for the molecule under its intrinsic symmetry are lost.

It should be noted that the degree to which the vibrational spectrum of a molecule in a site of low symmetry will deviate *observably* from the behavior of the free molecule depends on just how strongly the molecule interacts with its surroundings in the crystal. Symmetry considerations alone can of course tell us nothing about this, and the degree of deviation varies from one case to another. A few examples now to be considered will give some indication of the magnitude of such effects in these typical cases.

Qualitatively, the effects of low site symmetry are of two general types: (1) changes in selection rules and (2) splitting of degeneracies. A nondegenerate vibration may be inactive in the high symmetry of the free molecule but active in the symmetry of one or more subgroups of the same molecule. For example, the A_1' mode of the carbonate ion (totally symmetric C—O stretching, Figure 10.1) is not infrared-active under the full D_{3h} symmetry of CO_3^{2-}. The compound $CaCO_3$ occurs in two crystallographically different forms, calcite and aragonite. In the former, the site symmetry of the CO_3^{2-} ion is D_3, in the latter, C_s. Reference to the character tables for these two groups show that in D_3 the totally symmetric mode is still not infrared-active, whereas in the group C_s totally symmetric vibrations are infrared-active. In agreement with these expectations, the symmetric C—O stretching mode of CO_3^{2-} (known from the Raman spectrum of solutions of carbonates) is not observed in calcite but appears weakly in aragonite.

The effect of low site symmetry in splitting degeneracies is also nicely demonstrated in the forms of $CaCO_3$. The E' representation of the group D_{3h} correlates with the E representation in D_3. Hence, in calcite, both v_3 and v_4 (Figure 10.1) are observed as single peaks. In the group C_s there are no representations of order greater than 1; this means that the degenerate vibrations of CO_3^{2-} must be split by the C_s site symmetry of aragonite. Actually v_3 is still observed as a single peak, indicating that the magnitude of the splitting is too small to permit resolution or that one component has very low intensity, but v_4 is distinctly split into two peaks separated by 14 cm^{-1}.

As another example of the effect of low site symmetry in splitting degeneracy we may consider the thiocyanate ion in KNCS. The SCN$^-$ ion is linear and in the absence of perturbing influences has as one of its normal modes a doubly degenerate bending vibration. The degeneracy exists in the isolated ion because bending of the molecule in any given plane containing the molecular axis is entirely equivalent to bending in a plane perpendicular to the first one. However, if the site symmetry of the ion in a crystal is such that these two planes are not equivalent, then the frequencies of bending in the two planes need not be identical. In Figure 10.12, which shows the structure of KNCS, it can be seen that planes parallel to the plane of the drawing are not equivalent to planes perpendicular to the drawing. It was found that, while the bending vibration gives rise to a single peak at 470 cm^{-1} in the infrared spectrum of NCS$^-$ in solution, crystalline KNCS has two absorption peaks at 470 and 484 cm^{-1}.

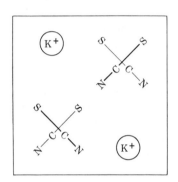

Figure 10.12   Sketch of the unit cell of KNCS.

The Correlation Field Approximation. In some cases it is not possible to explain experimental observations in terms of the site symmetry approximation, whereby the surroundings of a given molecule are treated as static. A clear example is provided by the crystalline form of the *trans* isomer of $[(C_5H_5)Fe(CO)_2]_2$, which has the centrosymmetric structure and the infrared spectrum shown in Figure 10.13. The *trans*

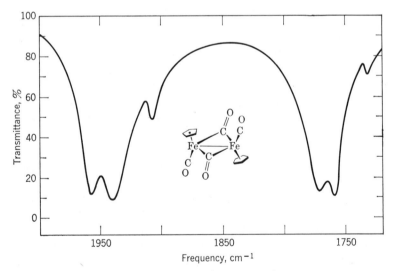

Figure 10.13   The structure and infrared spectrum of *trans*-$[C_5H_5Fe(CO)_2]_2$. (The two weaker bands are ^{13}CO satellites.)

molecule (other isomers exist) has inherent C_{2h} symmetry when rotational orientation of the C_5H_5 rings about their C_5 axes is ignored. With this centric form of symmetry it turns out straightforwardly that the only infrared-active C—O stretching modes are those arising from antisymmetrically coupled vibrations of the terminal pair and the bridged pair.

In the crystalline form of *trans*-$[(C_5H_5)Fe(CO)_2]_2$ each molecule lies on a crystallographic center of symmetry; the site symmetry is C_i. In this site symmetry it should still be true that all vibrations of the molecule can be rigorously classified as symmetric and antisymmetric with respect to inversion and that only the antisymmetric ones can be infrared-active. Therefore, according to the site symmetry approximation, only the *two* antisymmetric C—O stretching vibrations should be observed in the crystal spectrum. As Figure 10.13 shows, however, *four* C—O stretching bands are actually observed.

This can be explained by treating the entire set of molecules in the unit cell (two, in this case) as the vibrating unit, subject to symmetry restrictions arising from the symmetry of the entire unit cell. Thus, we have a total of eight oscillators, four bridging and four terminal, arranged in four pairs. The members of each pair are related by a center of symmetry which lies between the molecules. Each pair of centrically related oscillators gives rise to a symmetrically and an antisymmetrically coupled mode, and the antisymmetric ones are infrared-active. This is illustrated schematically for the bridging CO groups in Figure 10.14. Within each molecule there are symmetric and antisymmetric modes, v_s, v_{as}, v_s', v_{as}'. These can then couple *inter*molecularly as follows:

$$\left.\begin{array}{l} v_s + v_s' \\ v_{as} + v_{as}' \end{array}\right\} \text{Symmetric}$$

$$\left.\begin{array}{l} v_s - v_s' \\ v_{as} - v_{as}' \end{array}\right\} \text{Antisymmetic}$$

This way of expressing the overall modes for the pair of molecular units is only approximate, and it assumes that intramolecular coupling exceeds intermolecular coupling. The frequency difference between the two antisymmetric modes arising in the pair of molecules jointly will depend on both the intra- and intermolecular interaction force constants. Obviously the algebraic details are a bit complicated, but the idea of intermolecular coupling subject to the symmetry restrictions based on the symmetry of the entire unit cell is a simple and powerful one. It is this symmetry-restricted intermolecular correlation of the molecular vibrational modes which causes the *correlation field* splittings.

The site symmetry method is most likely to fail in just such a case as we have been discussing, where the crystal contains neutral molecules which lie close to each other and which also contain highly dipole-active oscillators (CO groups are among the most strongly dipole-active of all oscillators). In the several $CaCO_3$ crystals and other ionic compounds, the polyatomic

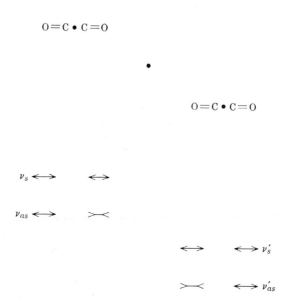

Figure 10.14   *Upper*: two pairs of CO groups centrically related both within each pair and between the pairs. • denotes center of inversion. *Lower*: symmetric and antisymmetric modes of vibration within each pair.

ions are separated from each other by intervening counter ions; since the correlation field effects depend on dipole-dipole interactions, which vary as r^{-3}, where r is the separation of the dipoles, it is reasonable that effects which amount to ~ 10 cm^{-1} for adjacent molecules are negligible when the separation is about 2-3 times as great: 10 cm$^{-1} \times 2.5^{-3} = 10/15$ cm$^{-1} = 0.7$ cm^{-1}! Correlation field effects have been generally observed in crystalline metal carbonyl compounds.*

The preceding discussion of correlation field effects is mathematically very superficial. A more rigorous treatment would require some discussion of space groups, which are not covered in this book. The reader interested in further details should consult the paper by Vetter and Hornig cited in Appendix IX.

* See, for example, papers by H. J. Buttery et al., *J. Chem. Soc. (A)*, **1970**, 471.

Exercises

10.1   Set up the symmetry coordinates for the five in-plane vibrations of the *trans*-N_2F_2 molecule and construct the F and G matrices.

10.2   Benzene is a good example of an important molecule with relatively high symmetry (D_{6h}). (*a*) Determine the symmetry types of the genuine vibrations. (*b*) Determine the activities of each type of vibration. (*c*) Classify all the in-plane vibrations as to whether they involve C—H stretching, C—C stretching, or angle bending. (*d*) What vibrations involve out-of-plane angle bending? (*e*) Derive the normalized expressions for the symmetry coordinates of the latter.

10.3   Determine the expected C—O stretching spectra in the infrared and the Raman for *cis*- and *trans*-$X_2M(CO)_4$ molecules, and discuss the feasibility of distinguishing the isomers by infrared alone, Raman alone, and both together.

Part III

Appendices

Crystallographic Point Groups,
Stereographic Projections, and
International (Hermann-Maguin) Notation

The Thirty-Two Crystallographic Point Groups

It can be shown that an infinite lattice, in which a motif is repeated indefinitely by translations, can contain only the following proper rotation axes: C_1, C_2, C_3, C_4, C_6. It is also possible for such an array to have centers of inversion. Systematic inspection shows that there are 32 distinct combinations of these, leading to the 32 crystallographic point groups. Table AI.1 lists them, giving in the second column the Schoenflies notation, which is used throughout the body of this text.

International Notation

In the Schoenflies system the improper rotation is a rotation-reflection. In the International system, used mainly by crystallographers, the improper rotation is a rotation-inversion.

The International system uses the numbers 1, 2, 3, 4, 6 to represent the proper rotation axes, C_1, C_2, C_3, C_4, C_6, respectively. A bar is placed over a number to denote the improper rotation. Thus we have $\bar{1}$, $\bar{2}$, $\bar{3}$, $\bar{4}$, $\bar{6}$. Clearly $\bar{1} \equiv i$, the simple inversion center, and $\bar{2} = C_2 \times i = i \times C_2 = \sigma_h$. The other equivalences to Schoenflies' symmetry elements are as follows:

$$\bar{3} = S_6, \quad \bar{4} = S_4, \quad \bar{6} = S_3$$

Finally, the International system uses the symbol m to represent a mirror plane. When a mirror plane, m, and a proper axis, X, coexist there are two possibilities. (1) The plane contains the axis: we write Xm. (2) The mirror plane is perpendicular to the axis: we write X/m.

The International notation aims to be as compact and nonredundant as possible. Thus, in most cases only the minimum number of symmetry elements required to specify completely the point group is written. For instance,

Table AI.1 The Thirty-Two Crystallographic Point Groups

NUMBER	SCHOENFLIES SYMBOL	INTERNATIONAL SYMBOL	CRYSTAL SYSTEM
1	C_1	1	Triclinic
2	C_i	$\bar{1}$	
3	C_s	m	Monoclinic
4	C_2	2	
5	C_{2h}	$2/m$	
6	C_{2v}	mm	Orthorhombic
7	D_2	222	
8	D_{2h}	mmm	
9	C_4	4	Tetragonal
10	S_4	$\bar{4}$	
11	C_{4h}	$4/m$	
12	C_{4v}	$4mm$	
13	D_{2d}	$\bar{4}2m$	
14	D_4	422	
15	D_{4h}	$4/mmm$	
16	C_3	3	Rhombohedral
17	S_6	$\bar{3}$	
18	C_{3v}	$3m$	
19	D_3	32	
20	D_{3d}	$\bar{3}m$	
21	C_{3h}	$\bar{6}$	Hexagonal
22	C_6	6	
23	C_{6h}	$6/m$	
24	D_{3h}	$\bar{6}m2$	
25	C_{6v}	$6mm$	
26	D_6	622	
27	D_{6h}	$6/mmm$	
28	T	23	Cubic
29	T_h	$m3$	
30	T_d	$\bar{4}3m$	
31	O	432	
32	O_h	$m3m$	

the group C_{3v}, which requires the presence of the three-fold axis (3) and three vertical planes of symmetry, is denoted $3m$. It is not necessary to indicate explicitly the existence of the other two mirror planes. Again, the group D_{2h} is written as *mmm*. It is true that along the line of intersection of each pair of planes there is, necessarily, a twofold axis, and a complete symbol would be $2/m, 2/m, 2/m$, but this is redundant and is simplified to the nonredundant *mmm*. In the cases of C_{4v} and C_{6v}, the usual symbols, 4*mm* and 6*mm*, are redundant, in that the presence of one mirror plane implies the existence of the entire set of four or six such planes. However, the symbol m is repeated because there are two *kinds* of planes (cf. Section 3.7), generating operations which are in different classes.

There are only two cases in which the meaning of the International notation is perhaps not entirely self-evident. These are the groups T and D_3, denoted 23 and 32, respectively.

The International symbols for each of the 32 crystallographic symmetry groups are given in Table AI.1.

Stereographic Projections

A convenient way to depict the symmetry represented by a point group (*any* point group, whether one of the 32 crystallographic ones or not) is to use a stereographic projection. These are employed chiefly for the crystallographic groups, and Table AI.2 shows them for these 32 cases. The rules for constructing them are as follows:

1. One imagines points arranged on a sphere. The sphere cuts an equatorial plane (the paper) to define a circle. The principal axis is perpendicular to this circle. Except in two very simple cases (1 and $\bar{1}$) two such spheres are employed: one to show the symmetry elements themselves, and the other to show the array of points on the sphere which would be obtained as the symmetry operations replicate some initial, completely general point.

2. The points are denoted by filled circles when they lie on the upper hemisphere and open circles when they lie on the lower hemisphere. A filled circle in an open circle simply means that there are two points lying on the same vertical line.

3. Vertical planes of symmetry are represented by straight lines crossing the circle diametrically. Tilted planes (found only in the cubic groups) are represented as curved lines which are the projections of their lines of intersection with the surface of the sphere. A horizontal plane, if it exists, is not represented directly, but its existence is evident since all points on the sphere will appear as filled circles within open circles.

	Triclinic	Monoclinic (1st setting)	Tetragonal
X	1	2	4
$\bar{X}$ (even)	—	$m(=\bar{2})$	$\bar{4}$
X (even) plus center and $\bar{X}$ (odd)	$\bar{1}$	$2/m$	$4/m$
	Monoclinic (2nd setting)	Orthorhombic	
$X2$	2	222	422
Xm	m	$mm2$	$4mm$
$\bar{X}2$ (even) or $\bar{X}m$ (even)	—	—	$\bar{4}2m$
$X2$ or Xm plus center and $\bar{X}m$ (odd)	$2/m$	mmm	$4/mmm$

346

Table AI.2 (continued)

Trigonal	Hexagonal	Cubic	
3	6	23	X
—	$\bar{6}$	—	$\bar{X}$ (even)
$\bar{3}$	$6/m$	$m3$	X (even) plus center and $\bar{X}$ (odd)
32	622	432	$X2$
$3m$	$6mm$	—	Xm
—	$\bar{6}m2$	$\bar{4}3m$	$\bar{X}2$ (even) or $\bar{X}m$ (even)
$\bar{3}m$	$6/mmm$	$m3m$	$X2$ or Xm plus center and $\bar{X}m$ (odd)

4. Proper and improper axes are represented by the following symbols:

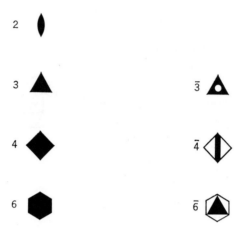

The symbol for each axis appears in the circle where its points of intersection with the sphere would project onto the equatorial plane.

Expansion of Determinants
and The Inverse of a Matrix

Expansion and Evaluation of Determinants

A determinant is a square array of numbers. The number of rows or columns is called the order, n, of the determinant, and it contains a total of n^2 elements.

A determinant, unlike a matrix, is a scalar quantity. Its value is given by the sum of $n!$ different products each containing n elements so chosen that each row and each column is represented but once. The sum of these products is called the expansion of the determinant. That there are $n!$ such products is easily shown. To form one of them we select an element from the first row, which may be done in n ways. An element from the second row cannot be chosen from the column to which the element from the first row belongs; thus for this choice we have only $n - 1$ possibilities. We can then choose an element from the third row in only $n - 2$ ways and so forth. Thus there are $n(n - 1)(n - 2)(n - 3) \cdots 2 \cdot 1 = n!$ different ways of composing a product.

In addition, each product is given a sign, $+$ or $-$. The choice of sign is determined as follows. We write down all of the factors in such a way that the row (column) indices run serially and count the number of transpositions (exchanges of neighbors) which are required to put the column (row) indices in serial order. If this number is even, the sign is $+$; if it is odd, the sign is $-$. Consider, for example, the following product, which would be obtained from a third-order determinant:

$$a_{31}a_{23}a_{12}$$

Written with the row indices in serial order, it is

$$a_{12}a_{23}a_{31}$$

To put the column indices in serial order we must perform two successive interchanges of adjacent factors, viz.,

$$a_{12}a_{23}a_{31} \longrightarrow a_{12}a_{31}a_{23}$$

$$a_{12}a_{31}a_{23} \longrightarrow a_{31}a_{12}a_{23}$$

Hence this product is given a positive sign.

The value of a determinant, $|A|$, of order 2 is given by

$$a_{11}a_{22} - a_{12}a_{21}$$

and that of a third-order determinant by

$$a_{11}a_{22}a_{33} + a_{12}a_{23}a_{31} + a_{13}a_{21}a_{32} - a_{11}a_{23}a_{32} - a_{12}a_{21}a_{33} - a_{13}a_{22}a_{31}$$

The results for these two simple and frequently occurring cases can easily be remembered. The positive terms are products obtained by selecting elements along diagonal lines running from upper left to lower right, and the negative terms are products of elements on lines running from upper right to lower left. For determinants of the fourth and higher orders the number of products (24 for $n = 4$) exceeds the number which may be enumerated in this way (8 for $n = 4$), and more labor is required to write down the complete expansion of the determinant.

Determinants of order ≥ 4 may conveniently be evaluated by the *method of cofactors*. Inspection of the list of six products whose algebraic sum is the value of a determinant of order 3 shows that we may rewrite it in the following way:

$$a_{11}(a_{22}a_{33} - a_{23}a_{32}) + a_{12}(a_{23}a_{31} - a_{21}a_{33}) + a_{13}(a_{21}a_{32} - a_{22}a_{31})$$

Each of the terms in parentheses is the expanded form of the determinant made up of the elements of the original determinant which remain after we strike from it the elements belonging to the row and column of the element in front of the parenthesis. It is given a $+$ sign if the sum of the indices of the element before it is even and a negative sign if the sum of these indices is odd. The terms in the parentheses are called the *cofactors* of the elements in front of the parentheses. Thus we see that the third-order determinant can be evaluated by finding the sum of the products of each element in the first row with its cofactor.

A little reflection will show that we could just as well have arranged the six terms in the expansion so as to have a sum of the products of *any* row or *any* column with their cofactors. For instance, choosing the second column, we can write:

$$a_{12}(a_{23}a_{31} - a_{21}a_{33}) + a_{22}(a_{11}a_{33} - a_{13}a_{31}) + a_{32}(a_{13}a_{21} - a_{11}a_{23})$$

It should also not be difficult to see that a similar process can be carried out on a determinant of any order. Thus, using A^{ij} to represent the cofactor of a_{ij}, we can write

$$|A| = \sum_i a_{ij} A^{ij} = \sum_j a_{ij} A^{ij}$$
$$\text{(for any } j) \qquad \text{(for any } i)$$

One further important property of any determinant is that *if any two rows or columns are identical the value of the determinant is zero*. This is easily proved. Suppose that the pth and rth rows are identical. Then for any term in the expansion, say

$$a_{11} a_{m2} a_{n3} \cdots a_{pi} a_{rj} \cdots$$

there must be another which is identical except that it will contain a_{ri} and a_{pj}. Now suppose that the column indices are arranged serially in the term shown above, that $p > r$, and that x transpositions are required to put the row indices in serial order. Then, if exactly the same x transpositions are carried out in the term

$$a_{11} a_{m2} a_{n3} \cdots a_{ri} a_{pj} \cdots$$

it will still be necessary to make an additional $2(p - r - 1) + 1$ transpositions to put a_{ri} and a_{pj} in their proper places, making $x + 2(p - r - 1) + 1$ transpositions in all. Thus, if x is even, $x + 2(p - r - 1) + 1$ must be odd and vice versa. It therefore follows that all the terms in the expansion will cancel out in a pairwise fashion. Obviously a similar argument could be made if we assume two columns to be identical.

The Adjoint Matrix

Before defining the adjoint matrix we must define the transpose of a matrix. This is a matrix of which the columns are the rows, and vice versa, of the original matrix. Symbolically, the transpose of the matrix $[a_{ij}]$ is $[a_{ji}]$. Now, the adjoint matrix of a matrix $[a_{ij}]$ is defined as follows:

$$\text{Adjoint of } [a_{ij}] = [A^{ji}]$$

That is, we treat the array of elements constituting $[a_{ij}]$ as a determinant, write the cofactor, A^{ij}, of each element in place of the element giving the matrix $[A^{ij}]$, and then make the transpose of $[A^{ij}]$. The matrix adjoint to $\mathscr{A}$ will be symbolized $\hat{\mathscr{A}}$.

The Inverse of a Matrix

The inverse $\mathscr{A}^{-1}$ of a matrix $\mathscr{A}$ is, by definition, such that

$$\mathscr{A}\mathscr{A}^{-1} = \mathscr{A}^{-1}\mathscr{A} = \mathscr{E}$$

Let us now look at the product $\mathscr{A}\hat{\mathscr{A}}$ for a square matrix of order 3. It is

$$\mathscr{A}\hat{\mathscr{A}} = \begin{bmatrix} a_{11} & a_{12} & a_{13} \\ a_{21} & a_{22} & a_{23} \\ a_{31} & a_{32} & a_{33} \end{bmatrix} \begin{bmatrix} A^{11} & A^{21} & A^{31} \\ A^{12} & A^{22} & A^{32} \\ A^{13} & A^{23} & A^{33} \end{bmatrix}$$

$$= \begin{bmatrix} a_{11}A^{11} + a_{12}A^{12} + a_{13}A^{13} & a_{11}A^{21} + a_{12}A^{22} + a_{13}A^{23} & a_{11}A^{31} + a_{12}A^{32} + a_{13}A^{33} \\ a_{21}A^{11} + a_{22}A^{12} + a_{23}A^{13} & a_{21}A^{21} + a_{22}A^{22} + a_{23}A^{23} & a_{21}A^{31} + a_{22}A^{32} + a_{23}A^{33} \\ a_{31}A^{11} + a_{32}A^{12} + a_{33}A^{13} & a_{31}A^{21} + a_{32}A^{22} + a_{33}A^{23} & a_{31}A^{31} + a_{32}A^{32} + a_{33}A^{33} \end{bmatrix}$$

We see that each diagonal element is the expansion of the determinant $|A|$ in terms of a row and its cofactors. On the other hand, each off-diagonal element is the sum of products of the elements of a certain row, say the ith, with the cofactors of the elements of some other row, say the jth. Such a sum is, in fact, the expansion in the elements of the ith row with their cofactors of a determinant in which the ith and jth rows are identical. Since we have already seen that the value of such a determinant must be zero, all off-diagonal elements of the product $\mathscr{A}\hat{\mathscr{A}}$ are zero. It is also easy to see that $\mathscr{A}\hat{\mathscr{A}} = \hat{\mathscr{A}}\mathscr{A}$. Thus we have the result

$$\mathscr{A}\hat{\mathscr{A}} = \hat{\mathscr{A}}\mathscr{A} = \begin{bmatrix} |A| & 0 & 0 & 0 & \cdots & 0 \\ 0 & |A| & 0 & 0 & \cdots & 0 \\ \vdots & & & & & \vdots \\ 0 & \cdot & \cdot & \cdot & \cdot & |A| \end{bmatrix}$$

$$= |A| \begin{bmatrix} 1 & 0 & 0 & 0 & \cdots & 0 \\ 0 & 1 & 0 & 0 & \cdots & 0 \\ \vdots & & & & & \vdots \\ 0 & \cdot & \cdot & \cdot & \cdot & 1 \end{bmatrix}$$

$$= |A|\mathscr{E}$$

Now, referring to the definition of $\mathscr{A}^{-1}$, we see that

$$\mathscr{A}^{-1} = \frac{\hat{\mathscr{A}}}{|A|}$$

That is, each element of $\mathscr{A}^{-1}$ is the element of $\hat{\mathscr{A}}$ divided by $|A|$. Since division by zero is not defined, only matrices for which the corresponding determinants are nonzero can have inverses. A matrix $\mathscr{A}$ such that $|A| = 0$ is said to be *singular* (no inverse), whereas matrices of which the corresponding determinants are nonzero are said to be *nonsingular*. Only nonsingular matrices can occur in representations of a group.

The working of a few examples, which the reader can make up *ad libitum*, will be helpful in developing a practical grasp of these results.

A. Character Tables for Chemically Important Symmetry Groups

1. The Nonaxial Groups

C_1	E
A	1

C_s	E	σ_h		
A'	1	1	x, y, R_z	$x^2, y^2,$ z^2, xy
A''	1	-1	z, R_x, R_y	yz, xz

C_i	E	i		
A_g	1	1	R_x, R_y, R_z	x^2, y^2, z^2 xy, xz, yz
A_u	1	-1	x, y, z	

2. The C_n Groups

C_2	E	C_2		
A	1	1	z, R_z	x^2, y^2, z^2, xy
B	1	-1	x, y, R_x, R_y	yz, xz

C_3	E	C_3	$C_3{}^2$		$\varepsilon = \exp(2\pi i/3)$
A	1	1	1	z, R_z	$x^2 + y^2, z^2$
E	$\begin{Bmatrix} 1 & \varepsilon & \varepsilon^* \\ 1 & \varepsilon^* & \varepsilon \end{Bmatrix}$			$(x, y)(R_x, R_y)$	$(x^2 - y^2, xy)(yz, xz)$

* Appendix IIIA is presented in two places: (1) here, in its proper location in the sequence of appendices, and (2) as a separate section in a pocket inside the back cover.

The C_n Groups (*continued*)

C_4	E	C_4	C_2	$C_4{}^3$		
A	1	1	1	1	z, R_z	x^2+y^2, z^2
B	1	-1	1	-1		x^2-y^2, xy
E	$\begin{cases}1 \\ 1\end{cases}$	$\begin{matrix}i \\ -i\end{matrix}$	$\begin{matrix}-1 \\ -1\end{matrix}$	$\begin{matrix}-i \\ i\end{matrix}$	$(x, y)(R_x, R_y)$	(yz, xz)

C_5	E	C_5	$C_5{}^2$	$C_5{}^3$	$C_5{}^4$			$\varepsilon = \exp(2\pi i/5)$
A	1	1	1	1	1	z, R_z		x^2+y^2, z^2
E_1	$\begin{cases}1 \\ 1\end{cases}$	$\begin{matrix}\varepsilon \\ \varepsilon^*\end{matrix}$	$\begin{matrix}\varepsilon^2 \\ \varepsilon^{2*}\end{matrix}$	$\begin{matrix}\varepsilon^{2*} \\ \varepsilon^2\end{matrix}$	$\begin{matrix}\varepsilon^* \\ \varepsilon\end{matrix}$	$(x, y)(R_x, R_y)$		(yz, xz)
E_2	$\begin{cases}1 \\ 1\end{cases}$	$\begin{matrix}\varepsilon^2 \\ \varepsilon^{2*}\end{matrix}$	$\begin{matrix}\varepsilon^* \\ \varepsilon\end{matrix}$	$\begin{matrix}\varepsilon \\ \varepsilon^*\end{matrix}$	$\begin{matrix}\varepsilon^{2*} \\ \varepsilon^2\end{matrix}$			(x^2-y^2, xy)

C_6	E	C_6	C_3	C_2	$C_3{}^2$	$C_6{}^5$			$\varepsilon = \exp(2\pi i/6)$
A	1	1	1	1	1	1	z, R_z		x^2+y^2, z^2
B	1	-1	1	-1	1	-1			
E_1	$\begin{cases}1 \\ 1\end{cases}$	$\begin{matrix}\varepsilon \\ \varepsilon^*\end{matrix}$	$\begin{matrix}-\varepsilon^* \\ -\varepsilon\end{matrix}$	$\begin{matrix}-1 \\ -1\end{matrix}$	$\begin{matrix}-\varepsilon \\ -\varepsilon^*\end{matrix}$	$\begin{matrix}\varepsilon^* \\ \varepsilon\end{matrix}$	$\begin{matrix}(x, y) \\ (R_x, R_y)\end{matrix}$		(xz, yz)
E_2	$\begin{cases}1 \\ 1\end{cases}$	$\begin{matrix}-\varepsilon^* \\ -\varepsilon\end{matrix}$	$\begin{matrix}-\varepsilon \\ -\varepsilon^*\end{matrix}$	$\begin{matrix}1 \\ 1\end{matrix}$	$\begin{matrix}-\varepsilon^* \\ -\varepsilon\end{matrix}$	$\begin{matrix}-\varepsilon \\ -\varepsilon^*\end{matrix}$			(x^2-y^2, xy)

C_7	E	C_7	$C_7{}^2$	$C_7{}^3$	$C_7{}^4$	$C_7{}^5$	$C_7{}^6$			$\varepsilon = \exp(2\pi i/7)$
A	1	1	1	1	1	1	1	z, R_z		x^2+y^2, z^2
E_1	$\begin{cases}1 \\ 1\end{cases}$	$\begin{matrix}\varepsilon \\ \varepsilon^*\end{matrix}$	$\begin{matrix}\varepsilon^2 \\ \varepsilon^{2*}\end{matrix}$	$\begin{matrix}\varepsilon^3 \\ \varepsilon^{3*}\end{matrix}$	$\begin{matrix}\varepsilon^{3*} \\ \varepsilon^3\end{matrix}$	$\begin{matrix}\varepsilon^{2*} \\ \varepsilon^2\end{matrix}$	$\begin{matrix}\varepsilon^* \\ \varepsilon\end{matrix}$	$\begin{matrix}(x, y) \\ (R_x, R_y)\end{matrix}$		(xz, yz)
E_2	$\begin{cases}1 \\ 1\end{cases}$	$\begin{matrix}\varepsilon^2 \\ \varepsilon^{2*}\end{matrix}$	$\begin{matrix}\varepsilon^{3*} \\ \varepsilon^3\end{matrix}$	$\begin{matrix}\varepsilon^* \\ \varepsilon\end{matrix}$	$\begin{matrix}\varepsilon \\ \varepsilon^*\end{matrix}$	$\begin{matrix}\varepsilon^3 \\ \varepsilon^{3*}\end{matrix}$	$\begin{matrix}\varepsilon^{2*} \\ \varepsilon^2\end{matrix}$			(x^2-y^2, xy)
E_3	$\begin{cases}1 \\ 1\end{cases}$	$\begin{matrix}\varepsilon^3 \\ \varepsilon^{3*}\end{matrix}$	$\begin{matrix}\varepsilon^* \\ \varepsilon\end{matrix}$	$\begin{matrix}\varepsilon^2 \\ \varepsilon^{2*}\end{matrix}$	$\begin{matrix}\varepsilon^{2*} \\ \varepsilon^2\end{matrix}$	$\begin{matrix}\varepsilon \\ \varepsilon^*\end{matrix}$	$\begin{matrix}\varepsilon^{3*} \\ \varepsilon^3\end{matrix}$			

C_8	E	C_8	C_4	C_2	$C_4{}^3$	$C_8{}^3$	$C_8{}^5$	$C_8{}^7$			$\varepsilon = \exp(2\pi i/8)$
A	1	1	1	1	1	1	1	1	z, R_z		x^2+y^2, z^2
B	1	-1	1	1	1	-1	-1	-1			
E_1	$\begin{cases}1 \\ 1\end{cases}$	$\begin{matrix}\varepsilon \\ \varepsilon^*\end{matrix}$	$\begin{matrix}i \\ -i\end{matrix}$	$\begin{matrix}-1 \\ -1\end{matrix}$	$\begin{matrix}-i \\ i\end{matrix}$	$\begin{matrix}-\varepsilon^* \\ -\varepsilon\end{matrix}$	$\begin{matrix}-\varepsilon \\ -\varepsilon^*\end{matrix}$	$\begin{matrix}\varepsilon^* \\ \varepsilon\end{matrix}$	$\begin{matrix}(x, y) \\ (R_x, R_y)\end{matrix}$		(xz, yz)
E_2	$\begin{cases}1 \\ 1\end{cases}$	$\begin{matrix}i \\ -i\end{matrix}$	$\begin{matrix}-1 \\ -1\end{matrix}$	$\begin{matrix}1 \\ 1\end{matrix}$	$\begin{matrix}-1 \\ -1\end{matrix}$	$\begin{matrix}-i \\ i\end{matrix}$	$\begin{matrix}i \\ -i\end{matrix}$	$\begin{matrix}-i \\ i\end{matrix}$			(x^2-y^2, xy)
E_3	$\begin{cases}1 \\ 1\end{cases}$	$\begin{matrix}-\varepsilon \\ -\varepsilon^*\end{matrix}$	$\begin{matrix}i \\ -i\end{matrix}$	$\begin{matrix}-1 \\ -1\end{matrix}$	$\begin{matrix}-i \\ i\end{matrix}$	$\begin{matrix}\varepsilon^* \\ \varepsilon\end{matrix}$	$\begin{matrix}\varepsilon \\ \varepsilon^*\end{matrix}$	$\begin{matrix}-\varepsilon^* \\ -\varepsilon\end{matrix}$			

3. The D_n Groups

D_2	E	$C_2(z)$	$C_2(y)$	$C_2(x)$		
A	1	1	1	1		x^2, y^2, z^2
B_1	1	1	−1	−1	z, R_z	xy
B_2	1	−1	1	−1	y, R_y	xz
B_3	1	−1	−1	1	x, R_x	yz

D_3	E	$2C_3$	$3C_2$		
A_1	1	1	1		$x^2 + y^2, z^2$
A_2	1	1	−1	z, R_z	
E	2	−1	0	$(x, y)(R_x, R_y)$	$(x^2 - y^2, xy)(xz, yz)$

D_4	E	$2C_4$	$C_2(=C_4{}^2)$	$2C_2'$	$2C_2''$		
A_1	1	1	1	1	1		$x^2 + y^2, z^2$
A_2	1	1	1	−1	−1	z, R_z	
B_1	1	−1	1	1	−1		$x^2 - y^2$
B_2	1	−1	1	−1	1		xy
E	2	0	−2	0	0	$(x, y)(R_x, R_y)$	(xz, yz)

D_5	E	$2C_5$	$2C_5{}^2$	$5C_2$		
A_1	1	1	1	1		$x^2 + y^2, z^2$
A_2	1	1	1	−1	z, R_z	
E_1	2	$2 \cos 72°$	$2 \cos 144°$	0	$(x, y)(R_x, R_y)$	(xz, yz)
E_2	2	$2 \cos 144°$	$2 \cos 72°$	0		$(x^2 - y^2, xy)$

D_6	E	$2C_6$	$2C_3$	C_2	$3C_2'$	$3C_2''$		
A_1	1	1	1	1	1	1		$x^2 + y^2, z^2$
A_2	1	1	1	1	−1	−1	z, R_z	
B_1	1	−1	1	−1	1	−1		
B_2	1	−1	1	−1	−1	1		
E_1	2	1	−1	−2	0	0	$(x, y)(R_x, R_y)$	(xz, yz)
E_2	2	−1	−1	2	0	0		$(x^2 - y^2, xy)$

4. The C_{nv} Groups

C_{2v}	E	C_2	$\sigma_v(xz)$	$\sigma_v'(yz)$		
A_1	1	1	1	1	z	x^2, y^2, z^2
A_2	1	1	-1	-1	R_z	xy
B_1	1	-1	1	-1	x, R_y	xz
B_2	1	-1	-1	1	y, R_x	yz

C_{3v}	E	$2C_3$	$3\sigma_v$		
A_1	1	1	1	z	$x^2 + y^2, z^2$
A_2	1	1	-1	R_z	
E	2	-1	0	$(x, y)(R_x, R_y)$	$(x^2 - y^2, xy)(xz, yz)$

C_{4v}	E	$2C_4$	C_2	$2\sigma_v$	$2\sigma_d$		
A_1	1	1	1	1	1	z	$x^2 + y^2, z^2$
A_2	1	1	1	-1	-1	R_z	
B_1	1	-1	1	1	-1		$x^2 - y^2$
B_2	1	-1	1	-1	1		xy
E	2	0	-2	0	0	$(x, y)(R_x, R_y)$	(xz, yz)

C_{5v}	E	$2C_5$	$2C_5{}^2$	$5\sigma_v$		
A_1	1	1	1	1	z	$x^2 + y^2, z^2$
A_2	1	1	1	-1	R_z	
E_1	2	$2\cos 72°$	$2\cos 144°$	0	$(x, y)(R_x, R_y)$	(xz, yz)
E_2	2	$2\cos 144°$	$2\cos 72°$	0		$(x^2 - y^2, xy)$

C_{6v}	E	$2C_6$	$2C_3$	C_2	$3\sigma_v$	$3\sigma_d$		
A_1	1	1	1	1	1	1	z	$x^2 + y^2, z^2$
A_2	1	1	1	1	-1	-1	R_z	
B_1	1	-1	1	-1	1	-1		
B_2	1	-1	1	-1	-1	1		
E_1	2	1	-1	-2	0	0	$(x, y)(R_x, R_y)$	(xz, yz)
E_2	2	-1	-1	2	0	0		$(x^2 - y^2, xy)$

5. The C_{nh} Groups

C_{2h}	E	C_2	i	σ_h		
A_g	1	1	1	1	R_z	x^2, y^2, z^2, xy
B_g	1	-1	1	-1	R_x, R_y	xz, yz
A_u	1	1	-1	-1	z	
B_u	1	-1	-1	1	x, y	

$\varepsilon = \exp(2\pi i/3)$

C_{3h}	E	C_3	C_3^2	σ_h	S_3	S_3^5		
A'	1	1	1	1	1	1	R_z	x^2+y^2, z^2
E'	$\begin{cases}1\\1\end{cases}$	$\begin{matrix}\varepsilon\\\varepsilon^*\end{matrix}$	$\begin{matrix}\varepsilon^*\\\varepsilon\end{matrix}$	$\begin{matrix}1\\1\end{matrix}$	$\begin{matrix}\varepsilon\\\varepsilon^*\end{matrix}$	$\begin{matrix}\varepsilon^*\\\varepsilon\end{matrix}$	(x,y)	(x^2-y^2, xy)
A''	1	1	1	-1	-1	-1	z	
E''	$\begin{cases}1\\1\end{cases}$	$\begin{matrix}\varepsilon\\\varepsilon^*\end{matrix}$	$\begin{matrix}\varepsilon^*\\\varepsilon\end{matrix}$	$\begin{matrix}-1\\-1\end{matrix}$	$\begin{matrix}-\varepsilon\\-\varepsilon^*\end{matrix}$	$\begin{matrix}-\varepsilon^*\\-\varepsilon\end{matrix}$	(R_x, R_y)	(xz, yz)

C_{4h}	E	C_4	C_2	C_4^3	i	S_4^3	σ_h	S_4		
A_g	1	1	1	1	1	1	1	1	R_z	x^2+y^2, z^2
B_g	1	-1	1	-1	1	-1	1	-1		x^2-y^2, xy
E_g	$\begin{cases}1\\1\end{cases}$	$\begin{matrix}i\\-i\end{matrix}$	$\begin{matrix}-1\\-1\end{matrix}$	$\begin{matrix}-i\\i\end{matrix}$	$\begin{matrix}1\\1\end{matrix}$	$\begin{matrix}i\\-i\end{matrix}$	$\begin{matrix}-1\\-1\end{matrix}$	$\begin{matrix}-i\\i\end{matrix}$	(R_x, R_y)	(xz, yz)
A_u	1	1	1	1	-1	-1	-1	-1	z	
B_u	1	-1	1	-1	-1	1	-1	1		
E_u	$\begin{cases}1\\1\end{cases}$	$\begin{matrix}i\\-i\end{matrix}$	$\begin{matrix}-1\\-1\end{matrix}$	$\begin{matrix}-i\\i\end{matrix}$	$\begin{matrix}-1\\-1\end{matrix}$	$\begin{matrix}-i\\i\end{matrix}$	$\begin{matrix}1\\1\end{matrix}$	$\begin{matrix}i\\-i\end{matrix}$	(x,y)	

$\varepsilon = \exp(2\pi i/5)$

C_{5h}	E	C_5	C_5^2	C_5^3	C_5^4	σ_h	S_5	S_5^7	S_5^3	S_5^9		
A'	1	1	1	1	1	1	1	1	1	1	R_z	x^2+y^2, z^2
E_1'	$\begin{cases}1\\1\end{cases}$	$\begin{matrix}\varepsilon\\\varepsilon^*\end{matrix}$	$\begin{matrix}\varepsilon^2\\\varepsilon^{2*}\end{matrix}$	$\begin{matrix}\varepsilon^{2*}\\\varepsilon^2\end{matrix}$	$\begin{matrix}\varepsilon^*\\\varepsilon\end{matrix}$	$\begin{matrix}1\\1\end{matrix}$	$\begin{matrix}\varepsilon\\\varepsilon^*\end{matrix}$	$\begin{matrix}\varepsilon^2\\\varepsilon^{2*}\end{matrix}$	$\begin{matrix}\varepsilon^{2*}\\\varepsilon^2\end{matrix}$	$\begin{matrix}\varepsilon^*\\\varepsilon\end{matrix}$	(x,y)	
E_2'	$\begin{cases}1\\1\end{cases}$	$\begin{matrix}\varepsilon^2\\\varepsilon^{2*}\end{matrix}$	$\begin{matrix}\varepsilon^*\\\varepsilon\end{matrix}$	$\begin{matrix}\varepsilon\\\varepsilon^*\end{matrix}$	$\begin{matrix}\varepsilon^{2*}\\\varepsilon^2\end{matrix}$	$\begin{matrix}1\\1\end{matrix}$	$\begin{matrix}\varepsilon^2\\\varepsilon^{2*}\end{matrix}$	$\begin{matrix}\varepsilon^*\\\varepsilon\end{matrix}$	$\begin{matrix}\varepsilon\\\varepsilon^*\end{matrix}$	$\begin{matrix}\varepsilon^{2*}\\\varepsilon^2\end{matrix}$		(x^2-y^2, xy)
A''	1	1	1	1	1	-1	-1	-1	-1	-1	z	
E_1''	$\begin{cases}1\\1\end{cases}$	$\begin{matrix}\varepsilon\\\varepsilon^*\end{matrix}$	$\begin{matrix}\varepsilon^2\\\varepsilon^{2*}\end{matrix}$	$\begin{matrix}\varepsilon^{2*}\\\varepsilon^2\end{matrix}$	$\begin{matrix}\varepsilon^*\\\varepsilon\end{matrix}$	$\begin{matrix}-1\\-1\end{matrix}$	$\begin{matrix}-\varepsilon\\-\varepsilon^*\end{matrix}$	$\begin{matrix}-\varepsilon^2\\-\varepsilon^{2*}\end{matrix}$	$\begin{matrix}-\varepsilon^{2*}\\-\varepsilon^2\end{matrix}$	$\begin{matrix}-\varepsilon^*\\-\varepsilon\end{matrix}$	(R_x, R_y)	(xz, yz)
E_2''	$\begin{cases}1\\1\end{cases}$	$\begin{matrix}\varepsilon^2\\\varepsilon^{2*}\end{matrix}$	$\begin{matrix}\varepsilon^*\\\varepsilon\end{matrix}$	$\begin{matrix}\varepsilon\\\varepsilon^*\end{matrix}$	$\begin{matrix}\varepsilon^{2*}\\\varepsilon^2\end{matrix}$	$\begin{matrix}-1\\-1\end{matrix}$	$\begin{matrix}-\varepsilon^2\\-\varepsilon^{2*}\end{matrix}$	$\begin{matrix}-\varepsilon^*\\-\varepsilon\end{matrix}$	$\begin{matrix}-\varepsilon\\-\varepsilon^*\end{matrix}$	$\begin{matrix}-\varepsilon^{2*}\\-\varepsilon^2\end{matrix}$		

$\varepsilon = \exp(2\pi i/6)$

C_{6h}	E	C_6	C_3	C_2	C_3^2	C_6^5	i	S_3^5	S_6^5	σ_h	S_6	S_3		
A_g	1	1	1	1	1	1	1	1	1	1	1	1	R_z	x^2+y^2, z^2
B_g	1	-1	1	-1	1	-1	1	-1	1	-1	1	-1		
E_{1g}	$\begin{cases}1\\1\end{cases}$	$\begin{matrix}\varepsilon\\\varepsilon^*\end{matrix}$	$\begin{matrix}-\varepsilon^*\\-\varepsilon\end{matrix}$	$\begin{matrix}-1\\-1\end{matrix}$	$\begin{matrix}-\varepsilon\\-\varepsilon^*\end{matrix}$	$\begin{matrix}\varepsilon^*\\\varepsilon\end{matrix}$	$\begin{matrix}1\\1\end{matrix}$	$\begin{matrix}\varepsilon\\\varepsilon^*\end{matrix}$	$\begin{matrix}-\varepsilon^*\\-\varepsilon\end{matrix}$	$\begin{matrix}-1\\-1\end{matrix}$	$\begin{matrix}-\varepsilon\\-\varepsilon^*\end{matrix}$	$\begin{matrix}\varepsilon^*\\\varepsilon\end{matrix}$	(R_x, R_y)	(xz, yz)
E_{2g}	$\begin{cases}1\\1\end{cases}$	$\begin{matrix}-\varepsilon^*\\-\varepsilon\end{matrix}$	$\begin{matrix}-\varepsilon\\-\varepsilon^*\end{matrix}$	$\begin{matrix}1\\1\end{matrix}$	$\begin{matrix}-\varepsilon^*\\-\varepsilon\end{matrix}$	$\begin{matrix}-\varepsilon\\-\varepsilon^*\end{matrix}$	$\begin{matrix}1\\1\end{matrix}$	$\begin{matrix}-\varepsilon^*\\-\varepsilon\end{matrix}$	$\begin{matrix}-\varepsilon\\-\varepsilon^*\end{matrix}$	$\begin{matrix}1\\1\end{matrix}$	$\begin{matrix}-\varepsilon^*\\-\varepsilon\end{matrix}$	$\begin{matrix}-\varepsilon\\-\varepsilon^*\end{matrix}$		(x^2-y^2, xy)
A_u	1	1	1	1	1	1	-1	-1	-1	-1	-1	-1	z	
B_u	1	-1	1	-1	1	-1	-1	1	-1	1	-1	1		
E_{1u}	$\begin{cases}1\\1\end{cases}$	$\begin{matrix}\varepsilon\\\varepsilon^*\end{matrix}$	$\begin{matrix}-\varepsilon^*\\-\varepsilon\end{matrix}$	$\begin{matrix}-1\\-1\end{matrix}$	$\begin{matrix}-\varepsilon\\-\varepsilon^*\end{matrix}$	$\begin{matrix}\varepsilon^*\\\varepsilon\end{matrix}$	$\begin{matrix}-1\\-1\end{matrix}$	$\begin{matrix}-\varepsilon\\-\varepsilon^*\end{matrix}$	$\begin{matrix}\varepsilon^*\\\varepsilon\end{matrix}$	$\begin{matrix}1\\1\end{matrix}$	$\begin{matrix}\varepsilon\\\varepsilon^*\end{matrix}$	$\begin{matrix}-\varepsilon^*\\-\varepsilon\end{matrix}$	(x,y)	
E_{2u}	$\begin{cases}1\\1\end{cases}$	$\begin{matrix}-\varepsilon^*\\-\varepsilon\end{matrix}$	$\begin{matrix}-\varepsilon\\-\varepsilon^*\end{matrix}$	$\begin{matrix}1\\1\end{matrix}$	$\begin{matrix}-\varepsilon^*\\-\varepsilon\end{matrix}$	$\begin{matrix}-\varepsilon\\-\varepsilon^*\end{matrix}$	$\begin{matrix}-1\\-1\end{matrix}$	$\begin{matrix}\varepsilon^*\\\varepsilon\end{matrix}$	$\begin{matrix}\varepsilon\\\varepsilon^*\end{matrix}$	$\begin{matrix}-1\\-1\end{matrix}$	$\begin{matrix}\varepsilon^*\\\varepsilon\end{matrix}$	$\begin{matrix}\varepsilon\\\varepsilon^*\end{matrix}$		

6. The D_{nh} Groups

D_{2h}	E	$C_2(z)$	$C_2(y)$	$C_2(x)$	i	$\sigma(xy)$	$\sigma(xz)$	$\sigma(yz)$		
A_g	1	1	1	1	1	1	1	1		x^2, y^2, z^2
B_{1g}	1	1	-1	-1	1	1	-1	-1	R_z	xy
B_{2g}	1	-1	1	-1	1	-1	1	-1	R_y	xz
B_{3g}	1	-1	-1	1	1	-1	-1	1	R_x	yz
A_u	1	1	1	1	-1	-1	-1	-1		
B_{1u}	1	1	-1	-1	-1	-1	1	1	z	
B_{2u}	1	-1	1	-1	-1	1	-1	1	y	
B_{3u}	1	-1	-1	1	-1	1	1	-1	x	

D_{3h}	E	$2C_3$	$3C_2$	σ_h	$2S_3$	$3\sigma_v$		
A_1'	1	1	1	1	1	1		x^2+y^2, z^2
A_2'	1	1	-1	1	1	-1	R_z	
E'	2	-1	0	2	-1	0	(x, y)	(x^2-y^2, xy)
A_1''	1	1	1	-1	-1	-1		
A_2''	1	1	-1	-1	-1	1	z	
E''	2	-1	0	-2	1	0	(R_x, R_y)	(xz, yz)

D_{4h}	E	$2C_4$	C_2	$2C_2'$	$2C_2''$	i	$2S_4$	σ_h	$2\sigma_v$	$2\sigma_d$		
A_{1g}	1	1	1	1	1	1	1	1	1	1		x^2+y^2, z^2
A_{2g}	1	1	1	-1	-1	1	1	1	-1	-1	R_z	
B_{1g}	1	-1	1	1	-1	1	-1	1	1	-1		x^2-y^2
B_{2g}	1	-1	1	-1	1	1	-1	1	-1	1		xy
E_g	2	0	-2	0	0	2	0	-2	0	0	(R_x, R_y)	(xz, yz)
A_{1u}	1	1	1	1	1	-1	-1	-1	-1	-1		
A_{2u}	1	1	1	-1	-1	-1	-1	-1	1	1	z	
B_{1u}	1	-1	1	1	-1	-1	1	-1	-1	1		
B_{2u}	1	-1	1	-1	1	-1	1	-1	1	-1		
E_u	2	0	-2	0	0	-2	0	2	0	0	(x, y)	

D_{5h}	E	$2C_5$	$2C_5^2$	$5C_2$	σ_h	$2S_5$	$2S_5^3$	$5\sigma_v$		
A_1'	1	1	1	1	1	1	1	1		x^2+y^2, z^2
A_2'	1	1	1	-1	1	1	1	-1	R_z	
E_1'	2	$2\cos 72°$	$2\cos 144°$	0	2	$2\cos 72°$	$2\cos 144°$	0	(x, y)	
E_2'	2	$2\cos 144°$	$2\cos 72°$	0	2	$2\cos 144°$	$2\cos 72°$	0		(x^2-y^2, xy)
A_1''	1	1	1	1	-1	-1	-1	-1		
A_2''	1	1	1	-1	-1	-1	-1	1	z	
E_1''	2	$2\cos 72°$	$2\cos 144°$	0	-2	$-2\cos 72°$	$-2\cos 144°$	0	(R_x, R_y)	(xz, yz)
E_2''	2	$2\cos 144°$	$2\cos 72°$	0	-2	$-2\cos 144°$	$-2\cos 72°$	0		

D_{6h}	E	$2C_6$	$2C_3$	C_2	$3C_2'$	$3C_2''$	i	$2S_3$	$2S_6$	σ_h	$3\sigma_d$	$3\sigma_v$		
A_{1g}	1	1	1	1	1	1	1	1	1	1	1	1		x^2+y^2, z^2
A_{2g}	1	1	1	1	-1	-1	1	1	1	1	-1	-1	R_z	
B_{1g}	1	-1	1	-1	1	-1	1	-1	1	-1	1	-1		
B_{2g}	1	-1	1	-1	-1	1	1	-1	1	-1	-1	1		
E_{1g}	2	1	-1	-2	0	0	2	1	-1	-2	0	0	(R_x, R_y)	(xz, yz)
E_{2g}	2	-1	-1	2	0	0	2	-1	-1	2	0	0		(x^2-y^2, xy)
A_{1u}	1	1	1	1	1	1	-1	-1	-1	-1	-1	-1		
A_{2u}	1	1	1	1	-1	-1	-1	-1	-1	-1	1	1	z	
B_{1u}	1	-1	1	-1	1	-1	-1	1	-1	1	-1	1		
B_{2u}	1	-1	1	-1	-1	1	-1	1	-1	1	1	-1		
E_{1u}	2	1	-1	-2	0	0	-2	-1	1	2	0	0	(x, y)	
E_{2u}	2	-1	-1	2	0	0	-2	1	1	-2	0	0		

D_{8h}	E	$2C_8$	$2C_8{}^3$	$2C_4$	C_2	$4C_2'$	$4C_2''$	i	$2S_8$	$2S_8{}^3$	$2S_4$	σ_h	$4\sigma_d$	$4\sigma_v$		
A_{1g}	1	1	1	1	1	1	1	1	1	1	1	1	1	1		x^2+y^2, z^2
A_{2g}	1	1	1	1	1	−1	−1	1	1	1	1	1	−1	−1	R_z	
B_{1g}	1	−1	−1	1	1	1	−1	1	−1	−1	1	1	1	−1		
B_{2g}	1	−1	−1	1	1	−1	1	1	−1	−1	1	1	−1	1		
E_{1g}	2	$\sqrt{2}$	$-\sqrt{2}$	0	−2	0	0	2	$\sqrt{2}$	$-\sqrt{2}$	0	−2	0	0	(R_x, R_y)	(xz, yz)
E_{2g}	2	0	0	−2	2	0	0	2	0	0	−2	2	0	0		(x^2-y^2, xy)
E_{3g}	2	$-\sqrt{2}$	$\sqrt{2}$	0	−2	0	0	2	$-\sqrt{2}$	$\sqrt{2}$	0	−2	0	0		
A_{1u}	1	1	1	1	1	1	1	−1	−1	−1	−1	−1	−1	−1		
A_{2u}	1	1	1	1	1	−1	−1	−1	−1	−1	−1	1	1	1	z	
B_{1u}	1	−1	−1	1	1	1	−1	−1	1	1	−1	−1	−1	1		
B_{2u}	1	−1	−1	1	1	−1	1	−1	1	1	−1	−1	1	−1		
E_{1u}	2	$\sqrt{2}$	$-\sqrt{2}$	0	−2	0	0	−2	$-\sqrt{2}$	$\sqrt{2}$	0	2	0	0	(x, y)	
E_{2u}	2	0	0	−2	2	0	0	−2	0	0	2	−2	0	0		
E_{3u}	2	$-\sqrt{2}$	$\sqrt{2}$	0	−2	0	0	−2	$\sqrt{2}$	$-\sqrt{2}$	0	2	0	0		

7. The D_{nd} Groups

D_{2d}	E	$2S_4$	C_2	$2C_2'$	$2\sigma_d$		
A_1	1	1	1	1	1		x^2+y^2, z^2
A_2	1	1	1	−1	−1	R_z	
B_1	1	−1	1	1	−1		x^2-y^2
B_2	1	−1	1	−1	1	z	xy
E	2	0	−2	0	0	(x, y); (R_x, R_y)	(xz, yz)

D_{3d}	E	$2C_3$	$3C_2$	i	$2S_6$	$3\sigma_d$		
A_{1g}	1	1	1	1	1	1		x^2+y^2, z^2
A_{2g}	1	1	−1	1	1	−1	R_z	
E_g	2	−1	0	2	−1	0	(R_x, R_y)	$(x^2-y^2, xy),$ (xz, yz)
A_{1u}	1	1	1	−1	−1	−1		
A_{2u}	1	1	−1	−1	−1	1	z	
E_u	2	−1	0	−2	1	0	(x, y)	

D_{4d}	E	$2S_8$	$2C_4$	$2S_8{}^3$	C_2	$4C_2'$	$4\sigma_d$		
A_1	1	1	1	1	1	1	1		x^2+y^2, z^2
A_2	1	1	1	1	1	−1	−1	R_z	
B_1	1	−1	1	−1	1	1	−1		
B_2	1	−1	1	−1	1	−1	1	z	
E_1	2	$\sqrt{2}$	0	$-\sqrt{2}$	−2	0	0	(x, y)	
E_2	2	0	−2	0	2	0	0		(x^2-y^2, xy)
E_3	2	$-\sqrt{2}$	0	$\sqrt{2}$	−2	0	0	(R_x, R_y)	(xz, yz)

D_{5d}	E	$2C_5$	$2C_5{}^2$	$5C_2$	i	$2S_{10}{}^3$	$2S_{10}$	$5\sigma_d$		
A_{1g}	1	1	1	1	1	1	1	1		x^2+y^2, z^2
A_{2g}	1	1	1	−1	1	1	1	−1	R_z	
E_{1g}	2	$2\cos 72°$	$2\cos 144°$	0	2	$2\cos 72°$	$2\cos 144°$	0	(R_x, R_y)	(xz, yz)
E_{2g}	2	$2\cos 144°$	$2\cos 72°$	0	2	$2\cos 144°$	$2\cos 72°$	0		(x^2-y^2, xy)
A_{1u}	1	1	1	1	−1	−1	−1	−1		
A_{2u}	1	1	1	−1	−1	−1	−1	1	z	
E_{1u}	2	$2\cos 72°$	$2\cos 144°$	0	−2	$-2\cos 72°$	$-2\cos 144°$	0	(x, y)	
E_{2u}	2	$2\cos 144°$	$2\cos 72°$	0	−2	$-2\cos 144°$	$-2\cos 72°$	0		

7. The D_{nd} Groups (Continued).

D_{6d}	E	$2S_{12}$	$2C_6$	$2S_4$	$2C_3$	$2S_{12}{}^5$	C_2	$6C_2'$	$6\sigma_d$		
A_1	1	1	1	1	1	1	1	1	1		x^2+y^2, z^2
A_2	1	1	1	1	1	1	1	-1	-1	R_z	
B_1	1	-1	1	-1	1	-1	1	1	-1		
B_2	1	-1	1	-1	1	-1	1	-1	1	z	
E_1	2	$\sqrt{3}$	1	0	-1	$-\sqrt{3}$	-2	0	0	(x, y)	
E_2	2	1	-1	-2	-1	1	2	0	0		(x^2-y^2, xy)
E_3	2	0	-2	0	2	0	-2	0	0		
E_4	2	-1	-1	2	-1	-1	2	0	0		
E_5	2	$-\sqrt{3}$	1	0	-1	$\sqrt{3}$	-2	0	0	(R_x, R_y)	(xz, yz)

8. The S_n Groups

S_4	E	S_4	C_2	$S_4{}^3$		
A	1	1	1	1	R_z	x^2+y^2, z^2
B	1	-1	1	-1	z	x^2-y^2, xy
E	$\begin{Bmatrix}1 \\ 1\end{Bmatrix}$	$\begin{matrix}i \\ -i\end{matrix}$	$\begin{matrix}-1 \\ -1\end{matrix}$	$\begin{matrix}-i \\ i\end{matrix}$	$(x, y); (R_x, R_y)$	(xz, yz)

S_6	E	C_3	$C_3{}^2$	i	$S_6{}^5$	S_6		$\varepsilon = \exp(2\pi i/3)$
A_g	1	1	1	1	1	1	R_z	x^2+y^2, z^2
E_g	$\begin{Bmatrix}1 \\ 1\end{Bmatrix}$	$\begin{matrix}\varepsilon \\ \varepsilon^*\end{matrix}$	$\begin{matrix}\varepsilon^* \\ \varepsilon\end{matrix}$	$\begin{matrix}1 \\ 1\end{matrix}$	$\begin{matrix}\varepsilon \\ \varepsilon^*\end{matrix}$	$\begin{matrix}\varepsilon^* \\ \varepsilon\end{matrix}$	(R_x, R_y)	$(x^2-y^2, xy);$ (xz, yz)
A_u	1	1	1	-1	-1	-1	z	
E_u	$\begin{Bmatrix}1 \\ 1\end{Bmatrix}$	$\begin{matrix}\varepsilon \\ \varepsilon^*\end{matrix}$	$\begin{matrix}\varepsilon^* \\ \varepsilon\end{matrix}$	$\begin{matrix}-1 \\ -1\end{matrix}$	$\begin{matrix}-\varepsilon \\ -\varepsilon^*\end{matrix}$	$\begin{matrix}-\varepsilon^* \\ -\varepsilon\end{matrix}$	(x, y)	

S_8	E	S_8	C_4	$S_8{}^3$	C_2	$S_8{}^5$	$C_4{}^3$	$S_8{}^7$		$\varepsilon = \exp(2\pi i/8)$
A	1	1	1	1	1	1	1	1	R_z	x^2+y^2, z^2
B	1	-1	1	-1	1	-1	1	-1	z	
E_1	$\begin{Bmatrix}1 \\ 1\end{Bmatrix}$	$\begin{matrix}\varepsilon \\ \varepsilon^*\end{matrix}$	$\begin{matrix}i \\ -i\end{matrix}$	$\begin{matrix}-\varepsilon^* \\ -\varepsilon\end{matrix}$	$\begin{matrix}-1 \\ -1\end{matrix}$	$\begin{matrix}-\varepsilon \\ -\varepsilon^*\end{matrix}$	$\begin{matrix}-i \\ i\end{matrix}$	$\begin{matrix}\varepsilon^* \\ \varepsilon\end{matrix}$	$(x, y);$ (R_x, R_y)	
E_2	$\begin{Bmatrix}1 \\ 1\end{Bmatrix}$	$\begin{matrix}i \\ -i\end{matrix}$	$\begin{matrix}-1 \\ -1\end{matrix}$	$\begin{matrix}-i \\ i\end{matrix}$	$\begin{matrix}1 \\ 1\end{matrix}$	$\begin{matrix}i \\ -i\end{matrix}$	$\begin{matrix}-1 \\ -1\end{matrix}$	$\begin{matrix}-i \\ i\end{matrix}$		(x^2-y^2, xy)
E_3	$\begin{Bmatrix}1 \\ 1\end{Bmatrix}$	$\begin{matrix}-\varepsilon^* \\ -\varepsilon\end{matrix}$	$\begin{matrix}-i \\ i\end{matrix}$	$\begin{matrix}\varepsilon \\ \varepsilon^*\end{matrix}$	$\begin{matrix}-1 \\ -1\end{matrix}$	$\begin{matrix}\varepsilon^* \\ \varepsilon\end{matrix}$	$\begin{matrix}i \\ -i\end{matrix}$	$\begin{matrix}-\varepsilon \\ -\varepsilon^*\end{matrix}$		(xz, yz)

9. The Cubic Groups

T	E	$4C_3$	$4C_3{}^2$	$3C_2$			$\varepsilon = \exp(2\pi i/3)$
A	1	1	1	1			$x^2+y^2+z^2$
E	$\begin{Bmatrix}1 \\ 1\end{Bmatrix}$	$\begin{matrix}\varepsilon \\ \varepsilon^*\end{matrix}$	$\begin{matrix}\varepsilon^* \\ \varepsilon\end{matrix}$	$\begin{matrix}1 \\ 1\end{matrix}$			$(2z^2-x^2-y^2,$ $x^2-y^2)$
T	3	0	0	-1	$(R_x, R_y, R_z); (x, y, z)$		(xy, xz, yz)

9. The Cubic Groups (Continued).

T_h	E	$4C_3$	$4C_3^2$	$3C_2$	i	$4S_6$	$4S_6^5$	$3\sigma_h$			$\varepsilon = \exp(2\pi i/3)$
A_g	1	1	1	1	1	1	1	1			$x^2+y^2+z^2$
A_u	1	1	1	1	-1	-1	-1	-1			
E_g	$\begin{cases}1\\1\end{cases}$	$\begin{matrix}\varepsilon\\\varepsilon^*\end{matrix}$	$\begin{matrix}\varepsilon^*\\\varepsilon\end{matrix}$	$\begin{matrix}1\\1\end{matrix}$	$\begin{matrix}1\\1\end{matrix}$	$\begin{matrix}\varepsilon\\\varepsilon^*\end{matrix}$	$\begin{matrix}\varepsilon^*\\\varepsilon\end{matrix}$	$\begin{matrix}1\\1\end{matrix}$			$(2z^2-x^2-y^2,$ $x^2-y^2)$
E_u	$\begin{cases}1\\1\end{cases}$	$\begin{matrix}\varepsilon\\\varepsilon^*\end{matrix}$	$\begin{matrix}\varepsilon^*\\\varepsilon\end{matrix}$	$\begin{matrix}1\\1\end{matrix}$	$\begin{matrix}-1\\-1\end{matrix}$	$\begin{matrix}-\varepsilon\\-\varepsilon^*\end{matrix}$	$\begin{matrix}-\varepsilon^*\\-\varepsilon\end{matrix}$	$\begin{matrix}-1\\-1\end{matrix}$			
T_g	3	0	0	-1	1	0	0	-1	(R_x, R_y, R_z)		(xz, yz, xy)
T_u	3	0	0	-1	-1	0	0	1	(x, y, z)		

T_d	E	$8C_3$	$3C_2$	$6S_4$	$6\sigma_d$			
A_1	1	1	1	1	1			$x^2+y^2+z^2$
A_2	1	1	1	-1	-1			
E	2	-1	2	0	0			$(2z^2-x^2-y^2,$ $x^2-y^2)$
T_1	3	0	-1	1	-1	(R_x, R_y, R_z)		
T_2	3	0	-1	-1	1	(x, y, z)		(xy, xz, yz)

O	E	$6C_4$	$3C_2(=C_4^2)$	$8C_3$	$6C_2$			
A_1	1	1	1	1	1			$x^2+y^2+z^2$
A_2	1	-1	1	1	-1			
E	2	0	2	-1	0			$(2z^2-x^2-y^2,$ $x^2-y^2)$
T_1	3	1	-1	0	-1	$(R_x, R_y, R_z); (x, y, z)$		
T_2	3	-1	-1	0	1			(xy, xz, yz)

O_h	E	$8C_3$	$6C_2$	$6C_4$	$3C_2(=C_4^2)$	i	$6S_4$	$8S_6$	$3\sigma_h$	$6\sigma_d$			
A_{1g}	1	1	1	1	1	1	1	1	1	1			$x^2+y^2+z^2$
A_{2g}	1	1	-1	-1	1	1	-1	1	1	-1			
E_g	2	-1	0	0	2	2	0	-1	2	0			$(2z^2-x^2-y^2,$ $x^2-y^2)$
T_{1g}	3	0	-1	1	-1	3	1	0	-1	-1	(R_x, R_y, R_z)		
T_{2g}	3	0	1	-1	-1	3	-1	0	-1	1			(xz, yz, xy)
A_{1u}	1	1	1	1	1	-1	-1	-1	-1	-1			
A_{2u}	1	1	-1	-1	1	-1	1	-1	-1	1			
E_u	2	-1	0	0	2	-2	0	1	-2	0			
T_{1u}	3	0	-1	1	-1	-3	-1	0	1	1	(x, y, z)		
T_{2u}	3	0	1	-1	-1	-3	1	0	1	-1			

10. The Groups $C_{\infty v}$ and $D_{\infty h}$ for Linear Molecules

$C_{\infty v}$	E	$2C_\infty^\Phi$	$\cdots$	$\infty\sigma_v$		
$A_1 \equiv \Sigma^+$	1	1	$\cdots$	1	z	x^2+y^2, z^2
$A_2 \equiv \Sigma^-$	1	1	$\cdots$	-1	R_z	
$E_1 \equiv \Pi$	2	$2\cos\Phi$	$\cdots$	0	$(x, y); (R_x, R_y)$	(xz, yz)
$E_2 \equiv \Delta$	2	$2\cos 2\Phi$	$\cdots$	0		(x^2-y^2, xy)
$E_3 \equiv \Phi$	2	$2\cos 3\Phi$	$\cdots$	0		
$\cdots$	$\cdots$	$\cdots$	$\cdots$	$\cdots$		

$D_{\infty h}$	E	$2C_\infty^\Phi$	$\cdots$	$\infty\sigma_v$	i	$2S_\infty^\Phi$	$\cdots$	∞C_2		
Σ_g^+	1	1	$\cdots$	1	1	1	$\cdots$	1		x^2+y^2, z^2
Σ_g^-	1	1	$\cdots$	-1	1	1	$\cdots$	-1	R_z	
Π_g	2	$2\cos\Phi$	$\cdots$	0	2	$-2\cos\Phi$	$\cdots$	0	(R_x, R_y)	(xz, yz)
Δ_g	2	$2\cos 2\Phi$	$\cdots$	0	2	$2\cos 2\Phi$	$\cdots$	0		(x^2-y^2, xy)
$\cdots$	$\cdots$	$\cdots$	$\cdots$	$\cdots$	$\cdots$	$\cdots$	$\cdots$	$\cdots$		
Σ_u^+	1	1	$\cdots$	1	-1	-1	$\cdots$	-1	z	
Σ_u^-	1	1	$\cdots$	-1	-1	-1	$\cdots$	1		
Π_u	2	$2\cos\Phi$	$\cdots$	0	-2	$2\cos\Phi$	$\cdots$	0	(x, y)	
Δ_u	2	$2\cos 2\Phi$	$\cdots$	0	-2	$-2\cos 2\Phi$	$\cdots$	0		
$\cdots$	$\cdots$	$\cdots$	$\cdots$	$\cdots$	$\cdots$	$\cdots$	$\cdots$	$\cdots$		

11. The Icosahedral Groups*

I_h	E	$12C_5$	$12C_5{}^2$	$20C_3$	$15C_2$	i	$12S_{10}$	$12S_{10}{}^3$	$20S_6$	15σ		
A_g	1	1	1	1	1	1	1	1	1	1		$x^2 + y^2 + z^2$
T_{1g}	3	$\frac{1}{2}(1+\sqrt{5})$	$\frac{1}{2}(1-\sqrt{5})$	0	-1	3	$\frac{1}{2}(1-\sqrt{5})$	$\frac{1}{2}(1+\sqrt{5})$	0	-1	(R_x, R_y, R_z)	
T_{2g}	3	$\frac{1}{2}(1-\sqrt{5})$	$\frac{1}{2}(1+\sqrt{5})$	0	-1	3	$\frac{1}{2}(1+\sqrt{5})$	$\frac{1}{2}(1-\sqrt{5})$	0	-1		
G_g	4	-1	-1	1	0	4	-1	-1	1	0		
H_g	5	0	0	-1	1	5	0	0	-1	1		$(2z^2 - x^2 - y^2,$ $x^2 - y^2,$ $xy, yz, zx)$
A_u	1	1	1	1	1	-1	-1	-1	-1	-1		
T_{1u}	3	$\frac{1}{2}(1+\sqrt{5})$	$\frac{1}{2}(1-\sqrt{5})$	0	-1	-3	$-\frac{1}{2}(1-\sqrt{5})$	$-\frac{1}{2}(1+\sqrt{5})$	0	1	(x, y, z)	
T_{2u}	3	$\frac{1}{2}(1-\sqrt{5})$	$\frac{1}{2}(1+\sqrt{5})$	0	-1	-3	$-\frac{1}{2}(1+\sqrt{5})$	$-\frac{1}{2}(1-\sqrt{5})$	0	1		
G_u	4	-1	-1	1	0	-4	1	1	-1	0		
H_u	5	0	0	-1	1	-5	0	0	1	-1		

* For the pure rotation group I, the outlined section in the upper left is the character table; the g subscripts should, of course, be dropped and (x, y, z) assigned to the T_1 representation.

B: Correlation Table for Group O_h

This table shows how the representations of group O_h are changed or decomposed into those of its subgroups when the symmetry is altered or lowered. This table covers only representations of use in dealing with the more common symmetries of complexes. A rather complete collection of correlation tables will be found as Table X-14 in *Molecular Vibrations* by E. B. Wilson, Jr., J. C. Decius, and P. C. Cross, McGraw-Hill, New York, 1955.

O_h	O	T_d	D_{4h}	D_{2d}	C_{4v}	C_{2v}	D_{3d}	D_3	C_{2h}
A_{1g}	A_1	A_1	A_{1g}	A_1	A_1	A_1	A_{1g}	A_1	A_g
A_{2g}	A_2	A_2	B_{1g}	B_1	B_1	A_2	A_{2g}	A_2	B_g
E_g	E	E	$A_{1g}+B_{1g}$	A_1+B_1	A_1+B_1	A_1+A_2	E_g	E	A_g+B_g
T_{1g}	T_1	T_1	$A_{2g}+E_g$	A_2+E	A_2+E	$A_2+B_1+B_2$	$A_{2g}+E_g$	A_2+E	A_g+2B_g
T_{2g}	T_2	T_2	$B_{2g}+E_g$	B_2+E	B_2+E	$A_1+B_1+B_2$	$A_{1g}+E_g$	A_1+E	$2A_g+B_g$
A_{1u}	A_1	A_2	A_{1u}	B_1	A_2	A_2	A_{1u}	A_1	A_u
A_{2u}	A_2	A_1	B_{1u}	A_1	B_2	A_1	A_{2u}	A_2	B_u
E_u	E	E	$A_{1u}+B_{1u}$	A_1+B_1	A_2+B_2	A_1+A_2	E_u	E	A_u+B_u
T_{1u}	T_1	T_2	$A_{2u}+E_u$	B_2+E	A_1+E	$A_1+B_1+B_2$	$A_{2u}+E_u$	A_2+E	A_u+2B_u
T_{2u}	T_2	T_1	$B_{2u}+E_u$	A_2+E	B_1+E	$A_2+B_1+B_2$	$A_{1u}+E_u$	A_1+E	$2A_u+B_u$

Appendix IV

A Caveat Concerning the Resonance Integral

On pages 137–139 we described a seemingly straightforward and obvious method for evaluating the integral β. This involved equating the so-called experimental or empirical resonance energy of benzene, defined as the energy difference between "real" benzene and Kekulé benzene, to a multiple of β. This procedure is widely used and is of at least empirical validity, as shown by the fact that it gives essentially the same value to β in various molecules. Some results illustrative of this point are given in the following table.

Table of Experimental Resonance Energies and the Derived Value of β

COMPOUND	OBSERVED RESONANCE ENERGY, KCAL/MOLE	THEORETICAL RESONANCE ENERGY, HÜCKEL APPROX.	$-\beta$
Benzene	36	2β	18
Diphenyl	70–80	4.38β	~ 17
Naphthalene	75–80	3.68β	~ 21
Anthracene	105–116	5.32β	~ 20
Phenanthrene	110–125	5.45β	~ 21

Vertical Resonance Energy

In the process just described we are actually considering not only the energy of delocalization but also the energy required to stretch and compress the C-C bonds in Kekulé benzene from the lengths 1.54 and 1.34 Å to the common length 1.39 Å found in real benzene. Thus the experimental resonance energy, R_{exp}, is related to the true delocalization energy, or vertical* resonance energy, R_v, and to the combined energies of bond stretching, bond compression, and other changes (as, for instance, in repulsive forces between nonbonded atoms) collectively denoted by E_C, by the following equation:

$$R_{exp} = R_v + E_C$$

Now it can be argued that the MO calculation actually applies to R_v and that in order to evaluate β we should equate the multiple of β given by theory to R_v and not R_{exp}. In order to do this, E_C must be calculated. For benzene† E_C has been estimated to be -37 kcal/mole. From this we estimate that $R_v = 36 + 37 = 73$ kcal/mole, and hence β must be about 37 kcal/mole, or about twice the value used in the calculations of experimental resonance energies. Thus, for estimating the actual separations of energy levels in the real molecule, this larger value of β should be more nearly correct.

The Spectroscopic β

One might reasonably observe that the Hückel assumption of negligible overlap between neighboring $p\pi$ orbitals seems physically unlikely, since just such overlap would be required to allow formation of multicenter molecular orbitals. This point has been considered in more detail by Mulliken, Rieke, and Brown,‡ and we summarize here their discussion.

Without the assumption that all $S_{ij} = \delta_{ij}$ our secular equations will take form (a) rather than form (b), which has been used in the Hückel approximation:

$$(a) \quad \begin{vmatrix} H_{11} - E & H_{12} - ES_{12} \\ H_{21} - ES_{21} & H_{22} - E \end{vmatrix} = 0$$

$$(b) \quad \begin{vmatrix} H_{11} - E & H_{12} \\ H_{21} & H_{22} - E \end{vmatrix} = 0$$

Thus the true value of β is not $H_{12} = H_{21}$ but $H_{12} - ES_{12} = H_{21} - ES_{21}$. It can be shown that the value of E (referred to the true zero of energy, which is the energy of an electron when completely separated from the molecule)

---

* The term vertical is used with respect to a diagram in which we plot energy on the ordinate against internuclear distances on the abscissa. Thus, if the internuclear distances are held the same in the hypothetical, nonresonating and the real, resonating molecule, the change in energy is a purely vertical change on the plot.

† See R. S. Mulliken and R. G. Parr, *J. Chem. Phys.*, **19**, 1271 (1951).

‡ R. S. Mulliken, C. A. Rieke, and W. G. Brown, *J. Am. Chem. Soc.*, **63**, 41 (1941).

is about 50 electron volts (1 eV = 23.06 kcal/mole). Moreover, S_{12} must have a value between 0.2 and 0.3 so that ES_{12} must be between 10 and 15 eV. Mulliken et al. note that the integral H_{12} should also be of about this magnitude, which is consistent with the fact that β turns out to be only 1–2 eV, that is, a small difference between two larger numbers.

It is therefore rather surprising that β remains substantially constant from one molecule to another, since relatively small variations in H_{12} or ES_{12} could cause relatively large variations in β. Empirically, however, this approximate constancy is observed.

On the other hand, we should perhaps not be too surprised if values of β derived from different kinds of measurements do not agree exactly, and indeed this is the case. We have seen above that β obtained from the vertical resonance energy of benzene is about 37 kcal/mole, whereas Platt* first showed that the best overall fit to the spectra of benzene and other unsaturated hydrocarbons was obtained in the framework of the Hückel approximation by taking β to be 55–60 kcal/mole. This high value, which has subsequently been widely adopted to estimate actual differences in energies between MO's, is generally known as the spectroscopic value of β.

* J. R. Platt, *J. Chem. Phys.*, **15**, 419 (1947).

Appendix V

The Shapes of f Orbitals

When the wave equation for a hydrogen-like atom is solved in the most direct way for orbitals with the angular momentum quantum number l equal to 3, the following results are obtained for the purely angular parts (i.e., omitting all numerical factors):

$$\psi_0: (5\cos^3\theta - 3\cos\theta) \qquad m_l = 0$$
$$\psi_{\pm 1}: \sin\theta(5\cos^2\theta - 1)e^{\pm i\phi} \qquad m_l = \pm 1$$
$$\psi_{\pm 2}: (\sin^2\theta\cos\theta)e^{\pm 2i\phi} \qquad m_l = \pm 2$$
$$\psi_{\pm 3}: (\sin^3\theta)e^{\pm 3i\phi} \qquad m_l = \pm 3$$

The seven functions are grouped into sets having projections of the orbital angular momentum on the z axis of 0, ± 1, ± 2, and ± 3. Each of the functions in the pairs with m_l equal to ± 1, ± 2, and ± 3 is complex as written above, but by taking linear combinations of each pair, for example,

$$\frac{1}{\sqrt{2}}(\psi_{+3} + \psi_{-3}) \quad \text{and} \quad \frac{1}{i\sqrt{2}}(\psi_{+3} - \psi_{-3})$$

the imaginary parts are eliminated. In this way the seven real, normalized (to unity) orbitals listed in Table AV.1 are obtained. As given in the table, the f orbitals are in a convenient form for problems involving only a single high-order symmetry axis. For instance, in treating bis(cyclooctatetraene)-metal compounds, where the point group is D_{8h}, we note that the orbitals are already grouped into sets belonging to the representations A_{2u}, E_{1u}, E_{2u}, and E_{3u}. However, for problems involving cubic symmetry, the functions given in Table AV.1 are awkward to use since they do not directly form triply degenerate sets, despite the fact that the entire set of f functions spans the representations A_{2u}, T_{1u}, and T_{2u} in the group O_h.

Table AV.1

TRUE POLYNOMIAL	SIMPLIFIED POLYNOMIAL	NORMALIZING FACTOR	ANGULAR FUNCTION
$z(5z^2 - 3r^2)$	z^3	$\dfrac{\sqrt{7/\pi}}{4}$	$5\cos^3\theta - 3\cos\theta$
$x(5z^2 - r^2)$	xz^2	$\dfrac{\sqrt{42/\pi}}{8}$	$\sin\theta(5\cos^2\theta - 1)\cos\phi$
$y(5z^2 - r^2)$	yz^2	$\dfrac{\sqrt{42/\pi}}{8}$	$\sin\theta(5\cos^2\theta - 1)\sin\phi$
$z(xy)$		$\dfrac{\sqrt{105/\pi}}{4}$	$\sin^2\theta\cos\theta\sin 2\phi$
$z(x^2 - y^2)$		$\dfrac{\sqrt{105/\pi}}{4}$	$\sin^2\theta\cos\theta\cos 2\phi$
$x(x^2 - 3y^2)$		$\dfrac{\sqrt{70/\pi}}{8}$	$\sin^3\theta\cos 3\phi$
$y(3x^2 - y^2)$		$\dfrac{\sqrt{70/\pi}}{8}$	$\sin^3\theta\sin 3\phi$

A set of functions directly useful in problems with cubic symmetry can be obtained by taking the following linear combinations of those in Table AV.1.

$$A_{2u}: \quad f_{xyz} = f_{xyz} \text{ (as before)}$$

$$F_{1u}: \begin{cases} f_{x^3} = -\tfrac{1}{4}[\sqrt{6}\,f_{xz^2} - \sqrt{10}\,f_{x(x^2-3y^2)}] \\ f_{y^3} = -\tfrac{1}{4}[\sqrt{6}\,f_{yz^2} + \sqrt{10}\,f_{y(3x^2-y^2)}] \\ f_{z^3} = f_{z^3} \text{ (as before)} \end{cases}$$

$$T_{2u}: \begin{cases} f_{z(x^2-y^2)} = f_{z(x^2-y^2)} \text{ (as before)} \\ f_{x(z^2-y^2)} = \tfrac{1}{4}[\sqrt{10}\,f_{xz^2} + \sqrt{6}\,f_{x(x^2-3y^2)}] \\ f_{y(z^2-x^2)} = \tfrac{1}{4}[\sqrt{10}\,f_{yz^2} - \sqrt{6}\,f_{y(3x^2-y^2)}] \end{cases}$$

It is these functions which are given in Table 8.1.

An Illustrative Semi-Empirical
Molecular Orbital Calculation:
BF_3 by the Extended Hückel Method*

In Section 8.6 the symmetry arguments used to set up a factored secular equation for BF_3 (or any similar AB_3 species of D_{3h} symmetry) were presented. A basis set of 16 atomic orbitals, consisting of the $2s$, $2p_x$, $2p_y$, and $2p_z$ orbitals of each atom, was used. This appendix presents an outline of the actual calculation, carried out on a digital computer by the so-called extended Hückel method, from which the orbital energy level diagram in Figure 8.11 was obtained.

In setting up the secular equation in the text, the local coordinate system on each F atom was oriented with its x axis directed toward the B atom and the z axis perpendicular to the molecular plane. This was done for conceptual simplicity, since the p_x, p_y, and p_z orbitals of each F atom are thus, directly, divided into $p\sigma$, $\|p\pi$, and $\perp p\pi$ sets. For computational purposes, however, it is more convenient to use the x and y coordinates shown below, with the z coordinates perpendicular to the molecular plane as before.

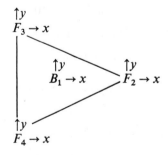

* These calculations were carried out using a program written by Professor Roald Hoffman of Cornell University.

Note also that the atoms are now numbered serially from 1 to 4 so that present $F_n \equiv$ old F_{n-1}. The new coordinates are related to the old ones by simple two-dimensional matrices, for example,

$$\begin{bmatrix} F_2^{new}x \\ F_2^{new}y \end{bmatrix} = \begin{bmatrix} -1 & 0 \\ 0 & 1 \end{bmatrix} \begin{bmatrix} F_1^{old}x \\ F_1^{old}y \end{bmatrix}$$

In the procedure used here, overlap of atomic orbitals is *not* neglected in computing the normalizing coefficients for the SALC's. Hence an *A*-type SALC, viz.,

$$\psi = N(F_{2s}^{(2)} + F_{2s}^{(3)} + F_{2s}^{(4)})$$

does not have $N = 1/\sqrt{3}$, but rather

$$N = (3 + 6S)^{-1/2}$$

where S is the overlap integral between the s orbitals on different F atoms. This integral was found to be 0.014 (see Table AVI.1) and hence $N = 0.568$ instead of $3^{-1/2} = 0.577$. This is obviously a small, virtually negligible, difference. However, in normalizing the actual MO wave functions, overlap is also included. Then it is often quite significant since the large overlaps between orbitals on bonded atoms are involved. It is for this reason that the sum of the squares of the coefficients in the MO's, as given below in Table AVI.3 is not equal to unity. It is actually the quantities

$$\sum_{i,j} c_i S_{ij} c_j$$

that equal unity. The sums

$$\sum_{i,j} c_i c_j$$

are generally <1 for bonding orbitals and >1 for antibonding orbitals, as the reader may verify.

Overlap of nonbonded atoms is also *not* neglected in evaluating the energies of the SALC's. Thus, the SALC just considered, viz.,

$$\psi = 0.568 \ (F_{2s}^{(2)} + F_{2s}^{(3)} + F_{2s}^{(4)})$$

is not assigned exactly the energy of a single F_{2s} orbital. Instead, the interaction between the three F_{2s} orbitals is computed, using a scheme analogous to the one now to be described for off-diagonal elements, and added (algebraically) to the single-orbital energy.

The off-diagonal elements are evaluated using the following expression:

$$H_{ij} = 1.75 \left(\frac{H_{ii} + H_{jj}}{2} \right) S_{ij}$$

Table AVI.1 The Atomic Orbital Overlap Matrix*

	2s B₁	2s F₂	2s F₃	2s F₄	2pₓ B₁	2pₓ F₂	2pₓ F₃	2pₓ F₄	2p_y B₁	2p_y F₂	2p_y F₃	2p_y F₄	2p_z B₁	2p_z F₂	2p_z F₃	2p_z F₄
2s B₁	1.000	0.355	0.355	0.355		$\overline{0.226}$	0.113	0.113			$\overline{0.196}$	0.196				
2s F₂		1.000	0.014	0.014	0.479		0.018	0.018			0.010	0.010				
2s F₃			1.000	0.014	$\overline{0.240}$	$\overline{0.018}$			0.415	0.010		$\overline{0.020}$				
2s F₄				1.000	$\overline{0.240}$	$\overline{0.018}$			0.415	0.010						
2pₓ B₁					1.000	$\overline{0.248}$	0.084	0.084			0.192	$\overline{0.192}$				
2pₓ F₂						1.000	0.084	0.021			0.014	0.014				
2pₓ F₃							1.000	0.004	0.192	0.014						
2pₓ F₄								1.000	0.192	0.014						
2p_y B₁									1.000	0.195	$\overline{0.138}$	$\overline{0.138}$				
2p_y F₂										1.000	0.004	0.004				
2p_y F₃											1.000	$\overline{0.029}$				
2p_y F₄												1.000				
2p_z B₁													1.000	0.195	0.195	0.195
2p_z F₂														1.000	0.004	0.004
2p_z F₃															1.000	0.004
2p_z F₄																1.000

* Blanks represent zero overlap; the complete matrix is symmetrical about the principal diagonal. A bar indicates a negative value.

372

H_{ii} and H_{jj} are diagonal elements, 1.75 is an empirically adjusted proportionality constant (values closer to 2.0 are sometimes used), and S_{ij} is the overlap integral between the overlapping SALC's (including the special case where one SALC is simply an atomic orbital on the central atom). These S_{ij}'s are obviously sums of overlaps between pairs of atomic orbitals, and are easily expressed as such using the expressions for the SALC's.

In view of the immediately preceding paragraphs, it is obvious that the first steps in a calculation such as this are the choice of a basis set of orbitals, interatomic distances in the molecule, and atomic orbital energies. In the present case these choices were as follows:

Orbital energies (electron volts):

	B	F
2s	−15.2	−40.0
2p	−8.5	−18.1

Interatomic distances (A):

B − F	1.300
F ⋯ F	2.252

Atomic orbitals (Slater type):

$$\phi = N \cdot A \cdot r \cdot e^{-\mu r}$$

where N is a normalizing factor, A is the angle-dependent wave function, r is the radius in atomic units (1 atomic unit = 0.5292 A), and μ is a parameter whose value is prescribed by an energy-fitting procedure of the type originally carried out by Slater. For the case at hand we have

$$\mu_{2s,\,2p}(B) = 1.300$$

$$\mu_{2s,\,2p}(F) = 2.425$$

Table AVI.2 Energies of the One-Electron MO's (in eV)

$1a_1'$	−41.489
$1e'$	−39.662
$2a_1'$	−19.112
$1a_1''$	−18.474
$2e'$	−18.467
e''	−18.043
a_2'	−17.814
$3e'$	−17.736
$2a_2''$	−5.713
$3a_1'$	21.084
$4e'$	23.640

Table AVI.3 The Molecular Orbitals*

	2s				$2p_x$				$2p_y$				$2p_z$			
	B_1	F_2	F_3	F_4	B_1	F_2	F_3	F_4	B_1	F_2	F_3	F_4	B_1	F_2	F_3	F_4
$1a_1'$	0.145	0.516	0.516	0.516	0.000	0.010	$\overline{0.005}$	$\overline{0.005}$	0.000	0.000	0.009	$\overline{0.009}$				
$1e'$	$\overline{0.000}$	$\overline{0.795}$	0.398	0.398	$\overline{0.055}$	$\overline{0.018}$	$\overline{0.004}$	0.004	0.000	0.000	0.008	0.008				
	$\overline{0.000}$	0.000	0.688	0.688	0.000	0.000	0.008	0.008	0.055	0.001	0.013	0.013				
$2a_1'$	0.152	0.097	0.097	0.097	0.000	0.530	0.265	0.265	0.000	0.000	0.459	0.459				
$1a_2''$													$\overline{0.152}$	0.540	$\overline{0.540}$	0.540
$2e'$	$\overline{0.000}$	$\overline{0.051}$	0.026	0.026	0.081	$\overline{0.430}$	0.385	0.385	0.000	0.000	0.470	$\overline{0.470}$				
	$\overline{0.000}$	0.000	$\overline{0.044}$	0.044	0.000	0.000	0.470	$\overline{0.470}$	0.081	0.656	0.158	0.158				
e''													0.000	0.818	$\overline{0.409}$	$\overline{0.409}$
													0.000	0.000	0.709	0.709
a_2'					0.000	0.000	0.505	$\overline{0.505}$	0.000	0.583	0.292	0.292				
$3e'$	$\overline{0.000}$	0.003	0.002	0.002	0.062	0.674	$\overline{0.520}$	0.520	0.000	0.000	0.089	0.089				
	$\overline{0.000}$	0.000	0.003	0.003	0.000	0.000	0.089	0.089	0.062	0.469	0.623	0.623				
$2a_2''$													1.051	0.285	$\overline{0.285}$	0.285
$3a_1'$	1.396	0.525	0.525	$\overline{0.525}$	0.000	0.349	0.175	0.175	0.000	0.000	0.302	0.302				
$4e'$	$\overline{0.000}$	0.720	0.360	0.360	1.409	0.422	0.098	0.098	0.000	0.000	0.301	0.301				
	$\overline{0.000}$	0.000	0.623	0.623	0.000	0.000	0.301	0.301	1.409	0.272	0.249	0.249				

* Blanks indicate coefficients which are identically zero. Bars denote negative values.

374

The next order of business is to compute the overlap matrix—an array of all the $(16^2 + 16)/2 = 136$ distinct atomic orbital overlaps. The overlap matrix for this problem is shown in Table AVI.1.

It is now possible to compute the values of all matrix elements and then solve the secular equations for the energies. These energies are presented numerically in Table AVI.2 and graphically in Figure 8.11 (page 221). With the energies in hand, the values of the coefficients of the atomic orbitals in each MO can be computed. These are given in Table AVI.3.

The computer program used also calculates an additional result, which can be obtained from the MO coefficients, namely, the charge distribution. It gives this first on an orbital by orbital basis and then for each atom as a whole. Some of the results for BF_3 are shown in Table AVI.4. It is well-

Table AVI.4
Calculated Charge
Distribution in BF_3

	B	F
σ Orbitals	$+2.362$	-0.787
π Orbitals	-0.142	$+0.047$
Net	$+2.220$	-0.740

known that charge separation or bond polarity is greatly exaggerated by a calculation of this type, chiefly because the initial atomic orbital energies used are too disparate. While the energies of the atomic orbitals are approximately correct for isolated atoms, there are interatomic penetration effects in the molecule which serve to lessen the differences, thus lessening the tendency of electrons to concentrate in the fluorine orbitals in this case.

Appendix VII

Character Tables for Some Double Groups

D_4'		E	R	C_4 $C_4^3 R$	C_4^3 $C_4 R$	C_2 $C_2 R$	$2C_2'$ $2C_2' R$	$2C_2''$ $2C_2'' R$
Γ_1	A_1'	1	1	1	1	1	1	1
Γ_2	A_2'	1	1	1	1	1	-1	-1
Γ_3	B_1'	1	1	-1	-1	1	1	-1
Γ_4	B_2'	1	1	-1	-1	1	-1	1
Γ_5	E_1'	2	2	0	0	-2	0	0
Γ_6	E_2'	2	-2	$\sqrt{2}$	$-\sqrt{2}$	0	0	0
Γ_7	E_3'	2	-2	$-\sqrt{2}$	$\sqrt{2}$	0	0	0

O'		E	R	$4C_3$ $4C_3^2 R$	$4C_3^2$ $4C_3 R$	$3C_2$ $3C_2 R$	$3C_4$ $3C_4^3 R$	$3C_4^3$ $3C_4 R$	$6C_2'$ $6C_2' R$
Γ_1	A_1'	1	1	1	1	1	1	1	1
Γ_2	A_2'	1	1	1	1	1	-1	-1	-1
Γ_3	E_1'	2	2	-1	-1	2	0	0	0
Γ_4	T_1'	3	3	0	0	-1	1	1	-1
Γ_5	T_2'	3	3	0	0	-1	-1	-1	1
Γ_6	E_2'	2	-2	1	-1	0	$\sqrt{2}$	$-\sqrt{2}$	0
Γ_7	E_3'	2	-2	1	-1	0	$-\sqrt{2}$	$\sqrt{2}$	0
Γ_8	G'	4	-4	-1	1	0	0	0	0

Elements of the g Matrix

Each element of a g matrix is identified by a double subscript which specifies the kinds of internal coordinates involved, using r for interatomic distances and ϕ for angles. Thus there will be three main classes of g matrix elements: $g_{rr}, g_{\phi\phi}, g_{r\phi}$. Within each of these classes subclasses arise based on the number of atoms which are shared by the internal coordinates; a superscript specifies this number of shared atoms. For example $g_{r\phi}^{2}$ is a matrix element involving a bond, r, and an angle, ϕ, with two atoms in common; the common atoms are those forming the bond, r, and one side of the angle. Even this notation is not entirely unequivocal, and some further distinctions must be made. The cases to be treated here are represented and assigned symbols in Fig. AVIII.1. In these sketches the atoms common to both coordinates are represented by double circles.

In some of the formulas for the g matrix elements it will be necessary to use two types of dihedral angles, as well as ϕ and r. In the expressions for $g_{\phi\phi}^{2}(1)$ and $g_{r\phi}^{1}(2)$ there occur dihedral angles denoted $\psi_{\alpha\beta\gamma}$, while in the expressions for $g_{\phi\phi'}^{2}(2)$, $g_{\phi\phi'}^{1}(1)$, $g_{\phi\phi'}^{1}(2)$, and $g_{r\phi}^{1}(1)$ a dihedral angle denoted τ will be found.

The ψ-type dihedral angles are defined by the following diagram:

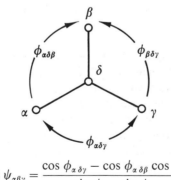

$$\cos \psi_{\alpha\beta\gamma} = \frac{\cos \phi_{\alpha\delta\gamma} - \cos \phi_{\alpha\delta\beta} \cos \phi_{\beta\delta\gamma}}{\sin \phi_{\alpha\delta\beta} \sin \phi_{\beta\delta\gamma}}$$

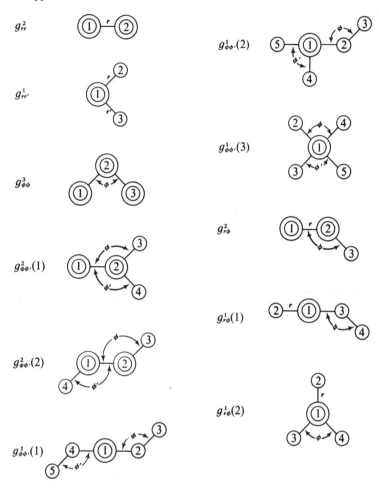

Figure AVIII.1   Diagrams showing how various common g matrix elements are defined.

The τ-type dihedral angle is defined as that between the planes formed by atoms 1, 2, 3 and 2, 3, 4 when atoms 1, 2, 3, 4 are bonded in sequence. When only two such planes are involved, as in $g_{\phi\phi'}^2(2)$ and $g_{r\phi}^1(1)$, τ without subscripts denotes the unique dihedral angle. In the other two cases, subscripts denote the first atom of the set defining the first plane and the last atom of the set defining the second plane. For example, in $g_{\phi\phi'}^2(1)$, τ_{25} denotes the angle between the planes defined by the overlapping atom triplets 2, 1, 4 and 1, 4, 5. It is necessary to have a sign convention for τ angles. These angles are restricted to the range $-\pi < \tau \leq \pi$ and τ_{14} is positive if, on viewing the atoms along the bond 2, 3 with 2 nearer the observer, the angle from the projection of 2, 1 to the projection of 3, 4 is measured clockwise.

In the following expressions μ_i represents the reciprocal mass (in atomic units) of the ith atom, and ρ_{ij} represents the reciprocal of the distance between the ith and jth atoms.

g_{rr}^2 $\quad \mu_1 + \mu_2$

g_{rr}^1 $\quad \mu_1 \cos \phi$

$g_{\phi\phi}^3$ $\quad \rho_{12}^2 \mu_1 + \rho_{23}^2 \mu_3 + (\rho_{12}^2 + \rho_{23}^2 - 2\rho_{12}\rho_{23} \cos \phi)\mu_2$

$g_{\phi\phi'}^2(1)$ $\quad (\rho_{12}^2 \cos \psi_{314})\mu_1 + [(\rho_{12} - \rho_{23} \cos \phi_{123} - \rho_{24} \cos \phi_{124})\rho_{12}$
$\qquad \cos \psi_{314} + (\sin \phi_{123} \sin \phi_{124} \sin^2 \psi_{314}$
$\qquad + \cos \phi_{324} \cos \psi_{314})\rho_{23}\rho_{24}]\mu_2$

$g_{\phi\phi'}^2(2)$ $\quad -\rho_{12} \cos \tau[(\rho_{12} - \rho_{14} \cos \phi_1)\mu_1 + (\rho_{12} - \rho_{23} \cos \phi_2)\mu_2]$

$g_{\phi\phi'}^1(1)$ $\quad -\rho_{12}\rho_{14}(\sin \tau_{25} \sin \tau_{34} + \cos \tau_{25} \cos \tau_{34} \cos \phi_1)\mu_1$

$g_{\phi\phi'}^1(2)$ $\quad \dfrac{\rho_{12}\mu_1}{\sin \phi_{415}} [(\sin \phi_{214} \cos \phi_{415} \cos \tau_{34} - \sin \phi_{215} \cos \tau_{35})\rho_{14}$
$\qquad + (\sin \phi_{215} \cos \phi_{415} \cos \tau_{35} - \sin \phi_{214} \cos \tau_{34})\rho_{15}]$

$g_{\phi\phi'}^1(3)$ $\quad \dfrac{\mu_1}{\sin \phi_{214} \sin \phi_{315}} [(\cos \phi_{415} - \cos \phi_{314} \cos \phi_{315} - \cos \phi_{214} \cos \phi_{215}$
$\qquad + \cos \phi_{213} \cos \phi_{214} \cos \phi_{315})\rho_{12}\rho_{13}$
$\qquad + (\cos \phi_{413} - \cos \phi_{514} \cos \phi_{513} - \cos \phi_{214} \cos \phi_{213}$
$\qquad + \cos \phi_{215} \cos \phi_{214} \cos \phi_{513})\rho_{12}\rho_{15}$
$\qquad + (\cos \phi_{215} - \cos \phi_{312} \cos \phi_{315} - \cos \phi_{412} \cos \phi_{415}$
$\qquad + \cos \phi_{413} \cos \phi_{412} \cos \phi_{315})\rho_{14}\rho_{13}$
$\qquad + (\cos \phi_{213} - \cos \phi_{512} \cos \phi_{513} - \cos \phi_{412} \cos \phi_{413}$
$\qquad + \cos \phi_{415} \cos \phi_{412} \cos \phi_{513})\rho_{14}\rho_{15}]$

$g_{r\phi}^2$ $\quad -\rho_{23}\mu_2 \sin \phi$

$g_{r\phi}^1(1)$ $\quad \rho_{13}\mu_1 \sin \phi_1 \cos \tau$

$g_{r\phi}^1$ $\quad -(\rho_{13} \sin \phi_{213} \cos \psi_{234} + \rho_{14} \sin \phi_{214} \cos \psi_{243})\mu_1$

Appendix IX

Reading List

Presented here is an annotated list of references to more complete or more advanced treatments of the topics covered in this book. References to basic theory relevant to all the chapters in Part I are given first, and then references pertinent to each of the areas of application covered in Part II are presented separately for each chapter.

Basic Theory

A legion of books on quantum mechanics presents the fundamentals as appropriate for chemical problems. Several which would provide a good basis to precede the study of this book are listed below. It is not necessary, however, to be conversant with all of the material in the more advanced of these books in order to study the present one. The last book listed provides perhaps the best-balanced contemporary treatment.

Valence, by C. A. Coulsen, Oxford University Press, Fair Lawn, N.J., 2nd Ed., 1961.
Introduction to Quantum Mechanics, by L. Pauling and E. B. Wilson, McGraw-Hill Book Company, New York, 1935.
Quantum Chemistry, by H. Eyring, J. Walter, and G. E. Kimball, John Wiley & Sons, New York, 1947.
Quantum Chemistry, by W. Kauzman, Academic Press, New York, 1957.
Valence Theory, 2nd Ed., by J. N. Murrell, S. F. A. Kettle, and J. M. Tedder, John Wiley & Sons, New York, 1970.

Many books cover the application of group theory to quantum mechanics at a more advanced and sophisticated level than this one, and the aspiring theoretician will find it necessary to consult some of these. A few of the best are listed below:

Group Theory and Its Physical Applications, by L. M. Falicov, The University of Chicago Press, Chicago, 1966.
Group Theory and Its Applications to Physical Problems, by H. Hammermesh, Addison-Wesley Publishing Company, Reading, Mass., 1962.

Group Theory and Quantum Mechanics, by M. Tinkham, McGraw-Hill Book Company, New York, 1964.

Group Theory, by E. P. Wigner, Academic Press, New York, 1960.

Symmetry: An Introduction to Group Theory and Its Applications, by R. McWeeny, Pergamon Press, London, 1963.

Chapter 7

Information on the details of various approximations besides the simple Hückel approximation for organic and hetero-organic molecules and on applications of these calculations will be found in:

Molecular Orbital Theory for Organic Chemists, by A. Streitwieser, Jr., John Wiley & Sons, New York, 1961.

The Molecular Orbital Theory of Organic Chemistry, by M. J. S. Dewar, McGraw-Hill Book Company, New York, 1969.

Approximate Molecular Orbital Theory, J. A. Pople and D. L. Beveredge, McGraw-Hill Book Company, New York, 1970.

The Molecular Orbital Theory of Conjugated Systems, by L. Salem, W. A. Benjamin, New York, 1966.

The Theory of the Electronic Spectra of Organic Molecules, by J. N. Murrell, Methuen and Company, London, 1963.

A tabulation of the results of Hückel calculations for the π systems of many organic molecules is available:

Dictionary of π-Electron Calculations, by C. A. Coulson and A. Streitwieser, Jr., W. H. Freeman and Company, San Francisco, 1965.

The account of symmetry-based selection rules for cyclizations and other rearrangements and reactions of organic molecules first published by Woodward and Hoffmann in *Angewandte Chemie* is now available in book form:

Conservation of Orbital Symmetry, by R. B. Woodward and R. Hoffmann, Academic Press, New York, 1970.

Chapter 8

Articles and research papers dealing with the quantitative aspects of MO calculations for AB_n molecules (including those suggested in Exercise 8.4 as practice problems) are:

H. B. Gray and C. J. Ballhausen, *J. Am. Chem. Soc.,* **85**, 260 (1963) (square $MX_4{}^{n-}$ complexes).

C. J. Ballhausen and H. B. Gray, *Inorg. Chem.,* **1**, 111 (1962) $[VO(H_2O)_5]^{2+}$.

R. S. Berry, M. Tamres, C. J. Ballhausen, and H. Johnson, *Acta Chem. Scand.,* **22**, 231 (1968) (AB_5 molecules).

R. F. Fenske, *Inorg. Chem.*, **4**, 33 (1965) (the TiF_6^{3-} ion).
L. L. Lohr and W. N. Lipscomb, *Inorg. Chem.*, **2**, 911 (1963) (An approach to Jahn-Teller effects via semiempirical MO theory).

Several semiempirical MO calculations on the electronic structures of $(C_5H_5)_2M$ and $(C_6H_6)_2Cr$ have been published:

R. D. Fischer, *Theoret. Chim. Acta*, **1**, 418 (1963).
E. M. Shustorovitch and M. E. Dyatkina, *Dokl. Akad. Nauk. SSSR*, **128**, 1234 (1959); *J. Strukt. Chem. (USSR)*, **1**, 98 (1960).
J. P. Dahl and C. J. Ballhausen, *Kong. Danske Vidensk. Selsk., Mat. Fys. Medd.*, **33**, No. 5 (1961).
J. H. Schachtschneider, R. Prins, and P. Ros, *Inorg. Chim. Acta*, **1**, 462 (1967).
M. F. Rettig and R. S. Drago, *J. Am. Chem. Soc.*, **91**, 3432 (1969).

An MO treatment of $C_6H_6Cr(CO)_3$ is reported by:

D. G. Carroll and S. P. McGlynn, *Inorg. Chem.*, **7**, 1285 (1968).

Chapter 9

The best general texts for chemists are:

Introduction to Ligand Field Theory, by C. J. Ballhausen, McGraw-Hill Book Company, New York, 1962.
Introduction to Ligand Fields, by B. N. Figgis, John Wiley & Sons, London, 1966.
Basic Principles of Ligand Field Theory, by H. L. Schlafer and G. Glieman, John Wiley & Sons-Interscience Publishers, New York, 1969.

A very sophisticated treatise is:

The Theory of Transition Metal Ions, by J. S. Griffith, Cambridge University Press, 1961.

Chapter 10

The definitive book on theory for discrete molecules is:

Molecular Vibrations, by E. B. Wilson, Jr., J. C. Decius, and P. C. Cross, McGraw-Hill Book Company, New York, 1955.

Two other books covering both theory and experimental data are:

Infrared and Raman Spectra, by G. Herzberg, D. Van Nostrand Co., Princeton, N.J., 1959.
Infrared Spectra of Inorganic and Coordination Compounds, 2nd Ed., by K. Nakamoto, John Wiley & Sons, New York, 1970.

The influence of a crystalline environment on the vibrational spectrum of a molecule is very lucidly covered by:

W. Vedder and D. F. Hornig in *Advances in Inorganic Chemistry*, Vol. II, H. W. Thompson, Ed., Interscience Publishers, New York, 1961, p. 189.

Index

Contents